The Green Evolution

How we can survive the global ecological collapse and
continue as a technological civilization

Jeffrey Ravage

Ravaged Books

Dedication:

I dedicate this book to my children, Emma, Ian, and Elias. And to your children, and all the children of the world. They are the ones who both understand and must live with the poor choices being made by their elders. This book is for them as well— the adults who understand climate change— and those who are still on the fence.

"Ecology as Technology"

Contents

PROLOGUE

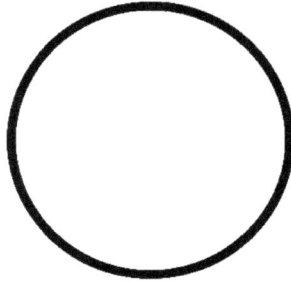

"People are like plants; they grow towards the light."
— ***Hope Jahren***

In the late 1970s, when I was a high school team debater, we often spoke about "The Green Revolution," both in and out of competitions. The Green Revolution was promoted as a "save the world" groundbreaking series of advancements in the sciences that promised to eliminate global hunger in our lifetimes. It was asserted to be a series of breakthroughs that would end food insecurity once and for all. And leaning reasonably liberal for kids growing up in Dick Cheney's Wyoming, we latched on, swallowing that hook deeply. As did the rest of society. With the promise of new methods of intensive agriculture and new strains of crops bred by leading universities, the pitchmen and pundits all aligned to promote the beginning of a golden age. Science would soon deliver on the promise of making the world a kinder and gentler place.

This revolution would shatter the Malthusian curve and ensure humanity's health and longevity for the foreseeable future. Manna wouldn't have to fall from the heavens, now it would rise from the ground. The sheer abundance would solve many social issues and make starvation a thing of the past. Once the developing world was freed from food insecurity, the proper business of elevating them to First World status could begin. Global agribusiness was poised to align with the farmers, large and small, and guide them to a healthy and prosperous future. Politicians of all stripes reached across the aisles and

joined their hands in a shared vision of a brave new world.
Kumbaya!

In retrospect, it didn't exactly turn out that way.

Thomas Malthus was a 19th-century scholar, cleric, and economist. He is still held in high regard, to this day, as an influential thinker of his time, on a par with Charles Darwin. His treatise: "An Essay on the Principle of Population," held a dire warning for humankind. He cautioned that while food production could only increase arithmetically, the population was growing exponentially. Meaning that every time food production was doubled, the number of hungry mouths would have increased by 4-fold. This was a significant concern for society's poorest, but it implicated the aristocracy in this march towards failure, as well. And since the productivity of any given plot of land was assumed to be fixed, the only known way to increase the number of bushels harvested would be to put more land under the plow. So, while food production could only increase by simple addition, the human population was expanding at an accelerating rate. There was a cliff, and they were headed, lemming-like, right towards it. In 1798 he predicted a worldwide famine was imminent, and his calculations seemed quite persuasive.

There was much debate over his provocative monograph. Sure, the world was seemingly becoming smaller and poorer, but Malthus was suggesting that everyone reduce their rates of reproduction, rich and poor alike. Stresses were being felt at all levels of society, but nearly everyone felt the course of civilization was on the right track. Great Britain was a growing world power, the rituals were occurring on schedule, and the aristocracy was well-kempt and coifed. Why change if your politicians and theologians felt that all was right with the world?

Economic systems have always had a strong protectionist bias towards their status quo. So there would be a more significant burden of proof set before Malthus in this squabble than for the ruling elite. Any attempt to save the common classes that might require some sacrifice from the wealthiest among them or even pose an inconvenience would not be well met. And while the governing powers did not like inconveniences, even less did they like being told not to procreate. Nor did the religions of the teeming masses. These influential denominations had perfected prohibitions against intercourse in the primary case, and the demand for unbridled reproduction once vows were sworn.

There was little patience for a reassessment of such values back then, much like today. Society's various decks were stacked against taking action. So the status quo would, in the end, do nothing. Even though they were required to pretend they were ever-ready to take the initiative, again, much like nowadays. The outcome of all the abundant posturing, oratory, and hand-wringing this book inspired, ultimately, was to kick the can down the road.

To punt, it seems, is one of humanity's primary reactions to nagging problems. So thankfully, Malthus was wrong. Pure luck averted Malthus' doom for reasons such as migration to the new world and innovations in supply distribution. More land was set aside for agriculture as well. Nor could he have foreseen the cholera epidemics that would begin less than a quarter-century after his publication. That slowed the rampant growth of humanity a tad. But his work left a nagging doubt in the collective consciousness of that status quo. Someday he might be right. The world was, after all, a finite place.

Status Quo: /stādəs ˈkwō/ *noun*

  1: the existing state of affairs, especially regarding social or political issues.
   Example: "They have a vested interest in maintaining the status quo."

In the mid-nineteen seventies, when we had our moment at the podium, the ghost of Malthus was rearing its ugly head again and his doom was threatening finally to come to pass. In his book "The Population Bomb," Paul Ehrlich argued essentially the same thing that Malthus had two centuries earlier, that humans were going to extinguish themselves by over breeding and starvation (or war). Westernization was spreading, and with it came the increasing exhaustion of scarce world resources. Consumption in developed countries represented a massive burden on the world's systems. But as long as most of the global population was in poverty, the camel's back would not break. However, if the developing world began to consume at the same rate as the "First World," disaster would be unavoidable. His predictions not only echoed the findings of Malthus; it was the excellent cleric's thesis: inequality in wealth distribution coupled with unrestrained rates of reproduction is "the human problem." But in the 1970s, the idea taken away from all this doom and gloom prophesy was instead: "the world needs to grow more food, and faster."

We were kids ourselves when the Green Revolution was occurring, and we wanted to save the world. So we "signed up" and got on board, as did many scientists and politicians. We promoted it in our speeches and considered it in our potential future careers. There was political willpower for work in this frontier, as well as grant funding. Television shows like "Face the Nation" invited scientists to come and talk about their breakthroughs, something every researcher loves to do. "Talk about my research? Don't mind if I do!"
But as clever as everyone was, they were totally in the dark as to what this revolution ultimately meant. In fairness, no one could have understood the implications as they do now, not the politicians, nor the scientists, and certainly not a bunch of high-schoolers from Wyoming. The critical thing to know was, seemingly, that science had found a solution to world hunger, and no one needed to worry about it anymore. Most of us still hold this faith, this prejudice, in our minds. Even though perhaps society should know better. Many think science will save us, regardless of how we treat it or its adherents. To be sure, Science is a tool that can better the world, but the more significant questions are: who wields this tool? And to what ends?
Our debate team was still just a bunch of kids back then, so at least we had an excuse. It did not sink in that:

The "Green Revolution" was petroleum-based farming.

The Green Revolution still is petro-agriculture. Massive crop productivity is realized via intensive mechanization and supported by the application of synthetic petroleum fertilizers. They replaced integrated crop and animal farms with concentrated specialization. One crop, or one animal, per facility, in what is called a monoculture. These intensive monocultures were "protected" by the applications of petroleum-derived insecticides and herbicides. Planting, husbandry, and harvest were all conducted by heavy machinery that ran on petroleum fuels. Only selectively bred crops could survive these toxic conditions and produce these increasingly heavy yields. Animals were placed in tightly confined quarters and given antibiotics to stave off the infections that thrive in such unnatural conditions. These same medicines had the side effect of stimulating rapid growth, a pleasant by-product from the point of view of the Green Revolution's proponents. Henry Ford's assembly line efficiencies had led to greater yields and lower prices in many markets, so why not in the open field? So this time it was Ehrlich's prediction of doom that was thwarted.

But at what cost?

In team debate, one team, the affirmative, proposes an idea, policy, or law. They must present a problem (preferably global) and then offer a plausible solution. This package of needs and solutions forms the "case." The negative team has one task only — to defeat the affirmative's case. There are a few restraints on how they may go about this nerd beat-down. They may not offer a counter plan. They may only argue that the greatest good is to do nothing to address the issue at all, or argue that the problem's solution was already in the pipeline of the status quo. The negative has the task of supporting business as usual.

The negative will attack the "need," the "plan," and present "disadvantages." The need can often be distilled down to just a point of view. "Do those who don't, or won't, help themselves really need another handout?" The plan, the nuts and bolts mechanisms required to achieve the affirmative's goals may be attacked as: "These systems may or may not already exist, or perhaps they will or won't work."

And then there are the disadvantages. Disadvantages are the unintended consequences. They are the horrible things that will happen if society were to enact the affirmatives' program. They are the doom that will befall us if anyone dares to challenge the status quo. For business as usual is given a presumption of fitness while the affirmative must prove its case. The process of debate is based on trial proceedings and how cases are argued in court. It is instructional to consider that the party of the status quo, the defenders of the desire not to change, is called the negative team.

> The goal of debate is not to find the truth; the goal is to *win* the argument.
> —This is the first lesson of debate.

One of our favorite disadvantages, when placed into this negative role, was nuclear war. Want to feed the poor in the inner city? It will lead to nuclear war. Want global peace? Sorry, but for all your good intentions, that will force a swift atomic retaliation. Do you wish to disarm the world's nuclear arsenals? Oh, that will definitely lead to human annihilation! How one gets to this fiery end was considered a challenging and worthwhile mental exercise. The more labyrinthine the links of cause and effect, the more convoluted the rhetoric, the better. Logical and grammatic gymnastics were encouraged. This type of intellectual exercise, divorced from any genuine respect for the topic, was the point, not the by-product of the debate.

Nuclear war was the demon that threatened all of us back then. It hovered over our lives, day and night, year after year. Since the 1950s, children had been taught to fear nuclear war and the possible eradication of humanity. Atomic bombs, nuclear proliferation, fission, fusion, these were the words that kept the children safely afraid. Our status quo was focused on a string that held an atomic hammer above our heads with a laser-like obsession. Such was this fear that our debate team could bandy it about and threaten, with cavalier insensitivity, everything of supposed value to humankind. All just to win an argument against some unfortunate kids from Lusk, Wyoming. We were very good at it, possessing the callousness of youth. So this was how the world could end: With a Bang!

Whether this was the primary threat to our society or not, people took it seriously and willingly changed their behaviors to accommodate it.

The environmental movement was powerful in the mid-seventies. Richard Nixon had fallen in disgrace, but not before creating the Environmental Protection Agency (EPA). Gerald Ford was our next President, and Jimmy Carter was on the horizon. The power of this EPA was growing. The Sierra Club was a national organization that would protect our forests and waters. There was still a "Cold War," but that was so much better than the other kind. Average citizens were fighting polluters and sometimes winning. Science was getting some respect, and many felt confident that this world would deliver a future worthy of an Arthur C. Clarke novel (one of the optimistic ones, anyway). It was a great time to grow up, for those in the mobile classes at least.

I read a lot. One of my favorite book series was the Time-Life Science library. In the 1970s, middle-class families could afford to buy educational reading for their children. There was no internet, as computers were still struggling to emerge from Texas Instruments calculators. Cars were big, and gasoline was cheap. I rode my bike around Laramie, Wyoming, a lot. I spent as much of my time in the forest as I could, cross-country skiing, hiking, and pushing over dead trees. The information I had learned from Time-Life brought the forest and nature alive for me. Sometimes when I was out and about with friends, I would tell them about something I had read. I liked to mention how Florida could disappear in the next century, and many of the world's great coastal cities might drown. The twenty-first century seemed so far away back then as to have been almost mythical. But even then, I knew, as did the editors of Time-Life, about a phenomenon called global warming.

It seems that the petroleum companies knew about it too.

But oil was the status quo.

A storm was brewing even back then. Humankind now stands on the edge of that cyclone, and policymakers and industrialists still enable denial about the seriousness of this situation. Unlike the paranoia that they evoked over nuclear war, this existential threat has received too little consideration. Quite the contrary, this knowledge was hidden, covered up beneath a monolith of silence. A lack of education about what was going on and the interdependence of the entire planet aided this nurtured ignorance, and many wish it would stay that way. Those who still debate global warming have a vested interest in pretending it away. And they question it as if it is something new that no one could've predicted. And as long as it is debatable, the first rule of debate is in full force. The pursuit becomes not one for truth but the discovery of the winning argument.

When economics leads discourse and commercial interests guide the search for "truth," inaction can be cultivated just as it was in the time of Malthus. Many in power have cast the seeds of doubt upon the ground and fertilized them with oil and our attachment to it. As a result, while most of us now know that the threat of global warming is real, many may feel powerless to do anything about it. Is this how the world really ends? Not with a bang, but in a puffy CO_2 laced cloud of denial?

It need not be. Together, we can do something about it. We have to stop it. And we can't wait for someone else to do it for us. Like the previous dark times, direct action could mitigate this catastrophe. But luck will not be on our side in today's scenario. Survival will be impossible if people choose our usual path of inaction. There are no new lands to expand to, no new sources of mucky tar to exploit. At least not in the quantities needed to continue this juggernaut called the Green Revolution. And it is crucial that everyone understand that oil is running out, it is never going to get substantially cheaper again, and we can't just pretend there's another wildlife refuge or sacred mountain we can harvest to keep this shindig rolling along a bit longer.

The climate changes brought on by our activities are a real and present danger. It is a by-product of our conventions and our lifestyles. It is an artifact of our fragmentation as a race and a post-truth world where societies can insulate themselves from reality with the sheer power they have wrested from it. Our problems can't be fixed with another "new" agricultural revolution, although we must have one. There will be no easy win by ending carbon emissions, though that must be achieved as well. Humanity can only survive this monumental challenge by changing the entire system we've built around ourselves because it operates in violation of a more powerful system's laws: the ecosystem. And that is unachievable if people hold on to current biases about who and what they are. Humankind needs to shed a part of their identity-the part that tells them they are "above" one another and everything else.

We are one human race, and together we have dodged every hurdle placed in our path to attain this fantastic civilization that we have built. Only now to meet the ultimate challenge of our existence. Finally, to discover that our potential dooms cause, and the greatest obstacle now before us is ourselves.

There is a system that has made life possible, that has made all life possible, and as a society, many know practically nothing about it. That's because most of us didn't care to learn about the living world. They only cared about how to make it work for their benefit. And often in the most narrow and short-sighted ways possible. The ecosystem holds the secrets to our long-term survival. But to learn from nature, we first have to learn about it. And that is the exercise we are about to undertake. By learning how life has survived, we will learn how we might persevere as well.

This book is a mea culpa. I am guilty of letting this horror unfold in my own little ways. And I have admitted to knowing about the climate crisis for most of my life. But you are guilty as well, and so is almost everyone else on the planet. And so are a handful of people circling above it right now. So while I will take the time to discuss how this situation got so out of hand, I will not dwell unduly on blame, and neither should you. There just isn't any time left to point fingers and shout "Ah-Hah!". And in terms of this climate crisis, there's plenty of blame to go around — more than we have time left to prosecute. We got into this situation one tiny action at a time, and we can get out of it in exactly the same way. We have to get out of it. There will be no last-minute salvation, no aliens coming to our rescue, no deity to pity us and hit the reset button. We have to avoid this peril the same way we met it. Together.

This book is a primer. It is here to help you get up to speed on where humanity has found itself, what's been done to this world, and what options we have available to fix it. This book is the story of this blue planet's natural history and how humanity's success has affected that long negotiated natural balance. Here is a tiny dollop of understanding that can help you recognize your place in relation to the natural world. And to help guide you to finding your proper employment within it. Little actions joined together can and will save us. They say you can't empty the oceans one drop at a time, yet that is precisely how they filled. Drop by dribble by dollop, tiny action by tiny action; everything adds up. It's a trickle-up world. Together we will see how humans created our current crisis one tiny step at a time, one little opinion, one meager fib, stacked upon the last. And together we will see how those actions can be undone, both by human effort and by the forces of nature herself. We can't wait for someone else to do it. And even if these tasks aren't easy, they need to happen. Accomplishing a simple task grants one little reward. The achievement of difficult things makes us proud and strong and defines what it means to be human. So if anyone tells you they can save you or that the answers to our problems are simple, they are lying. It's that cut and dry. Complex issues, by definition, do not have simple solutions. Good news! That means they do have solutions.

This book is an instruction manual for surviving the extinction our appetites have initiated. It is our point of departure, not the destination. This book will renew our understandings of how this synergistic world interacts and where things fit in, including ourselves, because everything fits in by definition. And that is your first tidbit of knowledge about how ecosystems work. Every part has meaning and a task to perform. Humans are here because we won our seat at the table. All of us have significance, and also a duty to discharge. Human expertise and imagination will get us out of this conundrum. And saving the world will save us all in every way imaginable.

So let's get started. The clock's ticking.

-Ravage- May Day, 2015

A Note on Conventions:

All measurements in this book will be given in metric system terms. The Imperial (American) system of measurement will follow in parentheses. For example: "a one-meter square (1.1 yd²) plot". The metric system is used by the vast majority of the world, just not in the United States of America (or Libya or Myanmar). It is the measurement system used by science, if for no other reason than because all values are expressed on a base10 decimal scale, making calculations easier and results more precise (you never liked fractions anyway, am I right?).
I will use standard rounding to eliminate decimal points unless those decimal values are essential. I will also occasionally use scientific notation for huge numbers. (ex: $1,000,000 = 1 \times 10^6$)
This text includes citations. The citations allow the reader to validate the facts as they are offered and look deeper into any piece of evidence if desired. If you find reading citations (ex: (Ravage, 2022)) annoying, just ignore them. Many

students do that very thing when reading journal articles for a class; it's okay. You can glide right over them or peel them open and take a more in-depth look if you wish. It's entirely up to you. Footnotes have additional information or sometimes little jokes[1]. They are below the text to avoid hijacking the train of thought and will occasionally offer the reader new rabbit holes to explore. I have taken great care to use accurate and up-to-date studies and values, but seeing as global warming is rapidly changing these calculations, some will be obsolete by the time of reading. When in doubt, assume a more catastrophic set of facts and figures confront you.

A good example would be the COVID-19 pandemic of 2020-2022. When I began writing this book, there was no hint that some virus would arise and start culling the planet's human population. Or that the United States would respond to it so inexpertly. But it was to be expected that something catastrophic was going to happen, and soon. Catastrophes galore await us as the ecosystem stutters and wobbles on its trajectory, thanks to our lifestyles. And there's a lot more to come.

This book aims to give you a sound basis in the sciences of the phenomena that make life possible. I will not bother to appear "balanced" on many topics because giving self-serving opinions the same credence as carefully researched data is absurd and should no longer be tolerated. The goal of this book is how we might best move forward in the healing of our broken planet and societies. Not whether we should even try.

---

[1] Jokes are just information coded in a surprisingly pleasant manner.

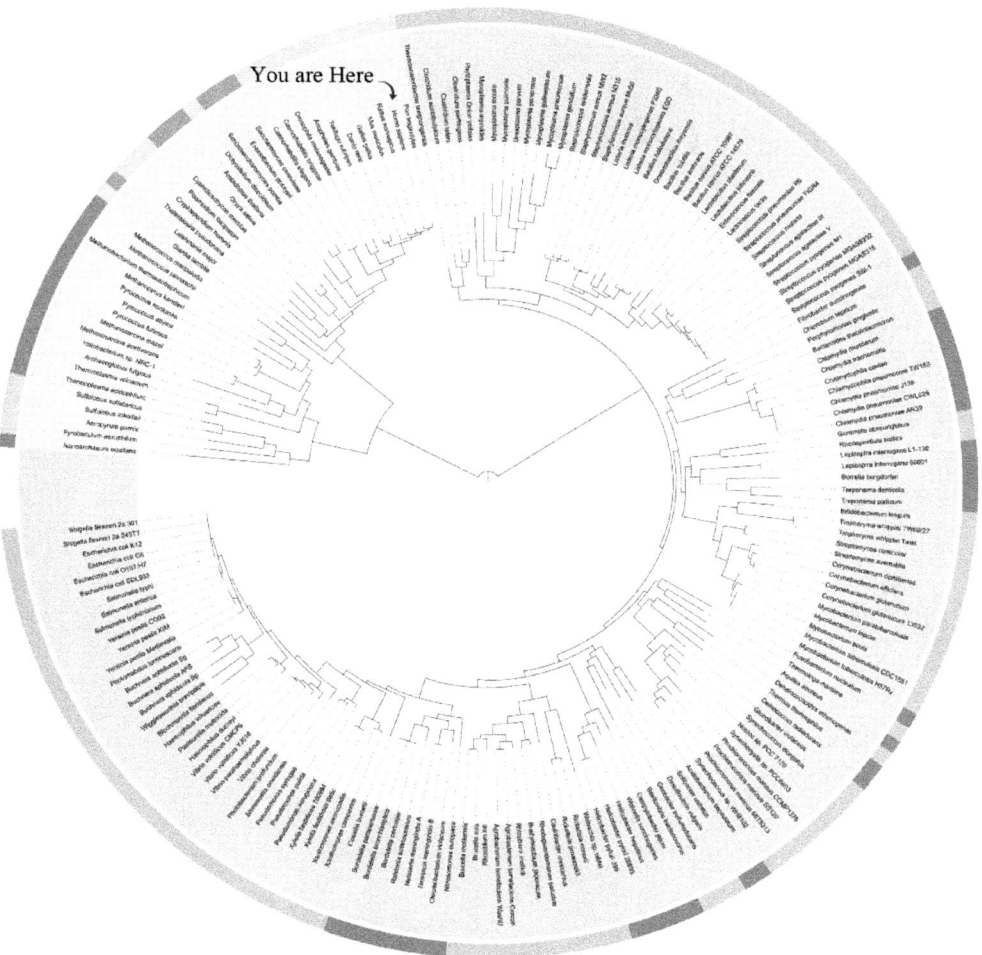

Tree of Life by: Ivica Letunic: Iletunic. Retraced by Mariana Ruiz Villarreal: (LadyofHats) Public Domain, all Rights granted by author. This is the direct connection between all of us as we depict it today.

"Science is not a philosophy. Science is a method of inquiry."

CHAPTER ONE: DISCOVERING THE GREEN EVOLUTION

"Nature cares nothing for logic, our human logic; she has her own, which we do not recognize and do not acknowledge until we are crushed under its wheel."
—Ivan Turgenev

On December 27th, 1831, a 22-year-old scientist set forth from Plymouth, England, on a Naval sloop named the HMS Beagle. This ship had been modified for its mission by another young man, Robert Fitzroy[2], captain of this globe circumnavigating voyage and only four years senior to naturalist Charles Darwin. Their ship was equipped for scientific discovery and the stresses of a sixty-four-thousand-kilometer (40,000 mi) voyage. It was astonishing that this prestigious undertaking would be commissioned to such a youthful crew in those days. Perhaps the high regard with which the social elite held Darwin and Fitzroy cinched this bargain. Or maybe it was their budding enthusiasm combined with the potential that this would be a one-way trip that closed the deal. Whatever the reason, commissions were signed, the hatches were battened, the mainsail set, and they bee-lined towards Bahia, Brazil, under fair winds and sky.

Thus began a scientific voyage to update sea currents, verify map coordinates, gather geological data, and collect plant/animal specimens. Darwin was credentialed in botany, biology, geology, and was a decent artist, which would come in handy as his drawings could capture more samples than the Beagle could carry in her holds. Fitzroy was on a further mission to find trading partners in Tierra del Fuego and return a trio of natives to their homes. These three men had previously been kidnapped, taken to England, and converted to Christianity as was the way in those days. The crew did not set out to turn modern society on its head. And while this voyage would do that, that drama wouldn't unfold for another 20 years.

After traveling around Cape Horn, and back up the South American Pacific Coast to Callao, Peru, they tracked west towards the Galapagos islands.

---

[2] Fitzroy was a prodigy himself, he soon achieved the rank of Vice Admiral in the Royal Navy. He later served as governor of New Zealand, and after retirement he created what we now know as: the art of weather forecasting.

Darwin's voyage before that fateful landfall had exposed him to vast fossil beds and other geological wonders. He had trekked into forests and gathered specimens to return to England for study. He had captured the many shapes of life with his detailed drawings and comprehensive notes. On February 20th, 1835, he witnessed an earthquake raze the City of Concepcion, Chile, and he noticed that the sea line rose several feet in just a matter of hours afterward. To him, the only rational explanation was that the land itself had slumped. But landmasses, being fixed in the 19th-century mind, didn't rise or fall, except by the act of a jealous God. However, Darwin was beginning to see the world in a new light. Perhaps this world was a system in flux and not the static, perfected domain of the Lord, which was the common understanding at the time.

And then he saw the iguanas, tortoises, and finches of the Galapagos. He saw forms and habits never seen before, for this refuge had separated from the mainland millions of years prior to his visit. There had been no communication with the continent since, nary a seagull or floating coconut. While the import of this discovery was not immediate, his carefully filled notebooks of illustrations and measurements captured a bombshell of truth. Birds that he collected from different islands initially seemed of multiple distinct species. Some had thin beaks, and some had fat vice-like ones. Some were dainty, and others brawny with the face of a grosbeak. It was only after he presented them to John Gould of the Zoological Society of London, on his return, that something strange became apparent about them. The renowned ornithologist correctly identified them all as finches and of completely unknown species. It would later become clear that these birds were either the same species or, at the very least, had all arisen from a common ancestor. This possibility, that these vastly different birds were the same bird, bordered on heresy. His was a time when the world was seen as an absolute reflection of a perfect design and a divinely inspired one, at that. All species were supposedly created by holy fiat, forever singular and distinct — never changing. And the entirety of life's diversity was purported to have occurred in just a matter of days. Something about that worldview now seemed to be amiss.

It would not be until 1859 that his work, those specimens, and the research of several free-thinking scientists combined would see the light of day as "The Origin of the Species- by means of natural selection." And what he had privately been calling "transmutation" gelled into the theory of evolution. The works of Malthus had greatly influenced Darwin. Therefore, he was keenly aware of the backlash he might experience from the clergy and the English establishment if his theories stepped too far out of line. He had planned for years, they say, to publish this work only after he had died, presumably to avoid the controversy that it was certain to cause. However, when he realized that this concept of divergent evolution of the species could be visualized as the branches on a tree, and like all trees, had a singular root, he knew that this work was too important to wait.

In 1858, Darwin received a letter from Alfred Russel Wallace, a fellow researcher coming to the same conclusions about life and its preferred methods of advancement and diversification. They joined forces and completed the manuscript within a year. These discoveries changed everything from a scientific perspective. But they changed social viewpoints only a little since nearly a century and a half later; cultural forces continue to assail and deny his findings.

The "Origin of the Species by Means of Natural Selection, or the Preservation of Favoured Races in the Struggle for Life" hit the streets in instant controversy. They did not intend it to be anything but a well-researched scientific text. But the subject manner, and the varied reactions to it, made it a best-seller in the common book market as well. People gladly exchanged their coppers for text so rapidly that in the United States, publishers quickly began the text's bootlegging (international copyrights having been less automatic back then). Darwin's masterpiece flew off the shelves. It had set off an international frenzy. And as feared, the clergy jumped upon the bandwagon, the podium, and the pulpit, claiming that the blasphemous tome attributed humanity to the descendants of monkeys. It mocked humanity, they claimed. Instead of humankind being presented as images of divine perfection, as the gospels promised, the book offered a picture of monkeys in spats and top-hats. And the press had a heyday producing just those sorts of images for the entertainment starved masses. In short, the controversy around this work of Darwin's elevated it beyond heresy; to utter blasphemy.

The book itself, of course, made no such sensationalistic claims about chimps and children. Instead, it was focused most conservatively on the morphology of finches, orchids, and evidence of the dramatic blossoming of diversity displayed on the Galapagos islands. It was the extrapolation of what these discoveries might mean that was the actual source of the outrage. But let's not let facts get in the way of a good debate.

Darwin was an accomplished geologist, so the book also mentioned the fossil evidence revealed by coastal erosion. Darwin put a geological date to the phenomena he witnessed in Chile. He estimated that 300 million years of rock strata slid seaward that afternoon. Such antiquity would be impossible in the lingering world of Bishop Ussher[3], who had so studiously dated the planet at less than 6,000 years old — using definitive biblical texts (and dubious maths). Many clergymen sarcastically lamented that this book's publication would keep them from meeting their friend, Mr. Darwin, in Heaven.

And so, he was met with both ridicule and accolades.

Charles Darwin's health was waning by the time stardom found him. His illness made him unable to debate these findings in public. A popular event of the day[4]. In time, and with further research, others would verify his observations repeatedly. Life is an ever-expanding collection of species that alter over time. As conditions change, life adapts, and any species that can't keep up with the change will perish or be transformed. So it is less unexpected that there would've been a row towards this shocking reevaluation of life on Earth in 1860 than that it would continue into the 21st century, and yet it does.

Evolution is simple. Species diversify through the twin processes of mutation and natural selection. This transformation leads to the beautiful variety of life displayed around us. Natural selection is also called survival of the fittest. But that doesn't mean that the strongest conquers the weak. It means that the species that can best capitalize on any given situation will tend to survive and even flourish in that environment. Sharks have been around, changing very little, for millions of years because their physiology and habits continue to succeed. The same goes for many a fragile flatworm. Darwin did not know that the recombination and interpretation of DNA via sexual reproduction were

---

[3] James Ussher was Archbishop of the Church of Ireland from 1625 to 1656 and dated the first day of genesis as: October 3rd, 4004 B.C. He was off by over 4 billion years. That should be awarded the Guiness book world record for rounding errors.

[4] Want to read more about the amazing Charles Darwin? Try "The Tree of Life" by Peter Sis, or "The Beak of the Finch" by Jonathan Wiener.

responsible for both mutation and stability within the genera. And that's just because no one knew about DNA back then.

DNA analysis can answer the question of evolution and answer it, it does. Incremental change, over great time spans (and sometimes not so great), selection by your relative fitness for your environment, and sometimes just dumb luck, leads to the "winners and losers" on the majestic tree of life. That many people can still argue against evolution 150 years post-publication is not only concerning but also dangerous. Today's scientific discoveries are often still assailed by prejudice and ignorance, as we shall soon see. But humanity does not have another century and a half to figure this one out; it may only have a handful of decades.

Evolution is the mechanism of life that explains its ability to diversify and spread over every inch of the planet's surface. Evolution is also the process by which social and academic thought changes gradually, over time (and sometimes quickly), often from simpler ideas to more complex ones. Or sometimes from complicated rationalizations into simple realizations.

Realizations such as: The changing climate and carbon pollution are soon going to end the human race if we don't do something about it.

In 2016, the Earth's atmosphere reached a sustained level of 400ppm [5] of carbon dioxide (CO_2), and it has continued to increase ever since. News media widely reported this fact. Scientists around the world declared a state of emergency. Unfortunately, declaring actual states of emergencies isn't up to scientists; it's up to politicians. Many of whom have so far declined to face the facts. What wasn't well communicated about this 400 ppm figure were all of its ramifications for the continuation of human life on this planet. That's because it takes a bit longer than a 30-second soundbite to explain them. The importance of this little statistic is three-fold:

First: CO_2 is a "greenhouse gas." Carbon dioxide acts analogously to the glass in a greenhouse. It allows heat from the sun to penetrate it and jointly reduces

---

[5] Ppm- Part Per Million: a measurement of concentration within a gas or a solution. And it's just what it seems: one molecule per million molecules.

the re-reflection of that heat back out into space. In a greenhouse, it's a remarkable phenomenon. A thin glass pane can insulate and warm the room beneath it. Unfortunately, it is not so great for a dominant species on a tiny planet orbiting a single sun in the seemingly endless void of space. The increased levels of CO_2 and other greenhouse gasses that our activities have put into the atmosphere with our activities are slowly warming our world and tampering with its life-bestowing systems.

The second point is: That the planet's atmosphere has not held this much carbon dioxide since the Pliocene era — 3.6 million years ago. (Williet, 2019) Before the industrial revolution, concentrations of CO_2 were steady at about 280 ppm. During the ice age, it was as low as 180 ppm. For around 800,000 years, the levels have only fluctuated around 100 ppm above or below the average. (E.S.R.L., 2013) Humankind now sees CO_2 levels from 100 ppm to over 200 ppm higher than our previous mean. So, what's the big deal here? The "deal" is that no humans were around during the Pliocene.
The human race has never before experienced the world now being created.
The earth may have been here before, but humans certainly have not. Our current prejudices tell us we are separate and above this world and its systems, but we most certainly are not. We are inadvertently terraforming[6] this planet, potentially for some life form other than ourselves or our cohabitants. This warming is primarily a threat to our continued existence, although it will probably take many other species with us.

And finally: The mechanisms by which this introduction of gasses occurs are, in fact, the very workings of our modern civilization. To "turn off" the flow of carbon dioxide would be akin to switching off our cherished ways of life. Our lifestyles are addicted to oil as an energy source, building material, and as foods and medicines. Each person in a developed country can expend a thousand times as much energy, day in and day out, as one did 150 years ago. Unfortunately, this expansion of productivity is accompanied by a similar thousand-fold increase in pollution and environmental degradation. The savvy

---

[6] Terraformation: The process of altering the climate and ecology of a heavenly body at a planetary scale. This discipline of study is based on the future potential of making moons, or other planets, habitable for humans. It's ironic that the first planet we may need to alter for our continued existence is our own.

broadcaster, or cautious scientist, knew the same cold calculus the moment the 400 ppm news was released: There's little chance in hell anyone was going to do anything about it. We're caught on our own flypaper.

Humanity is on the verge of extinction. There is no way to sugarcoat this. There are no opinions that can outweigh the facts before us. Our activities have altered the ecosystem to the near point of collapse. Our advanced cultures have leveraged our previous understandings to their absolute limits. What stands between us and our world's healing is not just the atmosphere, but our societies' very mechanisms and our manners of thinking. Humans have become masters of altering the environment for our practical benefit. They are also masters of kicking cans down roads and sweeping things under rugs. However, there are no more roads, and as we are about to see, there never were any suitable carpets. There is no "over there" and no "others" to sacrifice. We all live on one planet and share the same sun, air, land, and water. For those who decry "globalization" and claim that it is evil, it's too late. This is undeniably one world, and all are joined together now by a singular fate. It is time to wake up and face what has been done. That is the first step all must take to in order to fix things before it is too late.

As the globe warms, the systems that support human life are going to change along with it. Humankind grew up in a world with predictable seasons. People knew when to plant and when to reap. The rains would come at predictable times. They could expect the snowfall to arrive in winter and the snowpack to store that water for spring and the summer beyond. Our early cultures revolved around the seasons, as did our songs, ceremonies, and religions. They could count on the world to continue as it always did so that planning for our futures, rather than accepting them, became the order of the day.

Such is no longer the case. As the land and seas grow warmer, the systems that everyone relies upon are being altered. And in most cases, altered radically. The farmer used to have a grasp of what the seasons would hold. This is also no longer the case. We're facing uncertainty because of the utter complexity of the systems always taken for granted. The systems which support life. Many of which we have already disrupted, some of which have almost certainly been shattered.

Humankind now faces a future where many fertile lands will become deserts.

The snow will fall as rain on the mountains I call home, perhaps ending the alpine snowpack's annual accumulation. That snow storage is not just there for the skiing classes. It is needed to quench the thirst of millions of people and crops downstream. Even if the average amount of moisture that falls upon some areas remains the same, the end effect will be quite different if it falls as rain and not snow. Rain runoff comes and goes quickly. There used to be thousands of perennial snowfields, slowly releasing moisture and allowing it to soak into the soil in a measured manner. All life needs those snowbanks to keep the rivers flowing through the long summer growing season. Previously, as the icefields melted, which they did each spring for tens of thousands of years, they doled out their water in a measured fashion. With climate change, this might no longer occur. And glaciers are needed to do the same while also storing centuries of ice in reserve.

The polar ice caps are melting, and the oceans will rise as they fade. Cities will vanish beneath the waves, and millions of people will be displaced (Lonergan, 1998) (Rigaud, 2018). The loss of food production may cause millions more to starve. These disruptions will alter supply chains for all materials and threaten our continued success as a technological race. These disasters could end humanity simply because we have failed so utterly to understand our ecosystem.

Our future looks bleak, yet many continue to act like nothing untoward is happening. Our governments continue to bicker and sometimes war, to continue an illusion of supremacy and control. I speak primarily of my country, The United States of America, once a global leader in many things. Today, about the only thing the United States ranks first in is the per capita release of greenhouse gasses[7]. (Scientists, 2018) But Americans are not alone in this dismissal of reality, and every excuse minted to delay or reject our awakening to reality is another arrow in the negative debater's quiver. It seems all too similar to that old tale of the frog sitting in a slowly boiling pot of water and never noticing its peril until it is too late.

The story goes like this: If you place a frog in a pot of water and slowly increase the heat beneath it, the frog will fail to notice. The change, being too

---

[7] That, and COVID-19 deaths. Sigh.

slow to trigger the frogs' simple mind, will be ignored, and the frog will simply sit there until it boils to death. Fortunately for us, this little piece of "wisdom" is apocryphal[8], meaning it is completely bunk. The frog will figure it out. It will jump away, usually rather rapidly, after reaching what is called its critical thermal maxima[9]. (Seibel, 1970) It would be unethical to do this at home, so please don't try. But this is an excellent example of how concepts that may be taken for granted might simply be useless bits of misinformation. When faced with asserted "facts," we must demand evidence, even if we innately agree with them, especially if we are wont to agree to the point of accepting them without supporting evidence. Without proof, all claims are just more hot air in the wind, and there's enough of that already.

Critical thinking is not an innate characteristic of human consciousness; it needs to be cultivated. A frog, however, lacks our complex minds and our many layers of assumptions and preconceptions. So, our blessed froggy buddy will save itself without a moment's doubt. But we need a little more to get us hopping.

Many rightly look to science to help us as these various life-altering changes in the ecosystem advance, as new pandemics arise, as hurricanes endeavor to wipe our coasts clean. But it is critical to understand what science is and what it is not. Science is not magic and operates under the same assumptions and limits as all current human endeavors. Budgets, and inclusion at decision-making tables, are two notable limits on what science can do for us right now. Other limitations include anti-science biases and humanity's innate ability to pretend our problems away.

Science, in its infancy, suggested that humanity had a near-moral imperative to master nature. I shall call this overarching ethos Dominion. Dominion is a concept that came from our religions and would be familiar to both Darwin's supporters and his detractors. For, indeed, it is a creed that transcends religious sects, its adherents, including many a secularist and atheists as well. So I will capitalize its name to recognize its true primacy as the religion of our times. Its

---

[8] Apocryphal- Questionable, doubtful, spurious, or fictious- take your pick..

[9] Critical Thermal Maxima, the highest temperature that can be tolerated before action must be taken or death/damage will occur.

roots lie in a belief that everything was made for us, is owned by us, and should be exploited by us to all practical ends.

Many still cling to such ideas, where, if a scientific inquiry is not advancing our political goals, easing our overstimulated lives or enriching our chosen few, it is practically useless. People are encouraged to want more cool gadgets, and they don't necessarily want to hear from the people who create them. Scientists, these days, are often just corporate employees, contractually obligated to keep their mouths shut. Some of them may be searching for fame as well. Others may just need to keep their jobs so that their children can continue to attend school and visit the museum. In either case, it is economics that ultimately guides many of their endeavors and not the search for truth. Our thought leaders, those who actually call the shots, are often politicians or clergy. And they seem ready to open their mouths at a moment's notice. However, many of them may not have the simplest inkling of the sciences that study how the ecosystem that supports us all works. Too often, the individuals raised to positions of power have no other qualifications than their looks or charisma. Or perhaps they were born into power and wealth, an advantage for sure, but not a sign of competence by any means. But many of them do seem to share one thing in common, the desire to continue to consume and control an ever-shrinking abundance that is (was) this planet's bountiful resources. The desire to dominate nature is common across many ideologies and religious sects.

Dominion is a powerful driver of human thought and desire. Stalin promised to conquer nature for the good of the proletariat's revolution. Teddy Roosevelt pledged to stop all forest fires in the American West because the American spirit was more potent than the forces of physics. Chairman Mao declared war on nature and proclaimed that all lands should be put under the people's domination or the plow. This following quote, so often attributed to Sir Francis Bacon[10], a founder of scientific thought and natural philosophy, sums up such a dark mandate:

---

[10] **Francis Bacon**, 1st Viscount St Alban, PC QC was an English philosopher, statesman, scientist, jurist, orator, and author. 1561 CE-1626 CE.

"My only earthly wish is… to stretch the deplorably narrow limits of man's dominion over the universe to their promised bounds… [nature will be] bound into service, hounded in her wanderings and put on the rack and tortured for her secrets."

Disturbing stuff indeed. Fortunately, Sir Francis Bacon never actually said that. That quote is also apocryphal, meaning that while regularly stated, its actual provenance (origins) are doubtful, if not just purely invented out of thin air. In the common parlance, it would be called: bullcrap. But much like the amphibian in the saucepan story, one may accept such baseless assertions as fact because they may tend to agree with them. They validate presuppositions perhaps already incorporated into our standard operating principles. And as you are about to discover, many of our problems lie in our mindsets and beliefs, often created by misunderstandings and fictional tales about humanity and its unparalleled awesomeness.

What the first Viscount of Saint Albans, Sir Francis, actually believed and fervently stated was:

"We cannot command Nature — except by obeying her."

This would be quite the opposite of his previously assumed anti-nature position. So let's suppose that his position on the matter was complicated. But such antagonistic doctrines still rattle through the hearts of many a "master of the universe", and it gnaws at the rest of us because of our training. Early scientists may well have set out to command nature. And in many ways, they seem to have succeeded. Many self-proclaimed and self-serving scientists have proceeded intending to conquer nature, assuming that was the point of scientific inquiry. Some would see such as the philosophy of science, this goal of total human mastery of all aspects of existence. And in the developed world, humanity has certainly benefitted from the unbridled harvest of nature's resources. Advanced societies have altered the course of mighty rivers and bent the land to our desires, even affecting the planet's tilt on its axis. So when considered superficially, it might seem that humans have mastered nature.

But science is *not* a philosophy; science is not a goal or even a particularly potent driver of human ambition. Human thoughts, dreams, and desires are entirely separate from science, and they alone are responsible for the world in which all of us live. If there is a desire for more, ever faster, and with less effort, that seems the primary driver of human pursuit at the beginning of the 21st century. Then such a design would be a contraption created and debated by those who seemingly can't see their fingerprints on a collection of ideas entirely divorced from the actual discipline of science. Science, you see, is not a philosophy; it is a method of inquiry.

Our current state of statecraft, where so much opinion is paraded about to compete with facts, is based on a simple misunderstanding. Many believe that science is just another philosophy, like idealism or nihilism. And that therefore it can be debated like a philosophical construct, with assertions and logical acrobatics. But it is not, for one straightforward reason. Science begins with the premise that nothing is known for certain and then seeks to gather the evidence required to remedy that state. Science does not dictate facts; it shines ever brighter lights on them. Science is not a search for meaning; it is the pursuit of knowledge for its own sake. It is agnostic about data and will reject a hypothesis more quickly than accepting one. In this way, it has been extremely successful; because science has no underlying dogma. Pure science is beholden to no one or thing.

Science is not a philosophy; it is a toolbox. Science is a method of inquiry and nothing more. Science is a process for answering questions about the natural world. Those questions always have been and always will be ours. Science is a framework that allows us to isolate a problem and attempt to answer it. That answer must be repeatable and quantifiable. The simpler the question, the more universal that answer might be (and confoundingly more challenging to answer as well). Over time, systematic frameworks have been established by accumulating test results. These amassed answers may then streamline future investigations based on those past results. But such frameworks can be wrong, and good scientists are always aware of this. Science is self-correcting through repeated experiments. And with better techniques or instruments, our discoveries' accuracy becomes greater—our ability to see where past errors arose increases as well.

Science is so powerful because it acknowledges that *we don't know,* a near super-power as you shall see. Philosophies seek (or claim) to know truths, making their value questionable when designing a more efficient circuit or adapting to a shifting climate. Science is a discipline and a tool. And just like any tool, how it is used is up to the hand which wields it. Michelangelo could arguably use a chisel better than anyone else, yet it was still just a chisel, and others could, and did, use chisels. Anyone can use a chisel with some level of effect. But it was his perspective and skill that made the difference, and the same is true of science. Some will discover fundamental truths about the universe, and some will come up with cheaper fragrances for soap.

Scientists come in all stripes. There are great ones and average ones, and then there are hacks, just as in all other aspects of life. Einstein expanded our world view because he looked at the universe uniquely, not because he had special tools. He worked with mathematics and worked in physics, his toolset, and toolbox. He changed the way the world viewed itself because he could change the way he looked at it. But his science, his one question at a time, one answer revealed, used the same frameworks as all other scientists. And it gave us a better understanding of the universe, and our place within it. However, it did not rip the shroud of creation asunder and lay the essence of eternity before us. He only deepened the mysteries of existence. And that's how science, as a discipline, works. Each new answer leads to even more questions.

Some "scientists", such as those who work for the multi-national corporations that deny climate change are the hacks, beholden to the status quo, using their chisel to cut down and not to create. Perhaps they use scientific studies to claim this pesticide is totally safe. Or they maybe they use their titles to sell snake oil on television. But in essence, they are just using the chisel to cut off a chunk of the pie for themselves and their masters, seeking wealth and power and not strictly utility or knowledge. If Michelangelo was the master, then these are, perhaps, just chiselers. Pure scientific research and the profit incentive are rarely, if ever, in complete alignment. One need only remember the prolonged debate on the health benefits of cigarettes or the safety of DDT to realize that many a corporate MO[11] is to deceive, obfuscate, and misrepresent for yet another quarter of profit. While one might argue that this is customary and legal because it is in many cases. It is indeed unethical. Such

---

[11] Modus Operandi: basically; method of operation or business as usual.

behavior is also unnecessary and is a product of culture. It is not a feature of science or even of commerce, per se.

One of the primary and most demanding requirements of science is the removal of bias. You can't ask a question and look for evidence to support your preferred answer and only the evidence supporting it. Instead, you must begin each journey with the acceptance that you might be wrong — making science a tough bunch of balls to juggle. And that is precisely why our culture should support those individuals who can work in this manner. It requires almost Zen-like selflessness and rejection of personal biases to dedicate yourself to a lifetime of study that may end up proving itself incorrect or generate only a footnote by one's life's end. Of course, there is value in this, but it contradicts our social training to seek fame and glory.

A Buddhist might aspire to achieve a state of emptiness and a release from attachment. The discipline of science demands it. Zen Buddhists use koans as a method of self-examination. These mind-bending little riddles have answers meant to broaden the worldview of those who solve them. The answers are only clear once you find your biases and eliminate them. Remove the drives of the ego if you want to play this game. A common and well-known example is: "What is the sound of one hand clapping?"[12]
The emptiness of desire is the desire of pure researchers. They perform the balancing act of keeping an open mind and tempering it with skepticism. They seek only knowledge for knowledge's sake. Such pursuits are where our most remarkable discoveries have always come. It is then up to us to use our new understandings justly. And while this may sound like a philosophy, it is not. It is the complete opposite of philosophy, to be exact. So, if science is indeed an empty vessel, divorced from our social ideologies, the bride stripped bare, what good is it?

It is a tool that can now save us. And the only such tool we, the tool-using animal, have that might do for the job ahead. Science, properly applied, can save us. The critical point is that we'll have to wield it properly. We'll have to

---

[12] The answer to this Zen koan: "What is the sound of one hand clapping?" is: Nothing. You can't clap with one hand. It takes two to tango, and it's going to take everyone together to save the planet.

use it together with an aligned intent to save our environment and thereby save ourselves. We'll need to hold that chisel like a Michelangelo and cut away the dross that occludes our new future.

So if science is just a bunch of toolkits, what would be the proper use of them? Who or what can show us how to use such a power justly and with proper humility?

Nature.

Just like Sir Francis Bacon told us. "We cannot command Nature *except by obeying her.*"

Earlier, I pointed out that even if humans have never experienced a world such as this, nature has. Life has continued for billions of years; it has withstood extinction-level events and rebounded each time. Nature is the interconnected web of the living, and on it grows the tree of life. All humanity is just one leaf on a single branch of that tree. The forces of evolution are the processes of adaptation and the diversification of this life. Until now, these processes have acted at time scales beyond our comprehension. They have also acted indifferently, which has greatly benefited the human race. But now, the human world finds that our time in the spotlight is almost up. The human race is currently smack dab in the middle of a potentially final equation, and nature will solve for X unless we do it first.

There is scant time to act, and nature itself does not much care if humankind survives or not. It only cares if life survives. So here is the essential lack of bias needed to perform science in its right and proper manner. I will call this the will of no will.

In this current situation, our societies will need to find a course that leads to the survival of as many life forms as possible. We must save as many of the other lives that cohabit this world to secure our continued existence. The ecosphere is not a monoculture, and it cannot be made it into one. The ecosystem is a group function. All lives are related, and therefore must be saved together for human life to survive, or the default reset button will be pressed, and the process will start over.

Any particular biosphere is a group endeavor; all the species are intertwined and necessary for one another. They can't be unwound without unraveling the entire structure. Every time there is extinction, every time a dominant species or group is lost, the entire system changes radically.

Humankind won't, can't, continue if we allow the sacrifice our "lesser" fellows in our attempts to survive. Without affirmative action to save our contemporaries, the planet's flora, fauna, and funga, it will be us that the forces of evolution will select for removal. Life is not a Gordian knot[13] that some simpleton can just hack in two to dismantle. And you can't disentangle the complexity of an ecosystem to save only the apex predator. The choice is ours and ours alone. It'd sure be nice if together we got to continue along on this glorious road trip called life. And I believe we can. I want to show you how that can happen.

The mistake of the Green Revolution is a mistake people keep making over and over again. Humans are so awesome, so clever and resilient that we can psychologically project ourselves onto everything around us. Many mistake our motives for the motives of life itself. It can easily be believed that the ecosystem was created solely for us, and survives at our pleasure, when the reverse is true. This ecosystem created us, and we survive at its pleasure. There are few virtues our more selfish pursuits can add to the global web of life and its ways. There is, however, much that our understanding and compassion could contribute if humans could just pause our incessant pursuit of wealth and begin exercising them. Our world's societies must learn to look upon this planet with a lover's eyes, instead of with the lustful gaze of a conqueror.

This world made us to thrive within it. It did not make us to survive in the world our various obsessions are creating. Humans have effectively altered this planet, and the world that our practices is making is not one for us. It is a world birthed from our misunderstandings and socially inculcated wants and desires. Socially speaking, many don't always seem to know the difference between

---

[13] In another apocryphal tale: Alexander the Great becomes king of Asia by being the only person to untie the incredibly complex "Gordean Knot" He does so by slicing it in half with his sword. A supposed triumph of thinking outside of the box. Or perhaps, just another example of juvenile vandalism. The knot was not untied, it was destroyed. That wasn't the test.

their wants and needs. There are billions of us now, and all need clean water, fresh food, and room to stretch our arms. But some seem to want a world that will take whatever abuse is required so that they can have more toys, waste more, and feel self-justified while doing so. Many do this because they have been trained not to recognize any other benefit to living. Many would treat the world like an abusive husband treats his family, as props for his self-satisfaction and whipping posts for his disappointments.

People have tried to alter the laws of the environment for their convenience. Somehow, many have convinced themselves that there is an inexhaustible natural ability to adapt to our preoccupations built into this world. Perhaps it is felt that this life is just a game played upon an eternal and indestructible chessboard.

So while it may be believed that society was reacting to our needs, it was really just our wants that drove the petro-agricultural revolution.

The ecosystem is just that: a system — wheels within wheels within wheels, everything attached to the next by webs of interaction and interdependence. We'll examine these interrelations in depth in the sections of this book. For now, let's stay with the broader view of our predicament. Individually, most of us understand that the environment is essential. But as a culture, many of us don't know enough about how it works to form an opinion. Humanity has come a long way from those heady days of the conquest of plant biology that was the Green Revolution. But there's still quite a way to go.

The Green Revolution, it turns out, was akin to the subjugation of life for an expedient (and equally short-lived) benefit. It was the child of the Dominion theology and was based on exclusion. Exclude variety, remove pests with force, weed with chemicals that work by killing the target plant only slightly faster than the desired crop. Simplify, extract, and de-nature was the underlying ethos. How was it anyone thought they could rip chunks of existence apart and dispose of others without consequence? What hubris then possessed us? This disdain for the very forces that gave us life is an ancient human misconception. It resulted from centuries of training and now resides deeply in our consciousnesses. We'll need to remain aware of this tendency to scorn life because sweeping the awareness of our mistakes under the rug will only ensure that they will reemerge at some point, and stab us in the back.

I propose a Green Evolution to replace the Green Revolution. I will outline and suggest how our civilization might successfully use nature's frameworks to restore the balance that has been so disrupted. This book will explain not only how people fractured the cycles that had supported us for so long, but how those same systems might be used to restore balance once more. Technological systems can be altered to mimic natural systems, this is a must do task. Ideas such as biomimicry, bionics, and circular economies are gaining in acceptance, embrace them. Let's enact these schemes and do it quickly.

Life has been successful for trillions of generations because of some relatively simple qualities. All life, materials, and energy recycle themselves within the planetary construct. Birth, growth, and death are all necessary. Everything any life form takes must be returned. All actions must balance in the end.

Everything has a place, even if that place is transitory and fraught with peril. There are wonders and awe in every little action of life, and there is suffering and finally release as well. It makes a heck of a lot more sense to accept existence as it is than to try and force it to bend to our collective fantasies. That is the way of metaphysics, and it's an invitation to self-deception and failure. There is a path that has shown the ability to continue and thrive for countless eons. The other course, the one our world is currently on, is not looking very promising after only a handful of centuries. The timescales of ecological adaptation usually are far greater than our lifespans, greater even than our empires and dynasties. But this healing needs to be accomplished as rapidly as possible, so humans need to assist in adaptation and steward evolution. How can evolution help us back to the path of survival? Because we are farmers, and can help life itself adapt. Because we are shepherds, we can restore the ecosystems we have messed up so effectively. Because we can be stewards, there is hope this need we have created can be met.

A few billion trees can alter the atmosphere rather rapidly. A billion people, and their bodily waste, can rejuvenate millions of hectares of soil in relatively short order. You can take the organic refuse thrown away because it has so little economic value and compost it to restore our depleted fields. Human effort can convert wasteland back into forests and harvest our remaining forests sustainably to make for healthier and more resilient landscapes. Simple changes in personal consumption, amplified by all of us, can stop the waste that is a hallmark of our civilizations at this time and place. Together, people can end the toxification of our world. Doing so will make us all healthier and more resilient as well.

All life is one in a very meaningful way. Everything that "lives" on this planet has Deoxyribonucleic Acid (DNA) at its core[14]. DNA is the self-replicating molecule that stores the information and the order that is us, that fly buzzing around, the Blue Whale — everyone. The information stored in its bonds traces us to the first cell that rose from primeval Earth's organic soup. It is akin to computer memory in that data is stored in simple "bits." And if looked at as stored memory, then within us all is a history of the life that walked, slithered, or swam upon this planet. Included are the forms that existed when our world was like the one that is being haphazardly created. And not just their anatomies, but also the strategies those life forms used to survive.

Life is like a perpetual motion machine. Not that it will continue to run with no new input (energy), but in that, it will abide, adapt, and advance with no end in sight. It will seek and process its own energy. It will find the safest path in the face of uncertainty and seek shelter during calamity. Life has proven very difficult to exterminate, even though particular forms, such as dinosaurs, come and go. Life will march on, and that's a very comforting thought for those of us who are alive and wish to stay that way. We'll need that mojo to get us through this time of danger. Fortunately for us, the spirit of life is in every cell of our bodies; let's use it, shall we?

Life will survive, but it won't care if humanity fails. Life doesn't play favorites. Life is not like a Babylonian god that chooses one over the other and alters the course of fate for their benefit. The course of destiny is simple: it's just cause and effect. Our actions have changed the ecosystem's trajectory, unknowingly for sure, but there are no take-backsies and no supernatural intervention on the horizon. Our fate is dire. It will be a labor to save ourselves and a semblance of the world we know and love.

Life hedges its bets by having millions of species, many of which occupy more or less the same ecological niches in many differing climes and locales. This

---

[14] Okay so there's some lifeforms that use RNA as their main data store, but it's a pretty valid generalization, as generalizations go.

variety is insurance against cataclysmic events like meteorites or a species that might become too successful and lack basic self-control.

The diversity of species is more than an insurance policy, though; it is a strategy for success. Different beings have various jobs to do to advance the whole of life. The funga can do something nothing else can. They can digest rock, releasing minerals for other "higher" life forms. Plants can do something special as well. They can harness the sun's energy with photosynthesis and store it in chemical bonds (sugars) that can be used to power other life forms and themselves. Animals move around and can quickly colonize any available space or ecosystem. They return all the exotic chemistry they have accumulated to the fungus, bacteria, and plants that consume them when they die. Life shows us that diversity is strength, and recycling is renewal. Why is that so hard for us to understand?

I recommend replacing the Green Revolution with a Green Evolution. This evolution will use nature's inherent properties to recycle and restore what has been considered waste and treated like garbage. This evolution will also be a social one. Discarding the pretentious self-importance used to set ourselves apart from the world and each other. It requires a reexamination of who and what has been considered cultural garbage as well, for everyone has a place. Humans are children of this planet, and are all related to each other in the most fundamental ways.

Let's create an evolution of how we interact with one another and how (and who) our human systems serve. It will be an evolution in how we view ourselves and where we fit into the world. By definition, everyone fits, for they are of this world, and everyone's genuine needs cannot be anything but aligned with its rhythms. This little planet gave us everything. It's time to start treating it that way. Not above the world and its life forms. We are, in fact, subjects of the very ecosystem that gave us birth. And perhaps not subjects, but would-be partners in the continuation of this pyramid scheme we call life.

I implore discarding the Green Revolution for a Green Evolution. The most radical tenet of this proposition is that it is based on facts. With ourselves, our neighbors, and our planet, lies the one path that can lead us out of this dark nightmare of our own creation. Truth is what is real, not what one wishes were real. And this means it's time to give up many fables told for centuries. Stories that made us feel superior but also disconnected from everything else. What I'm suggesting is not a simple path; it's just the only viable one left. And that's a good thing because if there were another possibility, any other option, many would certainly take it if only to avoid facing the consequences of our actions.

Nature shows us the way forward. Choose to follow for a time. Let go of the world's metaphorical steering wheel and take a back seat to learn its processes. Humankind can work with nature to heal the damage done. There's no longer an option of continuing to pretend to master mother nature. There is simply no time left.

"We cannot command Nature — *except by obeying her.*" -Sir Francis Bacon.

Now it the time to get over our instilled sense of self-importance and relinquish the illusion that people are in control.

It is time to stop pretending humankind holds the steering wheel of creation and realize we are essentially just hitchhikers. Hitchhikers who can be booted off if we can't behave.

CHAPTER TWO: THE STATUS QUO

"The question is whether any civilization can wage relentless war on life without destroying itself, and without losing the right to be called civilized."
—Rachel Carson

Life exhibits interconnection in all of its ways and means. A simple yet profound example of this is the pelagic salmon. Many species have a very intriguing life cycle. They inhabit the upper layers of the open seas, traveling great distances and visiting many shores. But they need to breed and brood their young in freshwater streams, the streams where they themselves hatched. In what is called an anadromous migration, they will cross the oceans to find that very river or stream. They then swim upstream, en masse, to complete their ultimate task, that of reproducing. In a mating frenzy, they dig nests and spawn in the gravel of their hometowns. Bears, martins, and humans all greet them and harvest this bounty of giant fishes crowded into tiny streams. Once they have mated, the salmon die in an exhausted afterglow, returning their flesh to the currents and providing nutrients to the benthic insects and other fishes in what seems to us a morbid buffet. It doesn't stop with the waters, however. The bears take the fish onshore and eat only the fatty parts, strewing the protein they don't want about the woodland floor. This fish-emulsion fertilizer then feeds the grasses, forbs (flowers), and trees that make up that forest floor. Many Native American tribes would greet the annual salmon run with feasts and celebrations fitting for a visiting dignitary.

This noble fish's nutrients, spread far and wide, act as nourishment for thousands of creatures. Having collected and concentrated the bodies of other animals in the same way, they've become the backbone of a global web of sharing. A Korean shrimp might have its flesh redistributed to Alaska in this manner. Mayflys from the Nome River shoreline might make up the smolt that migrates across the sea and feeds a killer whale off Kamchatka. Life across the globe shares its flesh and energy through the many reciprocal processes of the food chain. And an interruption of even one small cycle can take the whole thing down.

One such disruption is our habit of building dams. Our desire to control the water system has broken the chain of salmon migration. If the salmon can't get home, everything else that depends on them will suffer, perhaps terminally.

Wildlife managers choose to deal with this problem by farming the salmon, giving them a population boost when they can't make it back home. They produce billions of hatchlings each year to "support" the native population and hopefully negate this fragmentation of their life cycle. This idea that humans can fix everything by imposing our version order is how our status quo currently behaves. And being based on the assumption that humanity is separate and above their cycle, above and separate from nature, it predictably fails. Our manner seems to be to break natural systems and then apply a band-aid. And it doesn't work. The more hatchlings released, the fewer salmon return. Were they not a desirable table fish, it might not even be given this half-hearted try.

This failure makes perfect sense. The problem is not with the salmon's reproduction. The problem is the rivers have been dammed. Our sensibility attempted to replace a wild animal with unadapted critters produced on farms[15]. And our concept of farming has gone so far off the rails, as you shall soon see, that our attempts at husbandry are turning pure gold into straw — a reverse Rumpelstiltskin effect. People repeat these simple errors because many have convinced themselves that it cannot be humanity who is mistaken. It must be the fish or the environment that is at fault. They must be the ones defeating our earnest attempts to offer aid. Our current views of who we are and how we relate to our ecosystems have led us to a cascade of tragic mistakes, perhaps the greatest of which now manifests as climate change.

Climate change is caused by global warming, which is in turn caused by the increasing amount of greenhouse gasses released into the atmosphere by human activity. I won't mince words to appear polite here. "The Greenhouse Effect" was first suggested in the 1820s and was given its name by Nils Gustaf Ekholm[16] in 1901, 14 years before Einstein's "General Theory of Relativity." They taught it in High School when I attended in the 1970s. It was considered a curiosity and not controversial in the least. The effects of it being so far off in the distant future as to warrant little more than a curious eyebrow lift. It was just automatically assumed someone would find a solution before it became a

---

[15] You can watch: "Artifishal" the Patagonia documentary on Salmon, executive produced by Yvon Chouinard, for their take on these practices.

[16] A Swedish meteorologist, who even in the 1890's knew what we were up to.

serious problem. The twenty-first century being seen as a remote and mythical time by general society back then as well.

Now, neither global warming nor climate change are taught in many American high schools, a scant half a century later. American politicians have called it a "hoax." And a majority of Christian evangelicals believe that it either doesn't exist or has nothing to do with humans. (Pew, 2015) As many corporations continue to obscure the data while simultaneously adapting their business models, it seems that complete acceptance of this issue is awaiting the time it becomes profitable.

The greenhouse effect is simple. Our atmosphere is a fragile shell of gasses surrounding the planet's surface. This mixture of nitrogen, oxygen, and trace gasses makes life possible. From the ground, the atmosphere seems enormous, voluminous, nearly infinite. But from space, the atmosphere is a frail, shimmering veil, slender as a fingernail, the original "thin blue line." This shell protects us from radiation and insulates us from the cold, dark vacuum of outer space. When radiation, much in the visible spectrum, from the Sun passes through the atmosphere, it travels to the planet's surface, warming it. Heat is then re-radiated from the ground as infra-red radiation (IR). But instead of traveling back into the dissipation of the cosmos, this re-radiation hits greenhouse gasses in the upper atmosphere, where it's reflected back towards the ground, as if by the partial mirroring of a pane of glass. This transmission/reflection property allows the world to warm up and maintain heat. The average temperature of the Earth's surface is 14 ° C (about 57° F). The temperature of space is -270° C (-455° F). This effect yields an average planetary warm-up of 266°C (398° F), an impressive amount of insulation for a membrane that is only 100 km (62 mi) thick and practically transparent. Without this functioning system, without the greenhouse effect, there would be no recognizable life on Earth.

Certain gasses contribute more to our insulation, and they are broadly called greenhouse gasses. Methane, nitrous oxide, and fluoridated gasses are all greenhouse gasses. However, the main greenhouse gas of concern is carbon dioxide (CO_2). Carbon dioxide results from the chemical combustion of carbon-based fuels. It is also the result of the metabolism of animals and fungi that inhale oxygen and exhale CO_2, for the two processes are analogous. This is essential to ponder and will come up again and again. Life can't stop breathing without dire consequence, so we must reduce combustion to curtail our carbon dioxide release.

Since the Nineteenth Century, when Ekholm first pondered the greenhouse effect, our species industrial revolutions have poured gigatons[17] of carbon dioxide into our tiny, albeit seemingly infinite, atmosphere. CO_2 has increased our atmosphere's insulative factor so much that the planet's surface is now heating faster and retaining that heat longer, thus creating the global warming that is so concerning to us. It's happening while I write these words, just as it is while you are reading them. It takes no breaks and is relentless in the incremental heating that speeds up chemical reactions and is drying out the planet's surface with increasing ill effects.

The planet's atmosphere has changed quite a bit in the billions of years it's been around. One could say it has evolved. Over the eons, the mixture of gasses has shifted. Our most recent combination of gasses is responsible for the moderate temperatures and climates all enjoy. Growing up together, so to speak, the flora and fauna of the planet's surface reflect the composition of the vapors around us and their degree of insulation from space.

This is the Holocene epoch, a tiny sliver of time that began a mere 12,000 years ago. This age is our species "Goldilocks" period, when the climate is neither too hot nor too cool. The ratios that compose our atmosphere's gaseous mixture are "just right," and that determines what life processes may occur and at what rates.

The metabolisms of life, in turn, alter the blend of gasses by the simple actions of respiration and decay. Every life breathes, everything dies, and these individually tiny actions add up into globe-altering effects. Life is a delicate balance, and changes once occurred over very long durations, much longer than a few thousand years. Stable change typically requires more than the couple of centuries human's have been jacking up our atmosphere. Slow adaptation allows all parts of the system of life to adapt and adjust with relative ease and minimal disruption. Our changes to the air have happened in a virtual blink of an eye in terms of these systems' previous time tables.

---

[17] One billion tonnes, in the metric system accounting (or as we calculate it in the U.S.A.: 1.1 billion tons)

American (and Multi-National) Petrochemical producers have known about this process since at least 1977 (Hall, 2015). I became aware of it around 1974, so if a product of the Albany County, Wyoming School District was aware of something that Exxon was not, that would be troubling for a whole different set of reasons. And now you know that it has been in the scientific lexicon since 1901, so let's assume they knew about it even sooner than 1977. But, as long as the vast majority of the public was unaware, nothing needed to be done from a status quo perspective. Once the masses began waking up to it, though, it was time to roll out a strategy. Facing the problem and fixing it was seemingly never in the cards. So, they concocted an active political messaging campaign to deny what they knew to be facts. The petrochemical industries chose to kick the climate change can down the road, proverbially speaking. As the understanding of these dangers began growing, new strategies were continually produced, tested, and implemented amongst a population previously kept in intentional ignorance. But, again, fixing the problem was not one of them. Now, many scientists fear conditions have deviated too far down the path of climate disruption to avoid catastrophe.

Some of the same voices responsible for stalling action currently suggest everyone may just have to "deal" with it. In much the same way, certain pundits suggested everybody just accept COVID-19 and pretend it's not a big of a deal. Such tactics come from a cold, cynical calculation made by some in power who think social disruption or the end of the world might work out to their benefit. Maybe they think they have enough money and power to insulate themselves from the consequences of their actions. Or perhaps they are old enough that they believe their time will have passed before this climate catastrophe hits the spinning fan blades of history. Whatever their reasoning, the future demands that we stop them. Maybe they won't have to deal with it, but we will, and so will our children. As will the plants and animals that cohabit this green world with us. All living species on the planet are mutualistic, to some extent or another. Meaning everything is inextricably interconnected and dependent on each other. With each loss of a tiny piece of the ecosystem, everything else comes closer to crossing the line of no return, the event horizon of global ecological collapse. Yet, as sentient beings, humans can still choose to do nothing, or possibly worse, choose to do the wrong things. Here is the situation society finds ourselves in by blindly following our status quo.

Hurricane Sandy laid waste to the Eastern seaboard in 2012, a Category 3 storm with a peak force that hit New Jersey on October 29th, with sustained winds of 185 Km/hr (115 mph). The storm surge flooded subways in New York City and killed over 200 people. It was perhaps the first storm that has been linked directly to global warming. (Lin, 2012) Sandy began as a tropical storm on October 22nd in the western Caribbean. It moved across Jamaica, over Cuba, and ran north up the US coast. Usually, such a storm would lose strength over the North Atlantic because the chilly waters beneath it would drain its energy. But sea temperatures were abnormally high in 2012, over 29° C (85° F), and this energy allowed the storm to rebuild and gain strength before landfall within the United States. In 2012, storm surges were higher than their historic averages because of rising sea levels due to icecap melt-off. John Boon of the Virginia Institute of Marine Science had noticed that sea levels were rising more rapidly along the Northern Atlantic since 1986 than points south of Norfolk, Va. (Boon, 2012). While it may be thought that the oceans would level off like a bathtub, it doesn't work that way on a global scale[18]. Even more confounding, where Sandy was concerned, was a "blocking" event where a high-pressure system over Greenland pinned this storm along the eastern seaboard, allowing it little chance to dissipate. Blocking events are becoming more common and are associated with several catastrophic weather events this century. These mechanisms, peaking together, gave us this initial taste of what the future of a rampantly warming world portends.

On October 29th, the needles of countless barometers dove off their scales as barometric pressure hit a low of 940 millibars (27.26" mercury). This was Hurricane Sandy's calling card as her initial force pummeled New York City. They measured the storm surge at Battery Park at 4.3 m (14 ft) as the Hudson River topped all barriers; flooding the streets and drowning the subway. This one flood trapped over 20,000 people in their homes. It took out the power for 8.5 million people. Her highest wave measured in New York Harbor was 10 m

---

[18] This is because of the planetary spin and seawater bulges created by the Coriolis effect, as well as molecular variabilities due to changes in salinity and water chemistry. Tiny differences build into measurable, sometimes enormous, effects.

(32.5 ft). No one would attempt to ride that wave. The city would soak under 88 mm (3.42") of rain over the next 24 hours. When the storms' focus hit New Jersey, tidal surges erased the boardwalks at Seaside harbor and Belmar. Atlantic City lost over half of its seaside walk. Hundreds of homes were destroyed, with 110 burned to the ground at Breezy Point, Queens. In the end, the damages totaled over $32 billion. And the vestiges of the storm continued, delivering 1 m (3 ft) of snow as far away as West Virginia. Fatalities over the entire footprint of the storm are now tallied to be 285 souls.

The phenomena that created hurricane Sandy are now gaining predominance globally. The warmer air associated with the greenhouse effect allows the atmosphere to hold more moisture. Relative humidity (Rh) is a measure of how much moisture air can hold and varies directly with air temperature. Air at 30°C (86°F) can hold more water than air at 15°C (59°F). So today's storms are carrying more water. One prediction of climate models is that storms may become fewer but greater in strength. Thus, a drought/drench cycle is expected with fewer torrential storms instead of a greater number of less violent weather events. Potentially, the dry areas of the planet may become drier and wet regions, wetter.

Many recent hurricanes are what are called "Superstorms." And this type of cyclone will undoubtedly continue to increase in frequency until the world gets our act together. These storms point to the great downfall of climate change deniers. These tempests are not going away, and no one will be able to pretend they aren't happening. Although it seems many will try for as long as possible, especially if there's money to be made from it.

Moving inland from the hurricane ravaged coast, you will encounter the opposite of drenching storms; our next catastrophe is the occurrence of drought. The National Oceanic and Atmospheric Administration (NOAA) has a drought tracker, which documents and predicts these increasingly devastating episodes. Droughts have accelerated in both strength and frequency since the beginning of this century. The droughts in California, which is a historically dry State, have fostered a catastrophic uptick in wildland fires. These fires have only grown more enormous and devastating as the years have advanced. The havoc wreaked by global dehydration will continue, as will its denial. During the 2016 US Presidential campaign, long-shot hopeful Donald Trump

declared that there was no drought in California during a debate. (Solis, 2016) Pitifully, a certain sector of society cheered despite knowing that those words were hollow lies. Hopefully, this particularly delusional form of reality rejection, for its own sake, will meet its extinction before the rest of us do. Weather events are apt to become a yo-yo of extremes as we careen towards the 22nd century. In 2019, the opposite of drought occurred as rains drenched the Midwest, washing away crops and their fertilizers. These cycles show the disastrous potential of cyclical crop failures, as either too much rain or too little can both lead to catastrophe. Worldwide, scientists predict cereal crop production to decrease by up to 13% in the coming decades (Wang, 2020). In America's grain belt, over-irrigation has drained the Ogallala aquifer by almost 30% of its volume since 1950. Withdrawals from this "water bank" are speeding up, and drought only increases the draw on this precious resource. As also does drench, when the water runs off faster than it can soak into the parched ground. Without continued and moderate moisture events, the actions of irrigators are all drawing and no filling. Over 2 million people rely on this water reserve across the Midwest.
Once drained, this aquifer will take 5,000—6,000 years to refill.

When Europeans first laid eyes on the Rocky Mountains, the average density of the forests was about 100- 250 trees per hectare (40-100 trees/acre). This can be confirmed by looking at an Alfred Bierstadt or George Catlin painting from the pre-migration period and comparing them to what is seen today. With the "urbanization" and "agriculturalization" of our mountains has come a public policy of fire suppression. Smokey the Bear has helped re-structure our forests to a density up to ten times greater than the historical average. Since fire has always been a part of the eco-system of montane and sub-alpine forests this overburden of vegetation, multiplied by the periods of extreme drought, yields a product of extreme fires.

In 2002, the Hayman fire in central Colorado burned 55,450 ha. (138,000 acres; 214 sq. miles). It was considered an uncharacteristically massive fire at the time. But by 2015, there were two fires twice that size in Washington State[19]. In 2017, in British Columbia, wildfires consumed 1,216,053 ha

[19] The Northstar Fire (83,205 Ha; 205,605 ac) and the Okanogan Complex (122,305 Ha; 302,224 ac)

(3,004,900 acres; 4,695 sq. miles). In 2018, the Camp Fire obliterated the town of Paradise in California, destroying 18,000 structures and killing 85 residents. In 2020, over 1.2 million hectares (3 million acres) on the west coast of the US burned. Also, in 2020, the August Fire broke a record when it reached 405,000 hectares (1 Million acres), creating a new term: the gigafire. This is our new normal, the present state of things: our status quo.

Within our forests, the tree species and fauna have adapted to a set of conditions that have changed little since the last ice age. Things such as average annual high temperatures and average lows, the soil configuration, and even the air's gaseous composition make up the typical conditions for these life forms to thrive. These aspects combined are called the: "expected (or standard) range of conditions" by ecologists; since there are slight natural variations to be expected. An example of a radical deviation from this standard range is the likelihood that the mountain treeline will rise. There is a point of elevation in the alpine environment above which the conditions are too harsh for timber to survive. This creates what is called the treeline. You can tell your elevation pretty accurately by looking at the tree species next to you as you hike up the mountains (or the lack thereof). The high Alpine regions are the domains of boulder, shrub, and summers that may only last a few weeks each year.

Standing on the great plains, cottonwoods line our rivers and deciduous trees intersperse with the native grasses. Moving up into the Rockies, you'll encounter aspen, Douglas fir, the limber, and ponderosa pines. Each occupies its preferred zone and hosts a unique panoply of life. As you reach the subalpine zone, great forests of lodgepole and white fir may confront you. Above that are the realms of the bristlecone and white-barked pines. Global warming may kill these slow-growing trees as warming climes invade their territory and allow other species to out-compete. Gone with them will be many other native species, including the adorable rabbit-like Pika. Drought in combination with the increasing warmth may well bring this change rapidly, as habitat-altering wildfire. Typically, the nights are very cool in the high elevations year-round. This chill has a suppressive effect on any fires that could occur in these alpine environments.

Fire managers keep track of fire frequency in a geographic calculation called: The Fire Return Interval (FRI). At these alpine elevations, that number is often

5,000 to 10,000 years. Practically speaking, it seldom to never burns up there. That is going to change. Fire needs three things to propagate and sustain itself: oxygen, fuel, and heat. Remove any of these sides of this "fire triangle," and the fire stops. Thus goes the first lesson in wildland fire training. Unfortunately, our activities are playing with these elements willy-nilly in forests worldwide, to disastrous ends. Fires are now occurring in the Arctic Circle. Fires are happening across Asian forests and within the Amazon basin.

Slight changes in conditions will have potentially dire outcomes for most forms of life on the planet. Humankind is now entering the Seventh Great Extinction. The Cretaceous-Tertiary extinction was the sixth great extinction[20], and it occurred 65 million years ago. Today, the Holocene extinction looms before us. It's also called the Anthropocene extinction, named after its cause: humans. While this extinction is not caused solely by global warming, it is the exclusive result of our activities. It is essential to accept this and the responsibility it confers. If we deny the cause of this crisis, there will be little chance of averting it.

Pesticides, slash and burn agriculture, pollution of waters, both fresh and sea, are just a few of the imbalances our ways of life have wrought. Global warming, however, is the elephant in our room, and fixing that may very well help resolve many of our other problems. They are all interconnected.

The same deviations from the expected range of conditions wreaking havoc on the trees lay the same waste on our animal friends. Rainforest amphibians, such as the colorful aposematic frogs[21], are the first to go. They are called indicator species, and the indications are dire. Our schemes are changing the planet too fast for our neighbors to keep up. Bees are the vital pollinators of our crops, and they are dropping like, well, flies. There is a combination of causes for this apian genocide (all pointing back to us), but one primary reason is that global warming makes for warmer winters. And that seems to allow

---

[20] Scientists at this time tend to refer to the 5 great extinctions (over 75% of life forms killed) and posit the Anthropocene extinction to be the 6th. But those lists do not include the Great Oxygenation Event over 2 billion years ago where up to 90% of all life forms bit it. I include it because, as you will see, it's a pretty important event.
[21] Aposematism is the use of bright "advertisement coloring" to warn of toxicity. The famous Poison Dart frogs are aposematic species.

Varroa mites (*Varroa destructor*) to survive a season that should knock them back with deadly, sustained, freezing weather. Bees huddle in their hives to over-winter, dancing and shivering to stave off the cold. As the globe warms, this adaptation may no longer be needed and may instead benefit the mites by allowing for easier transmission from bee to shining bee.

Along with pesticides, these mites are a leading cause of "colony collapse disorder" in Bees. If the Bees die, our crops will lose their primary pollinator, and food shortages will be imminent. About one in every three mouthfuls of food we eat relies on insect pollination; most of that is courtesy of the bees (USDA, 2019).

How could this all be happening? How could the human race, as a species, let matters get to this point? The answers are complex and rooted in our social orders, beliefs, and conventions. Rachel Carson predicted an ecological crisis in her groundbreaking book "Silent Spring" in 1962. Many cite her work as the beginning of the American environmental movement. One of the main toxins considered by her book was DDT[22], a pesticide claimed to be totally harmless to vertebrate creatures. Society initially celebrated it as a breakthrough insecticide that would end humankind's war on insects. They arranged elaborate demonstrations where DDT was sprayed over children in swimming pools and crowds of unsuspecting civilians. It was manufactured by over a dozen American chemical companies, including Monsanto and Ciba. It created hundreds, if not thousands, of jobs. But it wasn't harmless; it was a fat-soluble poison that would linger in the tissues of children and bald eagles alike. DDT caused nerve damage, eggshell thinning, and infant mortality in birds. It was suspected to be linked to cancer and was terribly stable in the environment, moving from site to site once released by the unsuspecting farmers and public health agencies. Not only that, but like all insecticides, it led to the emergence of DDT-tolerant insects, completely negating its primary purpose. "Silent Spring" led to a growing call from ordinary folks to end this and similar practices. This one book led to demonstrations in the streets and some very tough questions for polluters and politicians.

And so it also marked the beginning of the discrediting of that same movement by the powers that be. What seemed like a new dawn for the American

---

[22] Dichlorodiphenyltrichloroethane

consciousness slowly dwindled away beneath the steady drumbeat of denial and distraction that is our marketing-driven reality. Of course, no one (or very few) wanted to ignore the environment and clean air and water. But the grind of modern life, and the echoes of advertising and politics, wore down our attention until our culture effectively forget what initially concerned us, in this case, the planet's health. And this is what the status quo desired. As long as our attention stayed focused on human concerns, there would be fewer bothersome complaints about our society's unintended consequences.

But we have never been *here* before. We have never been on the verge of an extinction that now looms as a self-fulfilling prophecy of our rejection of nature. We could always delay the inevitable and pretend that everything was okay or that our "gods" were watching out for us. It is now time to stop playing our game of Dominion because that will lead only to our self-immolation as a species. But first: we need to see the game.

Our civilization is a mighty machine, and our economies work in ways similar to an ecosystem. The crucial difference is that with economies, people make up the rules. And they have a great capacity to ignore the consequences of our actions and change the rules when they become inconvenient. Billions are fed every day, and to do that, our systems place a heavy burden on the ecosystems that support all living activities. Few may even realize that the ecosphere is a part of our endeavors. Countries are continually building and hoarding the resources needed to advance our success. And at the center of our endeavors is a system of waste, where what is wanted is taken and the rest discarded, often thoughtlessly. Some accumulate great personal wealth while others are starving; for some reason. Most assume that all of this will have no impact on us or the planet because it is enormous, and life seems to function transparently in the background.

Humans have created advanced societies that thrive on disposable products. One-use plastics and highly processed foodstuffs are a specialty. These societies demand all goods in enormous quantities and don't consider the raw materials or post-use processes seemingly at all. Garbage goes to the dump, and sewage goes to the treatment plant. Nuclear waste goes somewhere. Certainly, someone is taking care of that problem for us?

Our developed societies rely on industrial systems to provide for all needs, so that they can indulge in the distracting things first-worlders do, such as texting and posting our status on social media. Many find ourselves consumed by petty annoyances with little mental capacity left to spare. Most of us have jobs that contribute to the machine that feeds, clothes, and keeps us safe, even if we hate them. We'll keep at it. It's easy to ignore the tiny shortcuts because that's what is needed to get the job done. At the end of the day, it can all be sloughed off like snakeskin, forgetting about what was done to make it to the time cards last punch. These little shortcuts together add up to massive waste, and our tiny fibs have built into one enormous lie.

Our politicians tell us what they believe we want to hear, and that is, generally, that everything is fine[23]. This seems to keep us are safe and free to focus on the essential things, such as castigating those different from us and demonizing other religions, the poor, or the wealthy. Our political world is currently filled with "social issues" such as homosexuality, feminism, liberalism, and how they are destroying our perfect world. These are just distractions designed to confuse our attention with empty prejudices instilled by our social training. Or perhaps the greatest threat is the terrorists, the foreign kind, not the home-grown type (there are "some very fine people" in that nationalistic gang). All of this occurs, oblivious to the fact that everything is not okay. It clogs the mind and freezes the heart, the way many can dismiss people in need. Inequality is not just rampant; it is an industry unto itself. And that's a problem because if someone can make a profit off of others barely subsisting in poverty, we've crossed an ugly threshold. But everyone is truly equal in how the collapse of the environment will affect them. This will be seen over and over again. Our current pursuits have no end game, no goal in sight. None.

Humans use animals as products. They use plants as products. And they have little use for fungi, except the few that taste good on salads, but products they must be. Societies use people as products. Even though they pretend otherwise. Around the globe, the growing homeless increase their ranks daily. And despite all evidence to the contrary, most prefer to blame this situation on them. It's not just "them".

---

[23] Except when it is to their benefit to say: "Everything is horrible- but I know who we can blame to make it better".

We can show little respect for anything but ourselves. And not in any broad or inclusive manner, just our singular selves alone. We've become fragmented, and many of us live in virtual worlds, not necessarily computer-created realms, but individual fantasy lands nonetheless.

We move the waters of the planet to our whims. We reap the forests and crush mountains to sift them for ores to run our machines. Pouring the waste products of our industries into our rivers and oceans, we then top it off with our excrement, untreated in many instances, thrown into the waterways, or spread anywhere we might pause.

We are addicted to energy, as modern societies. We use so much energy to power our cars, heat our homes, and calculate our bitcoins[24] that generations to come will run out of petroleum and natural gas. If, indeed, there are generations to come. Who cares? Drill baby drill, and let the future take care of itself. By all means, don't prepare for the future, don't look for newer, cleaner energy sources because we have so much! Don't plan for pandemics; that's a waste of money. Conserving is no longer a conservative value. We act as if everything is taken care of, and we are free to focus our attention on the most important business of humanity: the economy. We need the continued growth of wealth, and it must never end. Ignore the fact that unrestrained, perpetual expansion is the developmental pattern of cancer. Currently, that ignorance is bliss is our status quo.

A fine example of how humankind interacts with nature is the story of the Aral Sea. For millions of years, the rivers Amu Darya and Syr Darya fed the world's 4th largest inland sea. Situated in central Asia, this freshwater ocean was once a source of life over a wide area that was, without it, nearly uninhabitable. Humans lived on its shores for millennia, surviving off its bounty of Sturgeon, Carp, and the bottom-dwelling Barbel. This "sea" was an oasis in the desert. The Aral was once larger than Lake Superior in the United States. Then, around 1960, the USSR began damming the source waters to

---

[24] Most people don't realize the energy consumption required to calculate blockchains. Alex de Vries, calculated the annual draw at 22 Terawatt/hours/year, about the same energy use as the country of Ireland. Just to count some virtual booty. There are 4.9 million citizens of Ireland. There are 400,000 daily bitcoin users. When it all collapses, you can be forgiven for having a hearty laugh.

divert them for irrigation, slowly drying out the lake. By the 1980s, there remained only three smaller reservoirs in Aral's place. Once the inflow of freshwater was halted, evaporation concentrated the salts and minerals previously diluted by this shallow sea. As a result, these bodies of water became hyper-salinated. Meaning they became saltier than the oceans, far saltier. The hyper-salinity killed all the fish, invertebrates, and plant life that had once thrived there.

Gone was the bounty that had provided livelihoods for millions of people for thousands of years. Concentrations of petro-chemicals from fertilizers and pesticides also increased in the wake of the evaporation of those life-giving waters. Making them life-taking waters instead. Entire towns and harbors were simply abandoned. Disease and hunger visited those who did not leave. Perhaps the most brutally hit were the Karakalpaks, ancient Turkish peoples who had worked this land since the 16th century or even before. They had nowhere else to go. The people had no choice but to stand by and watch as their ancestral foundations crumbled into dust. Societies do such things because they can, not because they should.

Dominion views the world as ours to do with as we see fit, and how we see fit is often to treat it with indifference and disdain. People can act like children, and our leaders tend not to behave like parents but as the "aunt" or "uncle" who gives them too much candy, takes them to the bar, teaches them to smoke. It is a startling paradox that we currently demonize the term: liberal when we are all libertines in this developed world. Self-comfort, over mutual preservation, is a hallmark of our status quo.

Some of the United States still clings to a post-WWII era illusion of the perfect home, with a green lawn and a white picket fence surrounding it all. Papa smokes while watching a burning pile of leaves, keeping his kingdom neat and tidy like some 1950s Chesterfield advertisement. Junior has just mowed the lawn and bagged the clippings into plastic trash bags to go out with the garbage. Mom frets as bees land on the tart apple pie she left on the windowsill to cool. She can't shoo them all away lest she disturb her intricate crust. So she puts out insect traps to get rid of those nasty bugs. Secretly, she wonders why God would allow such pests to exist at all.

They use weed killer to eliminate the dandelions and turf builders because, for some strange reason, there are brown patches in their sea of green lawn. They use beauty products and strong laundry detergents to keep themselves clean, fragrant, and seemingly as perfect as their home. The same home they sanitize with commercial-grade disinfectants and scent with plug-in deodorizers. And when they are done, they throw the waste in the trash and flush everything left down the drain. Their lives may not be perfect, but they will go to any ends necessary to make it appear that way. This is the classic "American Dream," and it is an ecological nightmare. It is based on some very fundamental misunderstandings, perhaps the worst of which is the illusion of perfection.

Perfection is an alluring concept, and it is purely fictitious. The state of perfection suggests a stable finish, a place, or a time when everything is done and all needs are satisfied. This idea is the myth of a final destination. But as many a philosopher would tell you, it's the journey that is life, not the destination. Physicists will tell you that energy is motion and the cessation of motion is absolute zero- heat death. Our final terminal is death, and it always was death. Life is a verb, and it might never end, even though it always ends for the single participant. The physicist Edwin Hubble showed the universe is expanding and in a constant state of change, thus destroying the previous theological preconception that somehow it was static and perfected. Life is the tussle. Perfection is stagnation; finality is demise.

So, let's break down what the Cleavers are really up to: The grass is brown because the soil is dead. They killed it when they tried to eliminate the crabgrass. Just because a "targeted" weed killer doesn't kill the grass doesn't mean it doesn't poison the soil's microbes and fungi. They need to keep throwing that turf builder down because the nutrients are depleted, and the bioactive organisms are dead. Those same nutrients and bacteria were contained in the leaves Papa just burned and the clippings Junior threw out with the trash. The technical term for this is nutrient export. The plant tissues were made from the soil nutrients and, once removed, must be replaced, or a deficit is created. Humans are masters at creating deficits.

The bees on the pie weren't bees at all but flies with a color pattern that mimics bees. These are often hover flies that exhibit Batesian mimicry as a form of self-defense. Their larvae feed on aphids and are essential agents of natural control against those actual pests. But the bees do die because of the release of insecticides into the ecosystem. Bees pollinate most of our food crops, and they are rapidly approaching endangered status.

Those cleaning chemicals won't disappear when flushed. Instead, they will enter the wastewater stream, and many will survive the water treatment process. Then they will pollute the waterways, killing the invertebrates, bacteria, and algae that form the base of the food chain Papa's trout needs to survive. Or, the phosphates in their laundry detergent might cause oxygen destroying algal blooms. Maybe, depending on the year, both will happen. Humanity keeps assaulting the natural order, grinding life into the cobblestone with our hobnail boots, and it keeps coming back. Sure, each time it's a bit weaker and more disordered, but many think since it's coming back for more, they're just fine to continue such behavior.

Papa seems to remember that the fishing was better when he was a kid, but he's unsure why. Here's why: Mom and Pop unintentionally killed the aquatic environment with their pursuit of comfort and conformity. This illusion is perpetrated by commerce and the desire to sell more products as much as it is by anyone's misunderstandings of life and its cyclical nature.

And finally, the deodorants that make their home and armpits smell fresh contain artificial organic compounds, some of a type called endocrine disruptors, that they inhale and absorb directly into their skin. They may eventually affect the family's health and perhaps lead to annoyingly ambiguous maladies in the future. This image of perfection is disturbing when you stop to think about it. And we all do at least some of the things this family does. Often failing to consider the after-effects of our own actions, having been taught to believe that our place in the world is safe and divinely protected. Multiply the Cleavers by hundreds of millions (perhaps by billions), and you understand the scope of our problem. Mom and Pop would never act this way unless they had been taught to do so, and they have been. Armed with our growing understandings, we can begin to unlearn these behaviors and work with nature instead of against it.

This is our point of departure. The future looks a bit bleak from here, but it need not be so. We are on the edge of multiple revolutions: technological, social, and economical. Expanding and blooming around us is an explosion of pent-up desire and developing human ingenuity. What we need now is greater than a revolution; we need an evolution. A Green Evolution, where we come face to face with our proper place in this world and end the old illusions of superiority and omnipotence that blind our eyes and stall our action. The old world, where a few powerful groups hold reign over a captive audience, is ending, one way or another. They can't stop this change with all their dark money or fear-mongering because they can't block the planet's evolution with laws and prejudices. The difference between natural "laws" and human-made laws is that you can't break a law of nature. You can't exceed the speed of light, and you can't stop an ongoing global process with words, or threats, or money.

Truth is exploding around us, and it can set us free as surely as it can crush us. I am writing this to implore you to let it be the former. The world is collapsing in a frightening physical way, but we have everything we need to not only survive but make a better, fairer world for everyone. It can only be that way if we want to endure. This status quo is leading us relentlessly towards that lemming cliff. But this is a fate we can defy because it is an illusion. And there is another status rising, a "status si", if you will: a state of if.

As a whole, our societies are better educated than ever in human history, even if many still cling to many old prejudices and superstitions. There is a waking up to the horrors that this developed world lifestyle represents and the future it is creating with each discarded plastic water bottle and every beached whale. Environmental movements are gaining ground worldwide. But is it enough to care if you don't know what to do with that knowledge? Disinformation flies fast and free in this modern world, and concerted efforts have been underway for over a generation to reduce the quality of both information and discourse. Despite all the pretension, this is a battle over what is possible versus the limits those who have been in power for centuries prefer. Many of our thought leaders wish to delude us into believing that absolute power exists, and that it holds the reins of existence. Its imaginary omniscience has our best interests in mind, so all that is needed is to surrender to its (their) will and everything will work out in our favor. You can now see clearly that such ideas are not only incorrect, but increasingly dangerous. Life shows us that possibility is as nearly infinite as we can comprehend. And life is more than flexible enough to meet the needs of our survival.

Our submission to the status quo is made possible only by extended training. For example,many have been trained to react to certain words and ideas with knee-jerk negativity. Words like communist, homosexual, or feminist. The ability for simple words to stoke rage in a good percentage of a population is an excellent weapon against possibility and hope. The only way to combat this is with quality information and a populace willing to question, not rashly, but with measured contemplation.

IF we can learn to look for honest answers and judge the quality of those answers, we might be on our way to seeing what is truly important;
IF we can trust in each other's desire to survive and are willing to change our habits just a bit to adapt;
IF we can stop projecting our wants onto nature and learn from it instead;
IF we can then join together to heal the damage we have inflicted;
We will be just fine.
It all seems reasonable, right?

For the remainder of this book, I will examine, in a systematic and ecosystemic way, how our industrial and commercial practices have treated this planet that gives us life. We will also explore how the restorative processes of life can undo the damage we've wrought. But only if we join together to protect the planet and her systems from further insult. Now is the time to put away our adolescence and become adults, caring, nurturing parents to the life forms that share this world with us. Taking on this responsibility is only fitting because nature's structures, as we will see, gave birth to us. A common cycle in nature is that the child becomes the parent. This is one hopeful outcome of our current predicament.

There is no longer any time for debate, and attempts to paint ourselves and societies in a gentler light will only forestall our ability to get ahead of the ecological disaster that the Dominion mindset has brought down upon all. And time is one elemental force that is not on our side, nor is it within our powers to affect. In the time you take to read this book, more species will go extinct, more people will lose their homes and livelihoods. It is within our power to ease some changes ahead, but not to halt them. That horse has left the barn.

CHAPTER TWO: THE STATUS QUO

Change is upon us; whether that change arrives as the renewal of humanity or its demise is solely our choice.

Twisted power wants us to remain fractured and angry at each other so they can continue to slurp up what's left of the world's resources, like some pre-digested broth through a disposable plastic straw.

We can't let that happen. So instead, we can turn possibility into our new status quo. The first step is to understand how the ecosystem that sustains us all functions. Step two is to emulate it in order to live in balance with our homeworld's handling capacity. Our societies have taken so much for so long that we must find ways to give back, to repay our debt.

And we are about to learn how to do just that. A debt is owed to the future that can only be repaid by allowing it its chance to come into existence. The same deal every generation of every creature that lived before us extended to us.

CHAPTER THREE: EARTH

"What's the use of a fine house if you haven't got a tolerable planet to put it on?"
— **Henry David Thoreau**

In 2015, contractors working for the Environmental Protection Agency accidentally damaged the plug on a holding pond at the Gold King mine in Silverton, Colorado. As a result, eleven mega-liters (3 million gallons) of mine sludge poured directly into Cement Creek and the Animas River. This cocktail included: cadmium, lead, arsenic, beryllium, zinc, iron, and copper. It turned the Animas River a yellow-orange color and killed everything it touched. The Animas River runs a short 200 km (126 mi) jog outside Telluride, Co, to a confluence with the San Juan River in Farmington, New Mexico. All along that stretch, practically every inch of the river bed and bank withered as if by the touch of one Tolkien's Nazgul[25]. All fish and aquatic life perished. Plants and crops near the noxious flow withered and twisted. Any animals who drank from this stew suffered horribly before their deaths. The sight and smell of the river became monstrously twisted. Even the river's sounds were stolen as silence replaced the singing of birds and the click/buzz of insects. This flow of life was transmuted into a rambling torrent of silent death in the snap of a finger, with the flick of a joystick.

The Environmental Protection Agency took responsibility for the leak, which

---

[25] The Nazgul from the Lord of the Rings trilogy were powerful Kings, who, in their lust for greater Dominion took rings of immortality from the Dark Lord Sauron and subsequently withered away into shambling shadows of horror and death. But they were practically immortal- so they actually got what they bargained for.

meant that the US taxpayers would foot the bill for any clean-up and no more. They denied payments of any secondary damages because of sovereign immunity. So the EPA covered only the initial scoop up and haul off; all claims of consequential losses were swiftly rejected. While polluted sand can be removed and they can haul the corpses away, but there's no magic wand to set right such a calamity. Restoration, in current terms, often means relocation and entombment of contaminated objects. That's often the best option we've got. Only nature can heal such insults.

The mining company, which closed in 1991, had no culpability since the owner had followed all appropriate mining rules at the time of their exit. This tragedy is but one side effect of mining laws that are over a century out of date and promote the exploitation of natural resources over all other concerns. Located inside New Mexico's boundaries, the Navajo Nation was justifiably outraged when this environmental carnage poured into their lands. They could be excused if they saw it as just more rage unleashed at their reservation. This time as a kaleidoscopic flow of industrial annihilation. These people, bone-weary of watching what little is left for them again stolen, or desecrated, could only watch as their livelihoods were once again squandered by the machinery of our economies. Here was yet another slight in a seemingly endless catalog of wrongs perpetrated against our land's indigenous peoples by America's dominant culture. Another slap in the face, another poisoned blanket, yet more insult and injury against folks who only want to live in peace. Perhaps it can be argued that this one, at least, was an accident. An accidental by-product of how we treat the earth beneath our feet.

Many like to call this planet Gaia. By giving the Earth a personified name, they challenge us to see it as a living being, and in some fundamental ways, it is. Gaia is the only place we know for certain that life exists within the entire universe. Perhaps it is not, but there's a lack of concrete evidence for any other supposition. This planet's ecology is fully adapted to its geology and astronomical placement, long maintained and evolving in chorus. So I'll use this metaphor of the earth as a living being because it can help us understand our place upon it and within its systems. We can then use this analogy to comprehend the complexities of our circumstances, and for that reason, consider calling Earth by her creature name, Gaia. Whether this is technically true, using her name can help us acknowledge her value as our Mother and the only breast from which we all feed.

James Lovelock and Lynn Margulis developed the "Gaia hypothesis" in the late 1970s. And while it is not my strict intention to endorse their theory, everything I outline in this text might be seen to suggest a tacit acknowledgement. The Gaia hypothesis proposes that living organisms' complex interactions with the inorganic world create a self-sustaining singularity that is continuously in a state of change and, therefore, stable. Or, in simpler terms: "the goal of life is the perpetuation of life, and it does a damn good job of it." When viewed in these terms, concepts like "survival of the fittest" transform from a cynical dog eat dog competition into an awareness of complex interactions and, ultimately, cooperation. Life itself is both the experimenter and the experiment. What may sometimes seem cruel to our eyes is just another step towards a better outcome for all concerned. Life on Gaia is good, but it can be uncertain.

Earth is not only the currently accepted name for our world; it is the term used to describe the land — its soil, rock, and real estate. This earth that we walk upon is but a fraction of the planet herself, comprising only a scant 29% of the planet's surface area. The planet's diameter is 12,756 km (7,926 mi) and its circumference is 40,076 km (24,901 mi). The human race lives on 148 million square kilometers (57 million mi^2) of land. This number equates to 52 people per square kilometer (135 people/mi^2) at the current population of 7.7 billion. That's quite a bio-load of humanity, but it's also a bit deceiving. Somewhere around 17% of the land is uninhabitable. Inhospitable regions include high mountain peaks, deserts, and lands already made unlivable because of our activities. Humans tend to concentrate in urban areas, and 95% of the population lives on only 10% of the available land (JRC, 2008). Dhaka, Bangladesh's capital, has the greatest population density on this planet at 44,500 people/km^2 (115,210 people/mi^2). In Wyoming, the current population density is around 2.3 people/km^2 (6 people/mi^2).

Being social animals, people congregate together, leaving much of the land alone, as should be. But as our population continues to increase, humans push further into the frontiers, testing the ever-shrinking capacity of the environment to absorb our damages and provide for our needs. As our technology increases and our success escalates, many believe that they are somehow self-sufficient and search for ever-more remote areas to inhabit. Little do they realize that it is the interconnection of society that makes our prosperity possible, just as the web of the ecosystem makes all life possible. But off to the wild, they go. Some will slowly fade away, bears will eat some, and others still might just pollute new areas that had skirted humanity's onslaught up to that point. Wherever people settle, they take more than they give back. That has been our pattern. Our lifestyles displace other species. Our actions deplete fragile soils. Rarely is the land ever improved by our "improvements." That can change. And it must change.

This planet we call home formed from a consolidation of dust particles some 4.5 billion years ago. This dust was some of the matter left over from the Big Bang and elements ejected from the generations of exploding suns before ours. Suns are not only sources of light and heat; they are the forges of matter. Hydrogen is the primordial atom and is as simple as it gets: 1 proton, one neutron, and 1 electron, but in the fusion reaction of a sun, it transforms into helium[26]. And the building of matter doesn't stop there. As the original fuel of choice, hydrogen and its ions are consumed, secondary reactions occur, and heavier atoms are formed. In the final stages of a sun's life, the heavy elements, carbon and above, form quickly as the last of the star's fuel is exhausted. Finally, many stars explode and "seed" their local galaxies with the elements that will make up new planets, asteroids, moons, and us. We are all made of stars, which should be a transcendental enough realization for anyone.

As Earth formed, first as a ball of dust and then a collapsing sphere, gravity sorted the particles into layers based on density. Just as a centrifuge sorts material with its spin. Our sphere compacted, like a twirling dancer or ice skater. As her "arms" pulled in, the speed of her spin increased. With the mounting speed, the internal pressures increased equivalently. Heavy elements, many radioactive, ended up deep in the core, which is an excellent place for them to stay, as you shall soon discover. For the first billion years that you could call this world a planet, geology was the biggest game in town. While the crust was in upheaval, and the elements combined into mineral compounds

---

[26] 2 protons, two neutrons, and 2 electrons

and self-distributed, eruptions of startling immensity occurred over and over again. The planet was finding its equilibrium as it cooled and slowly compacted. This activity, which still occurs today (but to a much lesser extent), gave us deposits of metal ores, rare-earth minerals, and, yes, diamonds. Not willing to wait for things to calm down a tad, organic compounds began self-assembling in different ways. Organic scums bombarded by cosmic rays not yet filtered by an atmosphere organized into the first life forms. The ball of rock that was Earth birthed Gaia, the life that would eventually over-run its countenance. And life would become the dominant feature of this beautiful blue/green ball in space. Over 4 billion years later, this pyramid scheme called life led to us. So, it was actually humans that looked upon the face of the Earth and said: "This is good."

The geology of the planet is the inorganic world and one of our primary resources. What began with the gathering of, and obsession with, charming rocks led us to metallurgy. We are a race of boundless tinkerers, and this curiosity served us well. Early peoples forged copper and eventually discovered iron and then its cousin, steel. Once they became adept at smelting and smithing, our desire increased for more useful things. And to get the raw materials needed, they began mining. Humans started *seriously* mining. Now, giant machines consume mountains, tearing and crushing, sifting, and sorting an industrial world's raw materials. When there was limited technological capacity, the damage done was minimal, but that is no longer the case. With the ability to harvest massive quantities of ores comes the problem of gargantuan mountains of tailings. Mine tailings are all the materials that are not what the miners seek. Tailings are the discard, the trash of mineral extraction, mining's garbage heaps. Most ore deposits are not sitting directly on the ground, waiting for some grizzled but friendly old miner to pick up and take home in his pocket. Most deposits lie underground. To reach them, you have two choices: tunnel to it, or remove the ground layers above it, what they term the "overburden." Mining gave us the materials to create tools unimaginable to generations before and sped up the march of civilization. Industrial-scale mining became an obsession as societies accumulated evermore materials in a continuously expanding cycle of consumption.

In an open-pit mine, miners remove the overburden and pile it somewhere until they reach the ore "body." Where they stack the overburden is vitally important because there is no genuinely empty land anywhere. someone or something lives on almost every square millimeter of the limited terrain of Gaia. This mining process involves blasting, scraping, and sorting. While you are doing this, you are releasing massive amounts of dust. And in that dust, you may have substances like radioactive isotopes, potentially asbestos or its cohorts, riebeckite, and fluorine. None of these materials are anything you'd want to breathe in, and yet the miners and residents do just that. This mining technique has led to lung diseases and cancers worldwide (Ross, 2004). Once you have reached the ore, it's still not a pure product. That's why it's called ore and not, say, iron. Veins of ores or gold, or what have you, are seldom pure. The ore needs to be processed to remove the sought-after minerals. This distillation can be achieved for some ores by simple smelting. Heat the ore until the metal you seek melts out, and skim off the waste products of floating slag. Alternately, this is often done these days by crushing and leeching the ore in open fields. Gold, for instance, can be extracted from the ore with a cyanide solution. The MacArthur-Forrest process works by placing hill-sized piles of the ore in a shallow pit with a membrane beneath it. A cyanide suspension is sprayed over the top and allowed to percolate for a period of time. The collected solution goes through a couple more processes before electrophoresis[27] separates the gold from the cyanide stew. Sound tasty? Open-pit mining's environmental impacts are substantial, whether it be coal or gold or iron or copper produced. Pollution of ground and surface waters is also of significant concern. Sensitive habitats have been and probably will continue to be destroyed. Culturally significant sites have been plowed under and trashed for the minerals underneath. Government protections for either situation are neither adequate nor often even considered. When a toxic mess is created, the mining companies all too often get off scot-free, and it is the citizens who are left picking up the bill. Just as happened over on the Animas River.

Here in the United States, mineral rights are funny things (in the alternate, unhumorous meaning of the word). If you own some land, you do not

---

[27] Electrophoresis: the movement, or separation of particles in a solution under the effect of an electric field. Basically; an electrical filter.

automatically own its mineral rights. There is some statutory wiggle room, allowing continued access to a domicile by the "owner." But if industry finds strategic minerals under your home, you'd better move your house, or just move, period. Someone, somewhere, owns the mineral rights to your private property, and it's not you. So if you strike pay-dirt digging a foundation, you will often do best by keeping your mouth shut and kicking some dirt over the shine. Our entrenched ideas of security and prosperity seem to require that we exploit any mineral we can. Anywhere it is found. But that's just our training talking and entirely unnecessary for the continuation of the human enterprise. As shall be seen, the volume of minerals already extracted could likely suffice our advanced societies' needs (if everyone stopped throwing them into landfills).

On Federal lands, the right to make a mineral claim is still in full force. The federal mining act of 1872 grants ownership rights to anyone over 18 who stakes a claim on public lands and actively works to "improve" that claim. It's like a reverse eminent domain where an individual can take land from the public. The purpose back in 1872 was to stimulate the economy and incentivize settlements in the western lands. Back then, the idea was that westerners couldn't take the land from the indigenous peoples if they didn't occupy it.

Nowadays, if someone found gold under the oldest tree[28] in the nation and filed the proper paperwork, they could conceivably blast it down and take that gold as their personal property. Even though it is on land owned by "The American people" and is located in an area of ecological and cultural significance. Such an example is an extreme possibility, and they keep the exact location of that tree a secret. But this illustrates how our laws and customs continue to encourage destructive behavior.

Guerilla mining is a perennial concern, and what better example than the continuing search for gold? In the South Peruvian Amazon basin, large deposits of placer gold exist in the soils. Placer gold is a product of erosion and sedimentation. Placer mining is far different from hard rock mining, where deposits are removed from the solid parent rock. Instead, Placer gold is a

---

[28] This would be Methuselah, a 5,000 year old bristlecone pine located somewhere in the White mountains of California.

precious metal deposit distributed by erosion and found "in place." Placer mining requires removing and sifting soils or gravel. And that means you have to remove whatever is on top of the depositional layers — that dreaded overburden. Here, that means removing the rainforest.

Most of this mining is currently being done illegally. And as difficult as it may be to imagine, that the removal of thousands of hectares of forest and the heavy relocation and extraction of soil and gravel is occurring "under the radar", it is. And business is booming. To compound the issue, this isn't being done in a cautious manner whatsoever. Instead of cyanide, the most common gold extraction technique used in South America is mercury amalgamation. In another simple yet toxic process, the gold is dissolved in the mercury, and the resulting amalgam is then purified by boiling away the mercury solvent. This procedure leaves behind the tiny bits of gold initially mixed in with the soil. Mercury is one of the elements most toxic to life. And if you think these illegal miners are following safety protection rules, you'd be mistaken. So, this search for precious metal increases deforestation, accelerates erosion, and releases poisons onto the land. That's a three out of three for things not to do if you want a sustainable future.

So, why do we so covet Gold? Gold is a noble metal, which means it does not tarnish or rust. It is an excellent conductor of electricity and useful in making electrical contacts for long-term, reliable service. Wearing it on your finger or neck is simply a symbol of status, based on its rarity. This scarcity is the main reason it is so valuable. If you are wearing gold right now, look at it and ponder where it came from, what it means to you, and why. It has been around the planet for billions of years; it was made by a sun billions of years before that. And it will certainly outlast you, so who owns what?

So perhaps the tunneling method is a safer choice for grabbing our gold. Digging tunnels elicits fond memories of hard rock mining's early days where men (and mules, too) worked deep underground, chipping away at the solid mountain following that motherlode. The disruption of the scenery is a fraction of that caused by open pits. There are still health issues, but those dangers are isolated to the miners. And they are a stalwart bunch. Tunnel mines can

The Good Ol' Days of mining- This is what our "Traditions" looked like in reality.
Public Domain Image all uses permitted.

collapse[29]; they can explode from the accumulation of gasses because of poor
ventilation[30]. Miners suffer, and they die, predictably. However, in their
defense, the mine owners are required to install air filters and dust abatements
to protect their workers. Strict rules limit the time spent underground by any
given miner, and wellness checks ensure they stay in tip-top health. At least
that's how it looks on paper. In practice, miners are just another expendable

---

[29] Crandall Canyon Mine, 2007, 9 dead, including 3 rescue workers; Murray Energy
Corporation. For example...
[30] Upper Big branch Mine, 2010, 29 dead; Massey Energy Company. For example...

resource to most companies.

Underground mine safety rules have been under attack by the mining industry for years, and currently, the enforcement is poor. Tunnel mining is expensive for all the reasons mentioned above and seems to be falling from favor, primarily for economic reasons. It's very costly, especially if you have to follow the safety rules. And why dig a tunnel when you can scrape away the entire mountain?

Mountain top removal is open-pit mining on steroids and has led to catastrophic results in Appalachia and worldwide. The number of cancers and the decrease in life spans for the miners and their families are legion (Hendryx, 2011). Yet this seems just collateral damage from the viewpoint of industry. This ancillary wreckage is something that is rarely spoken of and hidden from the public as much as possible. Like the sausage factory, no one wants to gaze into the machinations that supply cheap electricity, copper pipes, or the wedding ring you place on the finger of your beloved.
It has become popular to criticize the environmental movement for exterminating the miner's "way of life." People condemn the loss of traditional values and mourn a lifestyle that has been around for, well, a while, at least[31]. They echo this outrage without really looking into just what that way of life entails. Rarely do many try to look into complex social interactions when it's easier just to find someone else to condemn. You'll see this tactic often repeated; when someone points out a crime perpetrated by industry, that someone themselves will be blamed.

Are good-paying jobs that are the equivalent of dollars per day of life lost worth it? Are there no alternatives?

Yes, there are alternatives, but those with a vested interest in keeping things just the way they are often stifle discussions of re-education or the development of clean energy and technology jobs. The argument that society

---

[31] Coal mining in America began in the 1740's in Virginia, but it wasn't a big deal until the Industrial revolution required megatons of it. It peaked in 1923 and has been declining ever since. That's a crest of only a bit over a century folks!

must preserve this way of life is just that — an argument. It doesn't hold much water when scrutinized. There will always be mining, but it could be needed much less, and in fewer areas, as we learn to save what we've found and recycle our materials. But the status quo has an economy based on mining and is quite happy to continue with its currently destructive practices because that's easier than changing; for those who profit heavily from these practices, anyway. A change could disrupt the steady flow of cash that goes into those miners' pockets but it mainly threatens the wallets of the few captains of this industry. And the anger instilled by that industry into the workers makes approaching them with a better offer risky at best. So they shun change in favor of a continuation of "business as usual."

Conflict mining is also an issue. Rare-earth minerals such as coltan[32] and wolframite are increasingly needed for computers and smart devices. Many mineral locations exist in parts of the world with unstable governments or territories, such as the Democratic Republic of the Congo (DRC). Warlords using slave labor in horrifying conditions may carry this mining out. Murder, rape, child enslavement, and extortion can and do occur (Kasinof, 2018). It is challenging to police this situation, and with the desire for these high-tech materials at ever-increasing levels, doing so would just be a big bummer. Conflict diamonds were the subject of the movie "Blood Diamond," a fictionalization of actual incidents in the 1990s in Sierra Leone. In the film, as in real life, slaves mined diamonds for warlords to fund more war. In real life, as in the movie, they are still doing it. Look in your hand or on any surface near you. How many lives per smartphone is acceptable?
Tech companies in the USA have taken the lead in attempting to rid their supply chains of conflict minerals. (Calloway, 2017) This vigilance occurred after a public outcry, which occurred after they (the public) found out about it. Before this general awareness, they did not see it as much of an issue. But better late than never, as they say. Then, in 2017, in an abrupt about-face, the Trump administration began efforts to eliminate conflict mineral rules in the name of cutting regulations. (Frankel, 2017) Why let little things like ethics or

---

[32] Columbite-tantalite

principles impede a good profit? Papa needs a bigger yacht. Daddy wants a wall[33].

Let's have the talk about these diamonds for a moment. While technically referred to as a form of "inorganic carbon," diamonds are not like any other mineral. The difference between organic carbon and inorganic carbon has to do with permanence, location, and whether it is available to fuel life processes. All organic compounds contain carbon, and most inorganic compounds do not. So while diamonds are considered an inorganic carbon, they are still composed of the same chemical element that makes up wood, chitin, and approximately 20% of you. Diamonds are the crystalline form of elemental carbon, and while produced naturally in rock (kimberlite), they are still just carbon and not rare at all. Diamonds burn, just like wood or coal. So while they are the hardest known substance, they possess a real Achilles' heel. They will ignite at 850° C (1562° F). So diamonds are only forever if you keep them away from a propane torch.

A geologist friend of mine once told me how she had to school her fiancé not to get her a diamond engagement ring: "I'm not wearing anything on my finger that burns!" And there are so many unique gemstones to choose from that are actual minerals[34].

Diamonds are relatively common, it turns out, and they make excellent drill bits and saws. The reason you may have paid a couple of months' salary for one is just a matter of marketing. A cut gem-quality diamond could run about $3000/carat (0.2 grams, 0.00044 lb.), while an industrial-grade diamond runs between $10 and 30 cents a carat, depending on the size of the crystal. That's at least a 3000% mark-up. The Huffington Post printed a now-infamous article entitled: "Diamonds are Bull$h!t," and it's worth a read and a chuckle. (Dhar, 2017) So, here is an example of a massive industrial mining complex with enormous energy costs, all for an end product that is as close to a scam as imaginable (except for the industrial diamonds, diamond saws are the bee's knees). This scheme exemplifies how our curated perceptions have led us to

---

[33] Building a wall on the southern U.S.A. border was, for some reason, a priority of US President #45's administration. Oh yeah- the reason was racism.

[34] Next time someone shows off their diamond engagement ring and uses the term: "Rock." Reply: "That's debatable", and see what hillarity ensues.

more destructive behaviors, with little need, except to fill an imaginary emotional hole. We can and will do better.

Our next type of mining may not seem like mining at all to some. For it uses those diamond drills to bore small holes into the earth's crust and then inserts steel straws into those holes to suck out petroleum, our cars' favorite drink. Petroleum, which is commonly called oil, is one of the fossil fuels and is foundational to our economy. It is considered an organic form of carbon despite actually being fossilized in mineral layers. It earns this title because of its molecular structure's preponderance of carbon/hydrogen bonds. And because when someone wants to draw arbitrary lines between "organic" and "inorganic" carbon, they have to make it somewhere.

This oil can be used to make gasoline, lubricating fluids, plastics, and soaps. Members of industrial societies ingest it as pharmaceuticals and vitamins. They make fertilizer for our food plants with it. The cases for our game consoles are made from it. We may even transport our dead wrapped inside it. Petroleum surrounds us, it infiltrates us, and we burn it. Boy, do we burn it! Forty-three percent of the annual global harvest of petroleum become fuels (Gasoline, Diesel, and fuel oils), so it is easy to understand why we have such a problem with carbon dioxide release. I'll save those calculations for a bit later, however.

In 2017, the United States consumed 913 million tonnes (1 billion tons) of petroleum. The rest of the world gobbled 4,621 million tonnes (5 billion tons), so that's a lot of straws sucking that black juice from the crustal layers of our planet. In the United States, people and industry consumed only 16% of the world's oil production that year (B.P., 2018). Americans may have started this game, but they're no longer the leader. Petroleum is measured in Barrels (BBL), containing 159 liters (42 gals) of muck. There are 196 BBL of oil per tonne (179 BBL/ton). Barrels are also the most prominent international use of the imperial measurement system. Why, you may ask, would that be? Because oil is America, Baby!

Colonel Edwin Drake drilled the first oil well in 1859 in Cowboyland[35], Pennsylvania. You can still find this "fact" in papers and libraries around the world (and on the internets). The fact that this isn't true is of no concern to

---

[35] Of course it was!

those who wish to glorify America's oil supremacy. Here is yet another apocryphal tale upon which to base our society's self-image (and found our bragging rights). The first wells were, in fact, hand-dug in Poland a couple of years prior to Col. Drake's endeavor. (Eig, 2017) Pointing that out may be just another big bummer for those who think such things matter in terms of power or legitimacy. But drilling did explode in the US of A, before it became a global obsession, and long before we discovered it was about to end life as we know it.

So how does oil extraction work?

Our crew Boss, Dirk Driller, pulls into an open field to find a survey marker left by the geology crew and begins his heavy equipment work. He bulldozes the ground to create a flat pad upon which they will set the drilling rig, tool shed, and a parking lot for the crew. First, they erect the derrick, a tall scaffolding that will guide the drilling and allow sections of pipe to be readily inserted as the bit drives into the ground. This rig is an enormous drill press. Large diesel engines run the rotary table, which gives the drill its spin. They attach a massive diamond bit to the pipes that form an expandable drill shank as the hole deepens. Once that hole is established, BOP's (Blow Out Prevention devices) are clamped around the bore, guiding the pipe and protecting the workers from the considerable forces with which they are meddling. This system can quickly cut through solid rock, and the hole may go hundreds of meters into the planet's crust. Next, the dust created by the drilling has to be removed, lest it dulls that diamond bit, or jams the whole mechanism deep underground. This removal is done by injecting mud or compressed air into the bore, ejecting the drill cuttings. Doing this also lubricates the cutting head. Once they hit the oil body, the well can kick with back pressure. So, they increase the mud pressure to match this geologic force and prevent a blowout. Once the well is completed, the drill can be removed, and a cap installed, awaiting the final procedures, cementing and constructing the wellhead, and installing a pump. At the end of a long, dirty days' work, the drill crew packs up and moves to their next conquest. Drilling is a non-stop endeavor in the United States of America and many other parts of the world.

This well may or may not be fenced, protecting it from the curious and keeping the wildlife away. When I worked as a roustabout[36] in Wyoming's oil fields, our crew had to fence around the dry ponds next to the wells. These pits would act as emergency impoundments if some oil leaked or the seals completely blew out. The lands on which we working had free-range cattle plodding about, and if they encountered a pool of oil, they'd take a drink. And then they'd die because crude oil is poison. A poison that, for some reason, tastes good to cows (who are we to question?) You can bring a horse to water, but you can't make him drink, as the saying goes. But if you let cattle near a puddle of oil, you can't stop them from drinking it. A suitable allegory for some of our fellows.

There are so many oil wells in the western United States that they disrupt migration paths for wildlife. There are currently over 3 million hectares (7.4 million acres) of land consumed by oil rigs or pads in the United States, the equivalent of 3 Yellowstone National Parks (Allred, 2015). That's a lot of wildlife non-refuges. Unfortunately, those fences don't stop birds, snakes, or small mammals from slipping in for a look-see or a guzzle. So, if there's a spill, there will be some mortality.

Oil wells leak, not all of them, but being a pressurized mechanical apparatus, left to operate without supervision, failures will occur. The methane emissions from these installations contribute directly to global warming, oil wells being responsible for over 30% of all methane releases worldwide. (Allison, 2018) The Environmental Defense Fund has been studying methane releases in the Permian basin in Texas. They've found the amount of this gas released into the atmosphere is up to 60% greater than stated by the EPA and could power 10 million homes annually. That's a tremendous amount of greenhouse pollution still escaping with impunity. (Lyon, 2016)

There were around 2,800 spills from drilling in Colorado, Wyoming, and Montana in 2018 alone (Wiess, 2018).[37] While often segregated from the cattle, these spills leak into the ground and pollute subsurface waters. Oil spills that occur on offshore drilling platforms are always catastrophic since they empty directly into our oceans. The list of environmental disasters, both big

---

[36] Basically- a grunt who wasn't allowed to touch anything moving or expensive.
[37] This is why it is essential to ban drilling in highly sensitive habitats, such as: the Artic Wildlife Refuge.

and small, created by oil drilling, can seem nearly endless. The only end will be when drilling for oil is reduced or even halted. Drastically curtailing these activities is possible and needs to occur as soon as possible.

This all begs the question: How much mining do we really need?

Currently, it appears that it is more economical to mine new materials than to recycle what is already around. But appearances can be deceiving. The base cost is lower for recycled materials than for freshly mined substances in terms of cash. In terms of the cost of labor, energy, and transportation. The disruption of the environment and release of pollution is definitely lesser for recycling. And yet, recycling still ends up costing more. How is this possible? Let's look at aluminum as an example. You could easily replace this example with glass, for both materials are practically endlessly renewable with virtually zero loss. The energy cost savings of reusing aluminum is substantial. The recycling of aluminum saves up to 90% of the energy of mining and processing the raw ore, but most people still keep throwing it away. Transportation costs, the lack of infrastructure, and the poor sorting by the recycling population all alter the price upwards towards—"let's just throw it in a landfill!" (Staff, 2012). Here, the onus falls squarely on the public. In the USA, there just isn't a recycling ethic. Consumers waste and don't even try to conserve in most instances. In some other cases, our attempts at recycling are half-hearted or poorly understood. Most recycling costs are spent on correctly resorting and cleaning the materials after they have been sent to the sorting facility. Most folks seem to think the recycling bin is just another trashcan, albeit one which allows us to feel a little better about ourselves. We could solve this issue simply by cleaning and sorting our recyclables correctly. And then by putting social pressure on businesses and governments to make recycling, if not mandatory, at least socially obligatory.
And in most cases, these entities already want this to happen. It just doesn't at present. And it's not the fault of the government or industry. It is apathy, plain and simple.
PLEASE RECYCLE, FOLKS!
Recycle and support public servants and organizations that support it as well. Wash your recycling and sort it properly; it's not that difficult. Promote cash deposits on containers that industry could easily reuse. Reuse containers

yourselves whenever and wherever possible, and reject products in elaborate mixed material containers, even if that product is marketed as "Green." Simple and reusable packaging is ideal. Standardized, interchangeable containers would be even better. The easiest way to recycle is to reuse, and currently, most packaging is nearly impossible to repurpose. The power of the pocketbook is indeed great, and you control when that wallet opens or stays shut. Recycling is the way of nature. Recycling is the path to our survival; this is not hyperbole. Nor is it, or should anyone interpret it as a political statement. It's a fact, Jack!

Trash is not natural in the natural world, where there is no such thing as a discard. However, humans have not only created the concept of waste, but we have nearly perfected it. In 2017, American citizens threw away 243 million tonnes (268M tons) of "garbage." That equates to over 2kg (4.5 lbs.) per man, woman, and child per day! (EPA, 2019) That is just about the same amount as an adult eats each day. And the waste is only increased when manufacturing the goods the public uses. For every 45k (100 lbs) of products industry manufactures, around 1450 k (3200 lbs) of waste materials are discarded, on average. (Hawken, 1997) This level of waste is mind boggling, and yet, here we are.

Twenty-two percent of what is thrown away is food waste, organic material that could be composted or used as a feedstock for agriculture. We waste, often without a thought or care. And as the rest of the developed world advances in recycling and trash reduction, in the USA, many still don't seem to "get it." The 745 kg (1,642 lbs.) of household trash each American throws away each year contains glass, metal, wood, and increasingly plastic. Plastic trash accounts for 19% of our annual waste stream, making it second only to scrapped food. We even wrap our trash in plastic bags before throwing them away. Worldwide, humans produce over 5 trillion plastic bags a year. That's 160,000 bags every second. And these are not even one-use products; they are often zero-use products. Garbage bags are unrolled, filled with waste, and then disposed of immediately. They have no use other than as discard. Here is waste consummated. This action represents a wealth that has been accumulated and then just tossed away. We are all litterbugs in the USA. Not that we throw our trash wantonly out the car window, but that we generate so much waste and throw it away at all. But we sure do toss it, on the land, into the seas, and into the air, as we shall discover[38].

But currently, where we throw our trash is into a landfill.

America's first landfill was established in Fresno, California, in 1937. It is now a Superfund site, with the EPA footing the bill to clean up its 59 hectares (145 ac.). Once again, the citizenry pays to clean up the waste and volatile organic compounds leaking into the groundwater and the methane still escaping into the atmosphere. Before this original "sanitary" landfill's construction, communities just dumped their trash in ditches, on the ground, in the river, anywhere really. People used to burn trash in their backyards while their kids danced in circles around the smelly black smoke. I know this because I remember doing it with the neighborhood kids.

It is essential to realize that humans used to produce far less trash than now. It is estimated there may be up to 500,000 landfills in the European Union. Currently, there are over 2,000 landfills across the USA, covering 485,000 hectares (1.2 million ac., 1875 mi²). The EPA doesn't have an updated statistic on the amount of land consumed by our landfills. So I gathered the raw data, rounded down, found an average, and then expanded it by the recorded number of landfills to find this number[39]. But the actual number is possibly higher.

Here is a figure all should ponder — 4,850 square kilometers (1875 mi²). This expanse represents nearly four Los Angeles, over 6 New York cities, and 46 Paris', all buried in rotting garbage, meters deep. What a joy!

Of the 8.3 billion tonnes (9.3 B tons) of plastic humanity has created since we began in the plastics business, 6.1 billion tonnes (6.7 B tons) has been buried in a trash dump. Every year, in the USA they throw 2.7 million tonnes of aluminum; 7.6 million tonnes of glass and 2.3 million tonnes of steel and iron into landfills. So little concern does our society have for actual material wealth that it is squandered and wasted with disdain. These are skyscrapers and

---

[38] And into space- let's not forget how much crap we've spewn into orbit around our planet.

[39] In 1997, a Denver, Colorado company, EnerQuest, mapped 18 standard landfills with LIDAR to an accuracy of greater than ±3 cm (<1 in). I used their findings to expand to the nationwide area consumed by landfills

aircraft carriers of resources tossed in with toxic chemicals, food wastes and whatever the next Picasso of the art of trash devises. More gold into straw.

Around 80% of landfills leak to one extent or another. They leak toxins either into the water or the air. Often, they pollute both simultaneously. The amount of leaked methane is colossal, 102 million cubic meters per day (3,589 million ft^3/day), every day, in the United States. Methane is a greenhouse gas over 30 times more damaging to the environment than CO_2. But it doesn't stop there. Landfill gasses are roughly composed of: methane, 30-60%; carbon dioxide, 30-60%; nitrogen, 2-5%; oxygen, 0.1-1.0%; ammonia, 0.1-1.0%; sulfides, disulfides, mercaptans, etc., 0-0.2%; hydrogen, 0-0.2%; carbon monoxide, 0-0.2% and trace constituents[40]. This abhorrent practice, our discarding and consolidating waste, is repeated, to some extent, around the globe. Lesser economically developed countries look to us as role models because we've told them to. And sadly, many of them believe us. If you need evidence that we have broken with nature and lost our way, here is it.

Repeatedly, we will see how our societies continually try to hide our messes and sweep our crimes under the soil or below the waves. But it is all still there, patiently waiting to come back to haunt us, returning home to roost. Because there is no there, there is nowhere else. There is only Gaia.

Soil is a mixture of the broken-down mineral by-products of erosion and organic materials, the results of decay. It varies significantly in quantity and quality from place to place around the globe. In 1900, surveyors recorded the topsoil depth in the US's great plains at 40cm (15.8 inches in Minot, North Dakota). As of 2017, 30 cm (11.8 inches, 75%) of that depth has been lost. (Arnason, 2017) This disappearance means only 10 cm (4 in., 25%) of this resource remains. This loss was caused by agricultural activities, the over-tilling of the land by our farming practices, and natural erosion. Add a little

---

[40] Which are: toluene, 34,907 ppV; dichloromethane, 25,694 ppV; ethyl benzene, 7,334 ppV; acetone, 6,838 ppV; vinyl acetate, 5,663 ppV; tetrachloroethylene, 5,244 ppV; vinyl chloride, 3,508 ppV; methyl ethyl ketone, 3,092 ppV; xylenes, 2,651 ppV; 1,1-dichloroethane, 2,801 ppV; trichloroethylene, 2,079 ppV and: benzene, 2,057 ppV. ppV= parts per billion. Data collected by: The Pennsylvania Alliance for Clean Energy (ACE)

drought and you get the "Dust Bowl" of the 1930s. This soil loss was, therefore, ultimately the result of human activities. And not just random human activities; this never happened under the original people's watch. This squandering of a vital resource resulted from a population with a particular set of beliefs and ethics. It directly results from the practice of the Dominion mindset.

This topsoil, which took millions of years to form, was originally the product of the sediment layers of ancient shallow seas. That, and erosion from the mountains to the west, mixed with thousands of years of grasses and forbs growing, fixing nitrogen, and passing away. These forces combined led to this dense blanket of life. Now, not only has there been a loss of soil, but with it comes a loss of the nutrients necessary for the production of those native grasses or our crops. The deep, rich soil the settlers encountered initially was perfect for agriculture, but their agriculture was "lossy." It took without replenishing the resources that made growth possible in the first place. It was like a swollen bank account that was continually drawn from, with no deposits to replenish it. To our settlers, it seemed like the land's fertility would always be there. That was a dangerous miscalculation because it won't be, if we keep acting towards our home like foreign cavaliers on holiday.

Grasses used to cover the Great Plains, far and wide, and their roots formed a formidable mat that held everything in place and intrinsically improved fertility. The wildlife that moved over and under the ground aerated and fortified the soil. These root mats were so dense that the construction of sod houses was not only possible; it was widespread. This interwoven layer made up a massive store of carbon in the form of living tissue. There is currently no economical method of replacing this topsoil or the carbon and nutrients it once contained. We've repeated this scenario across North America and around the world. Another very important type of soil, known for it's still pristine carbon storage, is the permafrost. Permafrost occurs above the artic circle around the northern hemisphere. Permafrost is defined as any soil that remains frozen for over two consecutive years, but that's misleading. Much of it has been frozen for thousands of years. It accounts for around 24% of the far northern soils and can be thousands of feet deep in Alaska and Greenland. Frozen inside of this soil mass are centuries of accumulated roots, grasses and trees, much like the great plains turf. As an added bonus, there's also some Wooly Mammoths, and the occasional giant sloth (*Megatherium*) frozen whole in some areas. All of this is currently sequestered carbon, and it is now at risk. Many of these soils have been frozen since the Pleistocene era.

As the permafrost melts, the ground buckles and massive sinkholes appear. This is already happening and in areas of Siberia, and the effects are severe. Buildings, roads, and cities can be, and are being, consumed by this expanding thaw. But worst of all is the release of carbon dioxide and methane released as the organic material rapidly rots. This load of dead tissue, once held safely in a deep freeze, will immediately begin to release greenhouse gasses. The world's permafrost is estimated to hold 1700 billion tonnes (1870 billion tons) of carbon. This is twice the amount presently held in the atmosphere. (UNEP, 2019)

Seventeen hundred billion (1.7×10^{12}) tonnes of carbon are twice the amount already held by the atmosphere. This is a disaster actually beyond comprehension. We have no idea what will happen if all of that carbon reaches our sky. The only thing that is certain is that if we continue to follow the status quo blindly, we're going to find out.

In the Colorado Mountains where I work, the "soil" is very different. At elevation, the actions of erosion allow much less time to develop anything recognizable as topsoil. Fine sediment particles are washed, or blown, downhill and deposited on the plains, so what we are left with is crudely degraded rock and gravel covered with a layer of often poorly decayed plant matter. We call this duff. Duff ranges from 20cm (<10") to 1cm (~1/2") thick, with shallow layers being far more common than deep ones. It is primarily composed of twigs and pine needles, with imperfect compost beneath. What we usually call topsoil is an intermix of organic and mineral materials, and it is practically non-existent in the high country. Needless to say, duff is not very fertile. Duff is extraordinarily fragile and held in place by a mycelial mat of different fungal species and tiny plant roots. Without this phyto/fungal framework, it would easily wash away when it rains, or be puffed away with the wind. And in times of drought or other ecosystem disturbances, this type of export is precisely what happens.

In a desert, almost no organic matter remains in the soil. Composed entirely of fine mineral particles, we recognize it as sand. Desertification occurs over time as climatic regimes create extended drought. All the organic material will have been churned up and blown away by the wind in the absence of moisture and the paucity of life. Desertification also occurs with startling frequency because of the actions of humanity. A recent study has suggested that the activities of humans might have created the Sahara Desert. The land, lush and fertile 10,000 years ago, changed as shepherds drove their livestock over it. Eventually, they eliminated the vegetation by forage and killed all the trees through firewood harvest. These activities spurred a change in the land's solar reflectivity, known as its albedo, slowly suppressing the region's monsoonal flow. Finally, the rain ceased altogether. (Wright, 2017) What happened next is a template for the damage we are doing worldwide.

The over-tilling of soil for agriculture and the exhaustion of natural organics can also convert fertile cropland into a desert in a matter of years. The diversion of water for industrial or urban use and energy production also contributes to desertification. Scientists predict this phenomenon to continue and even increase in the coming years under our current practices. Deforestation due to population movements and wildfires will cause deserts to form in the high mountains of many of our Western States. Most of Colorado's montane forests already exist within a high desert climatic regime. The only difference between them and Wyoming's treeless high deserts is just an atmospheric flip of a coin. We need healthy soils for life to survive, but we seem oblivious to this situation as a civilization. Now is the time to recognize all of this, and take stock of our activities.

Healthy soil is living soil. While its fundamental components are decomposed minerals and decaying organic matter, the real secret sauce is the vibrant ecosystem that thrives within. Beginning with bacteria that feed on organic particles, we encounter a pyramid of life. Earthworms glide through the ground, feasting on larger pieces of dead leaves and the excrement of surface animals. They churn and aerate the soil, improving its fertility with their agitation and castings. Fungi grow throughout, breaking down waste and minerals and freeing them for use by plants. Their mycelia hold everything together with between 1.6 and 5 kilometers of hyphal strands per 16 cubic centimeters (0.6-3 mi per1in^3). Isopods, those bugs we called roly-polies when we were kids, scavenge along with springtails (*Collembola*), who eat the fungi and help maintain balance. Balance is always of the utmost importance. It is of primary importance in the soil, as it is in all manifestations of life. Just ask an organic farmer; they'll go on for hours about their dirt because they know what their crops grow in will become the building blocks of your healthy body.

The funga are essential but often overlooked. Ancient beyond ancient, these life forms are not plants. The Funga[41] is a kingdom of life and one of the first (if not *the* first) to pioneer the land. Many possess the enzymes that can break down rock itself, freeing the minerals essential to every living being on, under, or above the ground. Lichen, the colorful coatings on rocks worldwide, is a symbiotic organism composed of a fungus and an alga[42], another terrestrial old-timer. The algae can photosynthesize and create sugars that feed both themselves and the fungus. The fungus produces a home for the couple and protects the fragile algae by preserving moisture and trading those digested minerals for a bit of sugar and a tad of carbon. Fungal companions in the soil work similarly, always in an intimate relationship with plants, either living or dead. Cooperation is a solid survival strategy, in what is called mutualism. Mycorrhizal fungi coat the roots of plants and perform these same services; they can protect the fragile root hairs from drying out and exchange minerals and carbon across the cell membranes, receiving sugars in return. Arbuscular mycorrhizae penetrate the cell walls of the plant's root hairs, blurring the lines between two interdependent organisms. Forests, as we know them, would not exist without this foundational relationship. The Kingdom Funga are keystone life forms and essential partners in our quest to survive this mess.

Healthy plants grow from living soil. When soil quality decreases, so will the vitality of the flora. You can't remove or replace any of the elements of living soil without causing a cascade of failures amongst all the others. When soil is in balance, the chemistry supports all the exchanges necessary for life; remove even one component and the tower of stacked block falls.

---

[41] There is currently a movement to rename the fungal kingdom Funga, instead of Fungi. This would make the three kingdoms: Flora, Fauna and Funga. Kind of makes sense? A tip of the hat to my good friend Giuliana Furci, PhD and ED of Fundacion Fungi de Chile, and a powerhouse behind this proposed change.

[42] Or a cyanobacteria, my colleagues would taunt me if I left them out!

Soil scientists talk about the Ph[43], the ORP[44], the balance of the nutrients: nitrogen (N), potassium (K), and phosphorus (P), as well as the presence or absence of the micronutrients. All of these components are needed for productive soil. And the interactions and transport of these substances by our micro zoo are required to keep it that way. Much of the soil's nutrient bank is stored in the bodies of the bacteria and bugs, with only some fraction floating freely about. Life binds nutrients and exchanges them through all its processes from birth to death. Life keeps its treasures held safely within.

Rachel Carson spent quite a bit of time explaining how our modern habits of using herbicides and pesticides were destroying our living soils (among other things, life itself included). Her masterpiece, "Silent Spring," should be required reading for all naturalists, ecologists, or anyone in the life sciences. It is considered a bit dated now, but the main tenants still hold. Sadly, its impacts on society were mostly transitory as the American public's gaze drifted away towards more mundane issues. It's easy to get distracted from tales of pollution, waste, and toxic chemicals. Our corporations changed many of the agricultural chemicals she highlighted as a direct result of her book. Unfortunately, the ill effects of many of their new chemicals remain about the same. The human race, specifically the developed nations, has effectively been waging war on this planet's natural balance for some time. Herbicides and pesticides kill the life within the soil, and not just at the point where these chemicals are applied. The natural actions of erosion and subsurface water flow spread them far and wide. In the next chapter, we'll discuss this as it applies to agriculture in depth.

The soil is a system of recycling, as is all of nature. Nitrogen, the essential nutrient for plants, is the coin of this realm. It is passed from root to mouth and back again. Nitrogen makes the leaves green and exists in forms that are both life-giving and deadly poisonous. The nitrogen cycle transforms this limiting

---

[43] PH: "Power of Hydrogen"- this is a measure of how acidic or basic a substrate is- different plants like different soil acidity, but within a narrow range- too acid or too alkaline is damaging to life.
[44] ORP- Oxidation Reduction Potential- this is a measurement of the ease at which electrons may be exchanged between atoms at a molecular level. The movement of electrons is the basis of life.

resource, turning the ammonia (NH_3) that animals excrete to nitrites (NO_2^-) and finally to nitrates (NO_3), which is the form that is a primary food for all plant life. Bacteria enable this cycle in the soil with the chemical properties of the minerals contained therein. The O in these chemical names is oxygen, and the presence of oxygen in soil is vital. The worms and moles and the action of roots and mycelia aerate the ground as they plow their way ever onward. These are all things that are either inadvertently or intentionally killed by our lifestyles.

Plants called legumes are vital because they work in conjunction with bacteria in root nodules to "fix" atmospheric nitrogen into organic forms that can become a part of the nitrogen dance. Beans, peas, and peanuts are all legumes, and plowing them in or leaving the root systems intact naturally fortifies the soil. When a plant is harvested and taken away from the spot it was grown, they remove all the nitrogen in its leaves and stalks from the local nutrient cycle. This transport was not a problem in the ancient order of things. If a deer ate some grass, it would deposit its ammonia nearby, in a similar habitat, and everything would equal out. Our production of massive crop yields, which are then hauled hundreds, even thousands of miles away, establishes a different order. Here is nutrient export on an industrial scale, and it is never actively balanced out by its perpetrators. That would negatively affect the bottom line. Over the years, the human race has depleted our soils of their original nutrients and have had to replace them. What we replace them with is of utmost importance to the soil community. Nitrogen fertilizer is not the same as organic compost, which is the product of a natural decomposition process. Yet chemical fertilizers are our current weapons of choice.

Composting is the process that used to break down plant-based "waste" on a large scale. Composters will mix fresh plant material called the green (nitrogen source) with woody material called the brown (carbon source). They sometimes add various minerals to buffer their interactions and complete the balance. Next, they build long piles called windrows and let the ever-present bacteria do their work. Micro fungi and actinomycetes break down the brown, and bacteria attack the green. As the bacteria eat the organics, the pile heats up. For the best results, they turn the rows several times as they mature. Not only does this dissipate the heat, which can eventually kill our bacteria buddies or lead to spontaneous combustion, but it folds fresh oxygen into the pile. There are multiple steps to the nitrogen cycle, one of the primary reactions running in our compost heap. Some stages are aerobic, meaning they require oxygen. Others are anaerobic, meaning they occur in the absence of oxygen. If the pile is not aerated, the chemical reactions will turn predominantly anaerobic, and the pile sours. This sour compost not only stinks but also contains mostly toxic nitrogen compounds and hydrogen sulfide, the rotten egg smell.

If our composting is carried out successfully, what is created is an artificial humus, the soil's organic component. It is called artificial because of how it is made and not because of what it is. It is a superior soil amendment and contains the primary nutrients required for plant life and the rich mix of micronutrients and minerals that were bound into the original materials put through this process of controlled decay. It also contains life. The bacteria and fungi that did the decomposition are still there and available to "seed" the soil into which they were introduced. Insects will have inevitably found our piles, as will isopods and annelids (worms). Compost is safe for all applications and well balanced. Add to the soil and grow your garden. Done!

But compost is labor intensive; just ask someone who had to turn a heavy pile with a pitchfork. It is also bulky, so transport costs are high. In a society that throws everything away, all mixed indiscriminately, into a landfill, we find ourselves lacking a steady source of the raw materials needed to replenish the land. It's in that organic 22% of the trash we squander daily. Returning compost (humus) to the soil is the only way to rebuild it, short of waiting a thousand years for natural forces to do it at its own speed. Returning compost is the *only* way we have to rebuild the soil in our lifetimes; this bears repeating. And it is possible, despite the efforts required. It also beats the chemistry based methods of returning fertility to the soil that the Green Revolution has saddled us with.

Terra preta (Dark earth) is a human-made soil found in the Amazon basin. Indigenous peoples created this rich black soil in layers up to 2 m (6.5 ft) deep between 450 BCE and 950 CE. In areas where the ordinary topsoil was only inches deep and lacking in nutrients, these black deposits are highly fertile even today. Humans built up these soils over centuries by adding manures, bone shards, bits of pottery, and a type of charcoal very close to what is now called biochar. If they did it, so can we.

Biochar gets its name from its source: life. Crews use wood chips as a starting point and then process them in a kiln to produce a valuable type of porous carbon. When they heat the wood to approximately 250° C (480° F) they begin a chemical process called torrefaction. This process is also what was involved in the high school chemistry demonstration of destructive distillation. Wood alcohols, esters, terpenes, and other volatile compounds are driven out of the chips and would ignite on contact with the atmosphere if they could. But they will not allow that to happen; when making biochar, they'll want to use those gasses as fuel for the next stage: pyrolysis. Using that fuel to up the heat the chips up to a blistering 600° C (1112° F), any remaining contaminants are driven out, isolating pure carbon in the oxygen-starved environment. If there were oxygen present at this temperature, it would immediately bind to the carbon and try to create a carbon oxide, but that's not the reaction they're looking for. Instead, they want to let our char mature in this kiln to allow the carbon to puff like popcorn. They're looking for pores to open, billions of pores that make microscopic tunnels in the carbon chunks, increasing the surface area by a million-fold. And viola' they've made biochar. In ancient Amazonia, the folks did this with clever inverted earthen kilns. Today, it is done in giant retorts with vacuums and computer controls, but the result is much the same (actually, ours is better). Biochar is a stable form of carbon that has many beneficial properties.

The billions of pure carbon tunnels can absorb large quantities of water, and organic compounds adhere to them at the molecular level. Adding biochar to the soil reduces the need for more fertilizers because nutrients stick to it. Agricultural plants require less watering because moisture is held by the pores where tiny root hairs can find it. And it sequesters carbon. Biochar, if properly formed, can last for centuries, thousands of years, as the Terra preta demonstrates. Moreover, biochar production has a low energy cost because producers can use the off-gasses to fuel the process; they just need to kick start the reaction. Soon we shall see the explosion of biochar use in the rebuilding our soils, just as we will need to expand composting to a worldwide scale.

The original people on this land, now called America, walked lightly on the ground and held it in spiritual esteem. They knew where their bacon (bison) came from and protected other life just as they harvested what they needed and very little more. The Great Plains were seemingly endless grasslands interspersed with wetlands and riparian cottonwood forests. Their stewardship helped keep it that way (Gill, 2009). Buffalo roamed freely, the game was plentiful, and birds filled the skies with color and sound. Life was ripe with diversity, and even ticks were understood to have a place within it. We can't return to that state of innocence, that Eden. But we can adopt their awe and respect, and we need to. We need to open our hearts to what was in the hearts of everyone at that place and time. We need a dose of humility, and we need to be content with just a bit less. The land can support us as it supported them if we let it.

When the first Europeans arrived on this continent, they couldn't help but drool over their linen blouses at this view that humankind had lost in their homelands. They laid eyes upon a virgin and untouched environment. The native peoples were in balance with their world. They were a part of life, not separate and not psychologically above it. But the horde that descended upon this continent with the promise of a fresh start and free land was conditioned differently. The land was seen as a possession, and people equated their status with its ownership. Europe was over-populated, and their cultures were stifled by religious and racial oppression. Here was the promise of a new start in a new world. The problem was that it wasn't a new world. It was very much a part of the same world, the only world. And they brought their prejudices and habits with them instead of freeing themselves from the bonds they were fleeing. Their education and apocryphal tales still held them enraptured. Try as they might, the new world became just a reestablishment of their old, dogmatic world. This land was a new and untapped area, to be sure. But it shared the same atmosphere. The same sun shone it upon. The same oceans bordered it. Their minds could not yet comprehend how closely they were still connected, and many today can't either. This land is but this tiny sliver on one small world orbiting an average sun in a massive universe. It is a very scarce resource upon which our very survival depends. Yet, many continue to act as if everything is just another object to be possessed, used, and cast off once we've had our way with it.

Those conditioned by that archaic European psychology had little choice but to plunder and horde. The patriarchal religions of the old world not only encouraged this, but seemingly demanded it. They unleashed carnage, the white population swelled, and psychopaths found gainful employment. The new "masters" of this land slaughtered the Buffalo in the most wasteful of manners; they sliced the land into bits, and a primary ethic was consumption and murder. When the immigrants first arrived, Bison herds numbered in the tens of millions and moved across the plains in vast herds. Buffalo Bill, an Army scout, led hunting expeditions for wealthy Americans and visiting royalty that were essentially bloody orgies of death. (Cody, 1894) But his antics were just a drop in the bucket as the Army used this Bison slaughter as a tool of Native American genocide. "Kill every Buffalo you can. Every dead Buffalo is an Indian gone" was an order attributed to an Army officer in 1867. (Merchant, 2007)

There was no incentive to conserve or steward the land and its seemingly endless expanse of resources, just the lust to dominate and possess it. While our desire to collect "stuff" may be partially genetically baked-in, just as crows and pack rats like their shiny objects. The real problem is that so many seem to have lost the ability to discriminate between having fewer nice things and having lots and lots of junk. We are sentient beings and are fully capable of exerting the discrimination and self-control required to make this choice. But too many are still distracted by their culturally induced pursuits. Marketing is the art of getting us to want something, whether or not it makes sense. Psychology makes this field a vibrant and lucrative industry.

By 1897, there were only around 1,000 Buffalo remaining in North America. According to the National Bison Range archives, just 85 of them were free-ranging in the wild. This type of utter disrespect for nature was actively encouraged, and by some, it still is to this day. Strangely, this ethic of wasteful consumption has become synonymous with the moniker "conservative[45]" in our culture, even though its primary ethos is to do anything but conserve. This twisted illusion that the planet is an endless resource for our unbridled exploitation is going to be the end of us. Westerners came to this land and stomped it down with spiked boots and heavy machines when those who came

---

[45] When I use the word "conservative" in quotes, it is meant to represent a false conservatism that values labels over actual definitions. -A social philosophy that promotes waste, greed, intolerance and power over all else.

before us walked it softly with bare feet and deerskin moccasins. They were the true conservatives.

Our childhood is over, and we need to all grow up and live within our means. We need to ground ourselves in the reality in which we now find ourselves. It always comes back to the ground, the finite limit upon which we stand. There is little more room to expand. There is no more time to waste. Over the next few decades, our actions will determine whether we thrive or return to the soil that once fed us. We can do this, and there is a map: the ways of nature, the history of life. We must stop looking just at our history like some enchanted mirror and look at the history of all life if we want to stay a part of this wondrous, continually unfolding tale. We need to stop looking for answers in our myths about ourselves, the ways we think life should be, and see it for what it is.

This Earth is not ours.
We are of it, and therefore:
If anything, Gaia owns us.

CHAPTER FOUR: AGRICULTURE

"... there is no solution to environmental problems without facing the problem of agriculture."
— **Robert Jensen**

The Book of Genesis tells the story of Adam and Eve, two archetypical humans living in a world that was a garden of perfection. Manna fell from the skies, and the climate was so perfect that clothing was optional. Life was leisurely, and the chief entertainments were watching the young fawns learn to stand and judging wild bird warbling contests. Then a colorful serpent tempted Eve to partake of the fruit of knowledge. That clever fork-tongued devil. She did so and offered some to Adam, who, feeling a bit peckish, takes a bite. Then all hell broke loose. The landlord flatted the tree and cursed the serpent. He revoked their lease and sent them packing without a prospect to their names. He condemned them to wander in pain and suffering for a term to be decided later. All because they finally had a thought of their own.

This preamble could've been the beginning of a tale explaining how humans were all helpless, hopeless mopes. For despite having gained knowledge, they were still clueless without the exterior guidance of "the One, True God". Let's just gloss over any culpability of the deity as a micromanaging boss, or negligent parent[46], and go with the assumption this was all the human's fault. The non-subtle subtext then becomes anti-intellectual. Knowledge will only ruin you. And then you'll deserve what you get.

---

[46] Seriously, who leaves the tree of knowledge just lying around unsupervised children?

Or it could be an allegory of humanity's cultural evolution from hunter-gatherer societies to agrarian cultures. The nomad's life may have been hard, but it was also simple. Go to where the fruit is ripe, the game is migrating, or where the weather is warm. The simplicity of plucking a meal off a branch and the few hours of labor required to meet your daily needs seemed proof that someone was watching out for you. Dalliance on the verge, the happy squawks of the children, and only the occasional saber-toothed tiger (*Smilodon*) were the gifts of this lifestyle. It would be easy to imagine that such a world was made for them. However, it would be nearly impossible for them to understand that instead, they were made for this world. But that's precisely the knowledge that we must now come to understand.

Once we set down roots, our farming civilizations had to acquire knowledge to feed their growing groups and co-exist within a new paradigm, one where they were no longer at the mercy of random fate. Civilization is indeed a bitter fruit in that they now needed to plan ahead. And such considerations required humanity to create the concept of ahead: the nasty concept of time. Moving from the simple interactions of innocence to adulthood's premeditated calculations and responsibilities was a significant jump indeed. And it was just a big drag.

Seen in this way, it could indeed have seemed to have been a fall from grace. So, understood no longer as the definitive telling of creation. Genesis could now be understood as a complex amalgam of ancient tales about the trials encountered along humanity's way. Once we divorce ourselves from the Usherian worldview and accept this tale as an allegory, it can be seen as a pretty decent metaphor for the end of humanity's innocence and the beginning of self-directed living. And our road to self-reliance began with our ability to grow our own food.

Agriculture was a milestone in human societal evolution. This newfound ability to control food stocks, plan for the winter, and store what would be required to survive it changed everything. Societies became fixed in place, no longer needing to move around like vagabonds with the seasons or food sources. The need to feed is primary. It even surpasses an organism's drive to reproduce because you can't breed if you are malnourished. Agriculture enriched our diets, increased fecundity, and eventually exploded our population size. The meaning of place evolved too; home became a concept fixed in space and time. It fostered the growth of the idea of ownership as well. My house. My food. My people.

Agriculture gave us the ability to have stable population centers, allowing us to diversify our pursuits. Some folks could now pursue the arts and rhetorical politics. Some could refine the language and copy it on stone tablets. It made knowledge and learning important. They passed skills from generation to generation. And discoveries were safe in the hands of many, not just one or two individuals. Thousands of years before science became a systematic pursuit, these undertakings planted its seeds, with the selective breeding of goats and the trials and errors of crop domestication. The need to keep stores fresh over the winter and safe from pests required learning and imagination. Staying one step ahead of the rodents is still an unfinished task. Thus, agriculture gave us both the desire, and the need for, ever-expanding knowledge.

In nature, plants and animals commingle with one another. You might find a field of nourishing grasses whose seeds could be ground into flour. And within that field would be many forbs, recycling and predatory insects, herbivorous animals, predators, etc. These biomes were balanced interactions that favored diversity and not necessarily the pure production of crops that humanity desired. So humankind flipped that script. They began tiling areas of land for the mass production that selective culture and intensive farming could give them. They fenced out the competing mouths and taught their domesticated dogs to keep predators at bay. Early farmers chose that one grass with its swollen seed heads and removed practically everything else. This system is still the practice most used today, more or less.

Ten thousand years ago, there wasn't as big a problem isolating crops into species-specific fields. Or perhaps it is more accurate to say there was less understanding of the problems it would create. But now, with massive factory farms, it is a huge issue. You could say this situation crept up on us. The more success these methods achieved, the further fresh problems arose. Locusts, drought, or strange wilts would appear, all of which could've been seen as a curse of some sort or another. When the land went fallow, or the soil turned salty, and no apparent framework existed to explain these phenomena, it was easy to attribute it to one of our old supernatural causes (A witch, a witch!). So, superstitions grew up side by side with our nascent Agro-scientific endeavors. It would take quite some time before microscopes were developed and an understanding of biology crystalized, but eventually, these advances came to pass. And with these new tools, hidden layers of understanding were exposed and new ways to see the world expanded. More accurate ways to see the world, as well.

It would take longer still until subsurface aquifers were mapped, and some farmers discovered their field lay on top of ancient ocean beds and their alkaline deposits. But despite finding the sources of our many plagues, many continue to entertain some of our ancient superstitions. For example, it is now known that ecological disruptions will lead to crop catastrophes. But rather than fix our behaviors, many will prefer to throw new layers of ecological disturbance at the problem (as you are about to see.) And many of us still prefer our old superstitions to the chore of continued discovery and its requirement that we continue to develop our ways of thinking.

Our current monocultural approach to farming improves efficiency via mechanization and the concentration of single crops into exclusive, massive fields. It reduces processing costs by excluding "contaminants," such as insects or different plant types that would need to be sorted out prior to our crop's storage, processing, and marketing. It makes it easier to fertilize because all the plants have the same nutritional needs, so there is a single course of enrichment. The farmer needs only one individual seed source, which is now processed by a tiller that is, by design, uncontaminated by and incompatible with seeds from other crops. It makes record keeping and reporting simple. These are all benefits in a world where farming now exists on a razor-thin profit margin that can only be achieved by mass production and computerized logistics. It is a factory on the ground, run by people with machines and chemicals. This framework of agriculture is now so entrenched; its practitioners rarely question it. And the multinational agribusinesses that call the shots and pocket most of the profits do not question it at all. This system is our Green Revolution.

But is this such a good idea? Single cropping leads to concentrations of pests that are species specific. A cornfield is susceptible to corn earworms, rootworms, and aphids, among other scourges. The spread of these pests will always be harder to control because the next plant over is inevitably more corn. Pestilence concentrations within such a field will always be exponentially higher than their occurrence in the wild, due solely to the monoculture itself.

Plant pests or weeds adapted to tilled fields also occur more frequently and exhibit similar interactions. They hide within the rows of crops, stealing moisture and robbing nutrients. And more importantly, contaminating the harvest with their bodies and sometimes their toxins. Then there are the plant pathogens: rusts and blights, smuts, and parasitic plants. These can exhibit ghastly abscesses and deformations on the cornstalks and ears. They can rob vitality and ruin the harvest. The presence of so many weeds and pests leads to the inevitable application of herbicides and pesticides.

Today, herbicides and pesticides present significant problems of their own. Monoculture demands their use, but what exactly are they?

Herbicides and pesticides are poisons, plain and simple. An herbicide is a discrete toxin designed to kill a broad spectrum of species, just hopefully not the main crop. Unfortunately, many of them are still toxic to that primary crop, so careful application is required. In our example, the corn can still be affected, with their leaves showing brown wounds from the herbicide application. However, if done correctly, the corn survives, the spots recede, and growth continues. The "weeds," however, die. Or at least they die back long enough for the corn to get ahead in the growth cycle. Then our main crop can again take control of the field, for a time at least. Eventually, there will need to be another application of toxins. For life adapts and is tenacious in its pursuits, it can not be fooled by simple trickery.

Herbicides, especially one called glyphosate, can now be found in (and on) many foods in our supermarkets. It is probably in your bloodstream right now. A recent analysis published in February 2019 found that glyphosate is linked to a 41% increase in cancer risk for people exposed to it (Zhang, 2019). This morbidity results from the direct contact to spray by those applying the herbicide, not casual contact, such as eating it on that lovely pear. But you might want to wash that thing, anyway.

Here is the pattern that repeats over and over again; Industry finds an herbicide or pesticide to be toxic. So a new one arrives to replace it that is "absolutely safe this time." Until it is found not to be, rinse and repeat. Take "Roundup®," for example. They marketed it as a "really, really" safe herbicide. Monsanto, the patent holder, swore up and down that it was utterly harmless: "if applied as recommended." That's the crux of this biscuit; because almost no one uses these substances: "as directed." The need for personal protective equipment (PPE) such as masks, gloves, and perhaps a Tyvek suit is too complicated and expensive for many to bother. Then there's the need to apply when the wind is still so that the toxin lays down in a nice, neat blanket. But if it's windy today, and crops are dying, what are you going to do?

You're going to use it anyway because your entire year's livelihood now relies on this one chemical application. So, in actual practice, it was and is not safe.

Another slick trick of agribusiness is confusing the questions about safety and causality. When done properly, these tactics confound the critics (and may even short-circuit lawsuits). For example, Monsanto spent millions of dollars publishing "ghostwritten" scientific papers supporting the safety of their herbicide that clouded the issue with a sheer onslaught of their peer-reviewed BS (McHenry, 2018) The con worked like this: they outnumbered the studies that showed toxicity, or other problems with their product, with multiple studies that showed none (because their paid hacks wrote them). Abracadabra! The question seemed answered in their favor. And it worked for quite some time, but eventually, reality crept back in, and ethical scientists began discovering the truth. So then it became time for a new "dodge," a new marketing campaign, and a new corporate image agency. Or, in this case, a new corporation entirely.

Monsanto sold "Roundup®" to the German mega-corporation Bayer just before the lawsuits could close in and saddled them with a $10 billion judgment. (Cohen, 2020) Ouch! Now that's some savvy business maneuvering! Sadly, these types of cruel, transparent, and cynical tricks are almost guaranteed to work because our system wants them to work. Glyphosate is now almost universally shunned by industry and has been replaced. However, Roundup® is still sold to homeowners to kill weeds on their lawns and driveways. The users least likely read the instructions and use: "as recommended." As long as it can be thought that we can eliminate our problems with chemical "magic wands", this type of chicanery will continue.

The next class of chemical toxins our food supply is marinated in are the Insecticides. Insecticides are used to control insect pests. The suffix "-cide" means, in practical terms, "killer." Plants create their own natural insecticides for self-preservation. And in the open fields of Gaia, these simple compounds work well enough. Nicotine is a good example. This is the tobacco plant's natural defense against being eaten alive by bugs. When someone ingests it, it gives them a little "lift" and a nice body "buzz." But take too much, and it could kill a human as well. Now that we understand it, it's fascinating that society would grow a major crop for centuries only to "huff" insecticide. But that's what it is, and that's what's been done.

Within our massive monocultures, something is needed that is much more potent than any plant's natural defensive weapons. Since we've laid out a multi-hectare buffet for these bugs, the chemical equivalent of a mass murderer is required to keep ahead of the insects. And that's exactly what Agro-industry has done with a veritable pandora's box of various chemicals for different crop/pest combinations. Like herbicides, these chemicals have to be applied widely and often. Insect life cycles can go fast, and what kills the bug might have no effect on the eggs and vice versa. "Spray early and spray often" is the mantra of our current system of agriculture. *Viva La Revolucion*!

So, what happens to all these toxins that are spread on our food crops? There are rules and sometimes even laws regarding when and where "-cides" can be used. Application has to stop a specific time before harvest to prevent injecting large amounts of toxins into our food streams. The FDA judges small amounts to be okay for the simple reason that the entire system would collapse without them. But where this noxious stew really goes is into the environment. Wind spreads particles far and wide. Rain ultimately washes excess chemicals away. And you aren't a very successful farmer without rain. The toxins then travel into Gaia's streams and rivers and, eventually, into the oceans. *There is almost no body of water on the planet without detectable amounts of agrochemicals.* (Giessen, 2015) (Md Wasim, 2009)

Our water treatment facilities are designed to deal with a specific and often narrow spectrum of pollutants. They are sediment (turbidity), bacteria, and common organic residues (soap scums). They were never designed to remove complex synthetic organic chemicals[47] or the fragments thereof. So, where do all the toxins released by agriculture go? Drink up; it's in your water! And it's still on the produce we eat, albeit in tiny amounts: Because they wash it, and that water goes... well, you get the picture.

Since monoculture concentrates single crops onto single fields, the nutrients drawn from the soils are predictable. Over seasons of single cropping, the ground becomes depleted of some specific set of nutrients. Again, the primary nutrients required by plants are nitrogen (N) and its compounds; potassium (K); phosphorus (P); and carbon (C). Carbon can come from the air via plant respiration or from the soil as digested organic residue. Plants inhale carbon dioxide (CO_2) and exhale oxygen (O_2), the opposite of animal and fungal respiration. In addition, a spectrum of other nutrients exists in the soil because of nature's cycles, which have deposited these compounds into the humus over many seasons. These micro-nutrients, once depleted, will have to be replaced. Replaced with petroleum-derived analogs.

Fertilizers supply nutrients, but not in the exact form that they exist in nature. Naturally occurring nutrients are a broad set of compounds, in different physical states, leftover from the decay of previous generations of plant and animal life through the processes of death and decomposition. Natural fertilizers are simply the nutrient load of the topsoil. Chemical fertilizers are single compounds that are synthesized for direct plant use. These are targeted nutrients and not the broad spectrum of compounds found in nature. In nature, there are multiple processes involved in nutrient uptake, with many steps and life forms taking part. Our chemical fertilizers are direct nutrient compounds, immediate and concentrated. Think of them as heroin, as opposed to the juice of a poppy flower. And that's precisely how powerful they are. If not metabolized by the plants, excess chemical fertilizers will end up washed out of the soil and, again, find their way into our water supply. Once loosed in the hydrological cycle, they can spread far and wide, causing dramatic effects on the ecosystem, almost none of which are good.

---

[47] Activated Charcoal is the main substance used to remove organic compounds, and it's fairly effective: until the filter is filled to capacity, and then its effectiveness is reduced to essentially zero.

The amount of contaminated water produced annually by agriculture exceeds the amount of water naturally present in all lakes, ponds, streams, and rivers (Giessen, 2015).

*The amount of contaminated water produced **annually** by agriculture exceeds the amount of water naturally present in all lakes, ponds, streams, and rivers (Giessen, 2015).*

That was not a typo — you need to let this sink in.

We find this seemingly incredulous statement to be true because of the constant recycling and overturn of the hydrological cycle. All life depends on fresh water, and of all the water on this planet, less than 1% of it is suitable for life to use. Humankind pollutes it all.

Our current agriculture system is highly efficient and supplies us (at least in the developed world) with cheap, plentiful produce, but is that enough? Another more significant problem with this system is that it's not sustainable. And it's on the verge of collapse.

Mechanized farming is based on tillage. Tillage is the use of disks or other machines to turn the soil. Turning the soil does several things: it breaks up clumps making it easier for young roots to penetrate. Tilling also turns young undesired plants upside down, effectively killing many of them. But it can also act as a weed multiplier by fragmenting plants that reproduce by root division, such as Canada thistle (*Cirsium arvense*) or field bindweed (*Convolvulus arvensis*). This accidental human-assisted reproduction leads to exponential amplification of the original pest plant, not just an arithmetic increase. This abundance of pest plants will lead to a similar exponential need for more herbicides down the road. Some will see this as a source of geometrically increasing profits. But at what actual cost?

By mixing the soil, a homogenous substrate for the application of our chemicals is created. So, in essence, much commercial agriculture is just a hydroponic[48] gardening system. When the soil becomes merely a medium to hold the plants upright, and all nutrients (along with water) are added, it is no longer strictly "farming." It is now high-tech synthetic horticulture at scale. It

---

[48] "labor of water" is a pretty direct translation— but in essence it means soilless crop production.

is tray culture where the tray is Gaia's belly. This cycle is a self-fulfilling prophecy for future crop failures. And not just failure, but complete collapse. It has recently been realized that tillage is almost entirely wrong. The Natural Resources Conservation Service is moving towards a policy of reducing it, hopefully towards zero. Tilling the soil reduces the potential for the ground to hold moisture. (NRCS, 2015) In a natural, untilled field, soil exists in "horizons," stratified layers of deposit, and decomposition. Composting occurs naturally within native soil filled with bacteria, fungi, and insects, all continually moving within it. Life is self-tilling, and this massage occurs without the damage of the plow[49]. This naturally loose soil will rapidly 'wick' and hold moisture, some of it within the bodies of the life forms themselves. Mechanically tilled soil often reaches a state where it actively rejects water (called hydrophobicity- basically: "repellant of water"), so tillage can cause water to run off more readily. These fields will require more water to achieve the same effect as natural soils. In extreme cases, the application of "wetting" agents (basically, soaps) will need to be applied on the field to compensate. If the soil can't wick moisture, runoff takes the nutrients away; so instead of fortifying the land's productivity, fertility is conveyed away. This system promotes an aqueous form of nutrient export.

In nature, nutrient supplies are self-improving simply by virtue of those nutrients' bio-availability to the soil's microfauna. In reality, the micro-organisms themselves contain much of the biologically available nutrient stores in their bodies. Fast life cycles can be seen as short-term loans of nutrients that are rapidly repaid on death. But these same micro faunae are now being extinguished by herbicide and pesticide use. And while that might seem like a return to the soil of nutrients from their bodies, in the absence of secondary recyclers, such as fungi, it's just waste. And the fungi are killed with our "-cides" as well. So we're inadvertently filling the soil with poisoned corpses that can't easily decay. That's just the way it is today. We have been pre-conditioned to think that bugs and fungus are disgusting- even evil. But they are the foundations of many cycles of life. For every harmful pest eliminated with poisons, many beneficial ones are also destroyed while we're at it.

---

[49] So, when someone suggests we "turn our swords into plowshares" remind them that a plow IS a sword— just one that is used against Gaia and not other people.

A startling indicator of the excessive presence of agrochemicals in our world is the honeybee's plight. Bees have been having a rough go these past few decades, and it's not getting any better for them. Hive collapse because of the *Varroa destructor* mite has spread worldwide since the 1960s. But beekeepers believe the problems with honeybee mortality are much more significant than just a single pest. A recent study has shown high concentrations of pesticides in bee pollen inside hives, pollen that came from non-agricultural plants. (Long, 2016). Let's consider that for a second, shall we? Bees are dying because of pesticides encountered from non-agricultural (wild) plants. So widespread has been the use of the "-cides" that they are now found on undomesticated plants. Much like we find them in our water supplies and our bloodstreams. Runoff, groundwater, and over-spray have bathed native plants in this noxious brew. Many toxic compounds that affect bees are called neonicotinoids: synthetic pesticides that are even applied to seeds, pre-planting, to protect them from predation. It seems there is no part of our current agricultural practice is not soaked in venom. These synthetic compounds are based on our old friend nicotine. They are now termed "neonics" by the media, perhaps for simplicity or possibly to avoid bringing up nicotine and all the conflicted emotions that its mention might conjure.

Our reliance on toxins as a means to combat other life forms is based on a very important misunderstanding. All life forms on Gaia are built to adapt and evolve. When they are assaulted with "-cides" they simply evolve immunity in the next generation. This is actually helped by eliminating competition from other species that are killed as "collateral damage" by the poisoning of Gaia's fields. There is no end in sight for our chemical wars against nature. Except, perhaps, for the one we have been inadvertently promoting with our combined assaults against her- the next great extinction.

The tilling activities of mechanized farming also create great dust clouds at both planting and harvest and can spread these compounds far and wide. This smog results from the massive disc machines, and not only does that dust spread the toxins, but it also exports what little topsoil may remain. According to UNESCO[50] the planet is losing about 24 billion metric tonnes (26.4B tons)

---

[50] United Nations Educational, Scientific and Cultural Organization

of topsoil per year, primarily because of these agricultural practices. This topsoil took centuries, if not millennia, for the planet to create, and it is difficult to imagine how it will be replaced at anywhere near the speed it is being lost. Nor within the time frames, the human race needs our soils replaced or restored. If we wish to continue eating, that is. Cynics would use this as an argument for continuing with the status quo. Technically, topsoil is no longer strictly needed for these industrial-scale hydroponic gardens.

The fertilizers, pesticides, and herbicides required for this type of agricultural production are almost exclusively derived from petroleum. The same black crud that is pumped out of the ground to fuel and lubricate our cars is the primary building block used for nearly all the produce we eat. Oil-based pesticides and herbicides are absorbed by the growing plants. The fertilizers become the plants' tissues. Even though protocols exist for the final application and waiting periods before harvest, traces of all these compounds remain on much of the food we consume. Sound appetizing? It really doesn't. We have passed peak oil, and the producers know it (Ahmed, 2013), yet we remain dependent on a toxic and increasingly scarce resource for our very food supply. If we "are what we eat," then, at least in the United States of America, we are "Sweet Light Crude."

Which brings us to meat production, a new group of contentious practices that has led to much outrage and nonsensical legislation in the US and abroad. Industry has used the same mindset with animals as when considering our plant crops. First: They deal with animals as if they are objects, with very little thought given to their needs as living creatures. Second: They have chosen to produce them in such crowded and potentially unhealthy conditions that the use of massive doses of medication is required just to keep them alive. And finally: They have convinced the rest of us that the "realities" of economics require them to produce in this manner, to crank out as much as possible, for as little money as possible, to remain competitive.

Factory farming of chickens and the feedlot finishing of beef place an absolute premium on space. Poultry producers may keep egg laying and meat chickens in cages of less than 0.03 m³ (one cubic foot) or crammed into feeding sheds where some may perish simply because they can't make it to their waterspouts. These sheds are more like massive barns, and if the power fails and the giant fans stop spinning, thousands of birds can die in hours of the heat and foul air. They are also excellent vectors for disease, and if an infection gets into such an operation, entire flocks may end up sacrificed. This mass euthanization can be done by shutting off those fans and letting them suffocate, as the US Chief Veterinary Officer suggested in 2015 when faced with an avian flu outbreak. The most intense fear a vertebrate animal can experience is that induced by a rapid increase in CO_2 levels, the suffocation reflex.

"No single federal law expressly governs the treatment of animals used for food while on farms in the United States." (AWI, 2007)

There are now over 60 billion chickens worldwide in poultry production facilities, large and small. In our American factory farms, the chickens' life begins with a rousing trip on a conveyor belt shortly after hatching. They are rapidly sexed and the hen's head to the de-beaker. As they mature in cramped confines with little to no stimulation, the hens tend to peck on each other. This pecking can even lead to death, but from a commercial perspective, this pecking leads to reduced production. So off the beaks come with a red hot wire. Who gives a crap if they have trouble learning to eat without a break. What's another 2%-3% mortality in the long run? This is mass production folks!

Into the next chute and down another conveyor they go! Ahead of them is a deafening, overcrowded life of food, poop, and egg-laying. When they reach their terminal age, they may be force molted. Force molting is a production enhancing practice where the older hens are starved for a period, and stressed, causing them to lose most if not all of their feathers. They may just die, but if they don't, a return to food and water will cause a flush of egg laying. This practice squeezes another dozen or so eggs out of the birds before they are sent to the slaughterhouse to become chicken sausages or burritos. Their bodies are a bit too thin and worn by their hard life to be sold as whole chickens in the meat department. A backyard hen can live happily and produce eggs for 6-7 years, under commercial conditions they rarely last longer than 2 years. Force molting is outlawed in Isreal and the EU, but is still a standard practice in many parts of the USA.

Broiler chickens are a hybrid that grows so fast their legs could give out from their unfathomable weight gain. The "hybrid" designation means that two distinct lines were crossed (two unique breeds) to produce the young who will live a hellish life of eating, growing, and suffering. They are designed to accumulate around 0.45 Kilos (1 lb) per week throughout their brief lives. They might double that amount of weight gain in their final week before slaughter or implosion. If they were to live beyond their 6-8 week "finish," their hearts might explode merely from the stress of this insane growth. Their legs may break under their heft and because the calcium can't lay down fast enough to support their bodies. In practice, these outcomes are not uncommon. You may have opened a pack of drumsticks to find one leg broken; ever wonder why[51]?

Broad-breasted Turkeys, another product of selective breeding, are incapable of self-sexual reproduction, reaching such obesity before sexual maturity that they couldn't "do it" if they wanted to. These birds are bred by sperm harvest from the males and then artificial insemination of the females. No one enjoys this, not the birds and not the people who do the harvesting and insemination. I discovered the hard way that this is not a good fact to bring up during Thanksgiving dinner.

Again, no federal laws in the United States of America address this cruelty to these animals. Each year 9 billion birds are raised for food in America, and over 305 million hens are caged and treated as egg-producing machines. These living creatures evolved to move as they grow, stretch their wings, and bask in the sun. Yet, in our wisdom, we, as a society have decided that none of that is required. This lack of basic compassion is why there is a PETA[52].

---

[51] Before the Lawyers get too far sharpening their pencils: The broken legs could also be a by-product of the processing machines. Or, just abuse from the also-abused farmworkers.

[52] People for the Ethical Treatment of Animals, you know them, and love them or hate them, they're are around for a reason.

Cattle are animals at least as intelligent as dogs or cats. Yet, in my experience, growing up around ranching, they are generally considered "stupid" by their caretakers. Sure, they'll drink oil, but you should see what some of our Cowpokes drink[53]. Most cattle out here in the American West have seemingly humane early lives. They wander pastures and range lands, usually in the herd they were born. They live off the grasses and forbs that grow on that land. The average amount of land they have to wander is often about 7-10 acres per head, a sweet gig indeed. But, of course, that's because out West, the forage is sparse, and as it turns out, most breeds of cattle don't digest the types of grasses present in North American efficiently.

Cattle were derived from the Aurochs of Europe, Asia, and Northern Africa, and their ancestral feedstuffs were far lusher than what grows in Colorado, Wyoming, Montana, and the Dakotas. Their four stomachs are required to get nutrition from poor-quality foods through re-mastication, fermentation, and microbial actions. However, "garbage in — garbage out," as they say. Their ancestors developed on more nutritious chow, so while they can use the local forage, the efficiency of grass to meat conversion is far lower here in the Americas. This digestive inefficiency presents a situation where the threat of overgrazing is an ever-present possibility. The wasteful conversion of feed to cow also leads to their constant burping and concurrent methane release. Methane is a potent greenhouse gas, 80 times more powerful than CO_2.[54] So while many a sniggering junior high-schooler will tell you that cow farts are causing global warming, it is actually their burps that are the most troublesome bodily discharge.

Cattle are often raised on the range to a certain age/weight, and then they are shipped off to feedlots for their "finish." Feedlots are generally rank-smelling places. High densities of animals mill about feed troughs, creating massive amounts of manure and equally massive amounts of stress (not just on the animals). Going from an open range with your family to a crowded chow and dung fest, surrounded by strangers, has got to be a frightening experience.

---

[53] I'm speaking of Energy drinks here, corrosive crap in a can.

[54] You caught it! I said earlier that methane was 30 times more damaging, and both are correct. It all depends upon how you measure the effect over time. The Global Warming Potential (GWP) of methane over 20 years is 86 times that of CO_2, The 100 year GWP is 34 times.

These are bovine "concentration camps." The intensive grain feeding creates toxic run-offs in the localities close to feedlots. Not only are massive amounts of ammonia and methane released into the atmosphere, but toxic organic chemicals, from the overuse of medication, taint the groundwater. With such a high population density, diseases can run rampant. And any cut occurring while standing in a pile of actual bull shit can quickly become infected. Therefore, continuous doses of antibiotics are given to the cattle during the final "beefing-up" process. An auspicious side-effect of antibiotic use is weight gain, so they are used universally as both a disease preventative and a production enhancement. These antibiotics not only leech out into the groundwater, affecting the feedlot soil's heavily taxed microbial community, but they can also lead to the creation of "super-bugs" — antibiotic-resistant bacterial strains.

The density of animals from differing locations also creates an excellent vector for viral spread and mutation. Add some stressed-out, over-caffeinated human workers to the mix, and you have created the perfect conditions for a novel virus to jump from species to species. Zoonosis is the process by which pathogenic diseases move from species to species. In this case, from food animals to humans. These feedlots could be the functional equivalent[55] of the "wet markets" in China that are believed to have been the novel coronavirus, COVID-19's, initiation points. There's much talk these days about the dangers of "Super-bugs," and much like the weather, it seems like there is very little to be done about them. (Hint: everyone can stop using antibiotics like tic-tacs, for a start.)

Hog farming is a mishmash of our chicken and cattle practices, combining the worst of each. Pigs are almost always ultimately confined in Concentrated Animal Feeding Operations (CAFOs). The current "American Model" has corporations owning the animals, supplying the feed and medications, and dictating exacting husbandry standards. Profits for the actual farmers are laughably tiny (but these farmers are not laughing). One farmer might breed the animals and send the piglets to the next farmer, who grows them out to mid-size and then to a third, who finishes them to slaughter weight. This business model's stress on the farmers often leads to animosity, for the

---

[55] They are the moral equivalent as well.

corporations for sure, but often directed outwards toward the animals. There is no debate that Pigs are very intelligent, smarter even than dogs. So, to be raised in these overcrowded conditions, sometimes mistreated by handlers, who are often shell-shocked themselves by the stress, and then injected with hormones and antibiotics, it's as close to Hell as you can imagine. Ask a hog in one of these operations what a monster looks like, and if it could, it would likely reply: "You."

It is wise to realize that most of the farmers themselves are not happy about these situations, either. No one gets into animal husbandry without caring about animals. Crop farmers may have come into a family business or found their way to fields for a love of the outdoors and a desire to work with the soil and plants. Then they become trapped in a vicious economic cycle where they are worn down and financially starved out by a system that values profits to an irrational extreme. This lifestyle used to be considered romantic, where one would rise and sleep with the sun. It was once seen as a life where children could grow up close to the land and nature. That way of life has been stolen from the farmers as surely as it has been from the crops and livestock. It is nearly impossible to get into the farming business these days and equally difficult to get out. They are bound by debt and, in many cases contractually obligated to do as they are told. Farmers and ranchers historically don't take too fondly to being told what to do. So, no wonder they hold such deep frustrations.

So how does the status quo deal with this inhumane system? By making it illegal to film, photograph, or otherwise expose these practices. I might be in hot water here in Colorado if I identified an organization or farm in this book. It's illegal to "slander" agriculture in Colorado, as it is in many of our states. (Stauber, 1997). This tactic is the "conservative" strategy for dealing with a systemic problem. I want to note again that what is currently called "conservative" is often anything but conservative. Their core desire being the continuation of mindless consumption and loyalty to a flawed status quo. There seems little inclination to conserve anything, as the slogan "Earth First — we'll mine the other planets later" plainly reveals. Perhaps the only things such a mindset wishes to see conserved (In the now rarely used, correct meaning of the word) are knowledge, creativity, and imagination. The very three things we so desperately need right now.

These practices done in our name serve no one. Well, almost no one; there are a few executives and stockholders who are pretty happy with this situation. But to make them pleased, we have allowed the sacrifice of the sustainability of our vital food supplies. There is no endgame here. If it all goes bankrupt, the investors will get a clean slate, but our society will starve. Unbalanced chaos will wreak havoc upon our cities and institutions. We have to fix this before it's too late. The list of dangerous activities: chemical agriculture, petroleum dependency, pollution, and poisons in our rivers and bloodstreams, the destruction of fertile topsoil by erosion, nutrient export, and exhaustion goes on and on. We can fix this. And that's good news because we have to fix this.

These problems have led many to tout the "organic" food market. And while there are plentiful claims that organic foods are more nutritious than factory-farmed produce, the science is still out on that. (Smith-Spangler, 2012) But maybe that is not the most crucial point. Organic produce is grown without chemical fertilizers, herbicides, or insecticides. Fertility is achieved by mimicking natural processes and is therefore inherently sustainable. Organic fertilizer comes from compost. And as discussed, compost is both natural and sustainable. Herbicides are replaced with the hand pulling of weeds or accepting their existence and a bit more post-processing to "weed" them out. Insects are dealt with using a minimal number of naturally occurring insecticides in low concentrations; or introducing predatory insects to counter them. One could also reduce the concentration of the host plants by turning away from monoculture. And finally, farmers can use hand removal of the insect infestations. This practice is very labor intensive, and it is a shame there isn't enough human power to achieve a safe and renewable food supply. Just kidding, there is if we wish to make it so. There are billions of us now, some sitting with their cellphones playing farming games right now.

Organic farming has a lower productivity rate regarding the food produced within a fixed land area. This deficit could be as great as one-third less production per hectare (2.47 acres). (Gilbert, 2012) Organic farming currently represents less than 1% of all agriculture in the United States of America. And while desirable, organic farming doesn't seem to be the "silver bullet" for feeding a growing world population. But maybe there is another way to use this knowledge and these techniques. Of course, keep in mind, that unsustainable is unsustainable, and the population will undoubtedly cease to grow if the entire system collapses. Or if widespread famine arrives, as Malthus (Malthus, 1798) feared. Civilization is painting ourselves into a corner in terms of food production. The mechanization and mass production of grains, vegetables, and meat products have followed the same frameworks used to create cars, refrigerators, and televisions[56]. Unlike those commodities, offshore sourcing is not a viable option for all of our food, nor is off-planet production, though some might try.

Looking into the various laws and lawsuits that have been pursued to blur the meaning of "organic food," one can see what agribusiness really desires. In 1990 Congress passed the Organic Foods Production Act. This law's enactment began a slew of lawsuits from manufacturers seeking to dilute the meaning of the term organic so that they could claim the label without actually needing to follow the rules. Many have argued that just the ability to claim that your food was produced without chemicals gives an unfair advantage in the marketplace. By manipulating our definitions of "natural" and "organic," the exploitation of our current agriculture system seeks a dark completion. Unfortunately, there is still a back and forth going on today, as business interests seek their profit; by scorched earth if necessary.

So, while organic farming solves many of our problems, it has its drawbacks too. Raw land is becoming scarcer and ever more precious. It is almost impossible for people to buy farmland and "work it" successfully. It just costs too dang much, loan interest makes payments outrageous, and prices for their farm products are too darn low. Factory farming has created unreasonable expectations of what food is worth, but just ask a starving person: What is food worth? And they will answer: "Food is worth everything."

---

[56] The last U.S. manufactured television rolled off the line in 1995, when Zenith sold out to Korean manufacturer LG. Let corporations have their way, and all food production might be similarly outsourced as well. Look for the brand: Soylent Green while you're at it.

Food is worth everything.

Governments can subsidize large corporations to produce cheap food that pollutes our world and bodies, but there is little help for those who want to do better. The destruction of autonomous family farming began as soon as the Green Revolution rolled off the assembly line. Agriculture is perhaps one of the noblest of human endeavors. It is the original science. But it has been transmuted it into the ranks of simple labor and low, menial labor at that. And that reclassification completes the exploitation of both the farmer and the farm. It may now seem that the Green Revolution was, in fact, a coup d'état.

But is there enough land to switch to a sustainable model even if we wanted to?

Enter: permaculture. Permaculture effectively uses poly-culture organic farming methods with multiple crops and animals cohabiting the same plot of land. Recycling within a more or less closed system. Polyculture is the exact opposite of monoculture. This system is agriculture that mimics nature, and it is not only possible; it is precisely what is needed today. In fact, this is very close to how farming used to be done. The solutions to our many problems can come from looking at nature and seeing how this tiny world sustains life. Understanding how diversity and interdependence make the ecosystem productive is step one. Mimicking it is step two. No step three is required unless it's pouring a tall glass of lemonade and watching the sunset.

> **Permaculture-** "is a system of agricultural and social design principles centered on simulating or directly utilizing the patterns and features observed in natural ecosystems[57]."

A primary tenet of permaculture is the restoration of ethics in all aspects of life. I challenge anyone to read the previous section on our current agricultural practices and suggest that our choices have been ethical. Whether by relinquishing ourselves to the inevitability of "the way things are" or choosing to accept or advance these practices consciously, the entire system has become

---

[57] Pretty much the generally accepted definition of permaculture.

unethical. When an imaginary commodity[58] may threaten the very system through which we feed ourselves and the world, this economy has become unethical. Of course, there will be those who will violently reject these suggestions, who will demand that we continue on this dangerous trajectory. And there will always be a rationale behind their reasoning that is corrupt. Humanity finds itself at this historic juncture because of a manifestation of "the One": one way, one right, one power of absolute and inviolate ownership and control. "One crop, or one species on each farm, separate and intensively produced for a maximum singular outcome" is the mantra, and it seems a little obsessive, doesn't it?

> "You shall keep my statutes. You shall not let your cattle breed with a different kind. You shall not sow your field with two kinds of seed, nor shall you wear a garment of cloth made of two kinds of material."
> - Leviticus 19:19

And here is seen the desires of Dominion as it relates to agriculture from the English Standard version of the King James bible. Very little room for mis-interpretation here. Just the absolute edict to segregate, scrutinize and judge harshly. There is an unabashed anti-nature sentiment reflected in these words, where the god of Abraham makes clear where his line is drawn. Fortunately, this is not taken literally by most anyone, anymore- especially that part about cattle. There are hundreds of breeds and counting. And aunt Celeste can continue to wear her paisley shawls and fancy peasant blouses as well.

Polyculture is planting multiple crops on the same plot of land, either in concert or in rotation. Polyculture is every natural field you've ever seen; it is the Rainforest and even the Arctic. It is how nature goes about her work. Polyculture tends to increase the fertility of the land. It increases the diversity of the harvest. And it also helps mitigate the problems of insect and weed infiltration. Do not think for one second that pests or plagues are not a danger to life in its wild state. Nor should you assume that diversity is not a direct reply to the risks an out-of-balance system poses. Polyculture may reduce the efficiency of machine harvest, but let's pretend for a moment that machine

---

[58] Hint: $

harvest is not the only way to reap what's been sown. Or, let's pretend that our engineers are capable of creating new mechanisms that can accomplish this feat of variable harvest. Or even suppose that skilled pickers can be paid well enough that they can raise a family and clothe, house, and feed themselves and their children. We won't have to allege this if we make the ethical choice to make farming work for the farmers and not just a few corporate titans. Polyculture is a multi-tiered approach to agricultural production, where the harvest is diverse and healthy at every stage of the process. When all of this is done with organic farming techniques and "Low-tillage," the problem of non-sustainability disappears into a puff of non-polluted air. Gaia has shown us that this is sustainable. The way out of our maze is right in front of us. Just follow the paw prints through the labyrinth.

Before the massive introduction of petroleum-based fertilizers, crop rotation was a primary method of sustaining soil fertility. Legumes are those plants that have symbiotic bacteria in their roots that fix atmospheric nitrogen into biologically active nitrogen forms. They essentially pre-fertilize the ground. Planting them between main crops, in a rotation, used to be standard practice. The rich mixture of bacteria and fungi found in healthy soil further increases this fertility naturally. Of course, under the present scenario, pesticides and herbicides kill those bacteria and fungi. But we're talking about doing better here.

Life is not uniform; some crops like full-sun (shade intolerant), and some like the shade (shade tolerant). By growing one over the other, practical productivity can be increased. Also, the need for shade nettings and other resource and time-intensive structures is reduced when we put plants where they want to be, not just where our habits want them. Everything has a place in nature, and we can use that to our advantage if we take the time to watch and listen to her rhythms.

Where permaculture really struts its stuff is that a farm is no longer technically required for certain production scales. Your yard, a city building roof, any place where you can put soil and let the sun strike it can be a "farm." Its greatest success will be in transforming the mega-scale applications on land that is now intensively factory farmed. But a desirable manifestation of this practice would be to help ease the city/country dichotomy. Farms will move into the city, and businesses and people who live in the city will move outward into the country, stabilizing our population's density, reducing the impact of our footprints. Changing our habits and synthesizing these seemingly opposing social structures through intermingling will reap greater harvests in our human endeavors as well. When you polyculture, the artificial separations of this activity here, that activity there, will slowly cease to be as important. Country morals and city ethics will merge on common ground. This synthesis is the natural and intended social design principle of permaculture; heal ourselves as we heal the land. Our societies can become more diverse in their cultural and spatial arrangements, helping break down our current social fragmentation. This transformation will also lead to better parties.

Right now, zoning regulations work on a "one thing/one place" rule. And while the logical example might be for our heavy industry's segregation into zones of tolerated pollution, even this caveat might no longer be necessary. Our industries have to become cleaner, and as industry moves away from an oil-drenched world economy, they will. This type of framework is not something new. It is actually closer to our agrarian roots than our modern factory farming. Since humankind learned husbandry from watching nature, it makes sense that farms used to be highly diverse. There were multiple crops, different animals, the waste of one becoming the food for the other in a constant cycle.

The cycle all live within, but have come to ignore as much as possible, is the rhythm of the seasons, the re-cycle of nature. There is a time to sow, a time to grow, a time to reap, and a time to lay fallow. Modern systems bypass this cycle by distributing cropped products across multiple countries and climatic zones. Our everything all the time philosophy results from economies that demand continual growth and ever-expanding profit. Monoculture is firmly rooted in the profit motive, where food is just a commodity to be turned into money. Gone are the motives of pride in a job well done and the satisfaction that you have given your neighbor the absolute best product available. In its place is a system of quantity over quality, form over function. It has given us strawberries that never ripen and pears that turn into tasteless grit (but at least they're cheap!)

Studies of polyculture have been promising and support what the small farmers who have been practicing it for the last couple of decades would tell you. Increasing diversity increases productivity while decreasing insect scourges and weeds. (Gliessman, 1982) (Picasso, 2008) These results make sense since the total bio-load of any given insect host plant is reduced per hectare. Other species planted nearby may also have a suppressive effect on troublesome insects as well. Increased biodiversity can allow these pest's natural predators to show up, taking care of the problem for us. Introduction of farm animals, such as chickens or guinea fowl, in scheduled periods, can eat the pests right off the vine as well. Humankind used to know all of this because we first learned to crop from the natural world. That wisdom was lost when we flooded our fields and pastures in black oil.

I want to mention quickly that chickens are omnivores; there is no such thing as a vegetarian chicken. If the health food store promotes "vegetarian-fed poultry" as healthy or healthier, they're just playing another marketing game. Chickens are not vegans, nor do they wish to be. If you saw how they run after a fat grub, you'd know that as well. Feeding chickens a solely vegetarian diet is unnatural, and perhaps mildly cruel. Be wary of ignorance portrayed as enlightenment in the marketplace. Be cautious of your emotional triggers being used to make you embrace nonsense for someone else's profit. Marketers don't care whether or not chickens are vegetarians. They're just looking for those things you hold dear, to exploit them and ultimately exploit you. If you're eating a chicken, you're not a vegetarian, so why would you care if the bird was one or not?

In sync with nature, another emerging practice is the use of fungal "pests" as insect predators that are species specific and environmentally neutral. *Cordyceps militaris* is a pathogenic fungus that feeds on living insects. It has been sensationalized as the "zombie fungus" because of its habit of growing inside an insect host and then taking control of its mind to make it climb obsessively. In the final stages of infection, the bug is compelled to go to the highest point it can find and latch on. Mycelium then rapidly grows out of every joint, anchoring the bug in place. And finally, in the more spectacular manifestations, a single mushroom explodes from the head and showers its spores onto the bug's unsuspecting fellows below. This macabre lifecycle could be the basis of an X-files episode if it happened to humans (it was).

These strains are species specific, making them very desirable for use in the field. The caveat is: releasing a nonspecific parasite of this type could most definitely lead to undesired consequences. Care is needed here, and "caring" doesn't seem to be a good descriptor of agribusiness today. But care will become the driving force of our new agricultural evolution, as our need to change changes everything.

Fungi are currently underrepresented in American farming as a food product as well. While US mushroom production has increased 20 fold in the past decade, it pales compared to Asian countries, where a meal can contain as much as 30% fungal material. Mushrooms are a fast-growing agricultural crop, and the waste involved is very low. (Chen, 1993) A ten-pound block of spawn can produce 3-4 pounds of fruit; the rest is a high-quality organic fertilizer; the mushrooms having broken down the wood or straw material into compost. Tempeh is an Indonesian food made of soybeans that is fermented with the mold *Rhizopus ogliosporus* and is eaten like a cake. In this instance, the product is 100% food with zero waste, as you essentially eat both the fungus and the medium on which it grew. Mushrooms are high in fiber, contain proteins, B-vitamins, and trace nutrients.

They can be and often are used as a substitute for meat. Quorn is a meat replacement originating in the UK and is based on fungal proteins from the mushroom *Fusarium venenatum* grown on a corn-based medium. Its popularity is growing worldwide, and their nuggets are considered just as tasty as the chicken-based kind, without the need for chemicals or cruelty.

Impossible Burgers, now a staple plant-based meat substitute, owes its flavor to fungally produced Heme, the protein in blood that gives meat its unique taste. Still works in progress; many of our engineered "meats" are highly processed preparations that may have different issues than just their sourcing. It will be found that simpler foods with lower energy costs will be essential in balancing our human population with the planet's nurturing capacity. There is little doubt that people will consume more manufactured foodstuffs in the future, but we still should try to keep them as lightly processed and straightforward as possible.

Americans can and will eat more fungal products soon, with a zero loss of taste and variety. In the USA, we are rapidly losing our strange repulsion to fungi as a society, and that's an excellent development. The fungus among us can help shepherd humanity to a long and prosperous future, just like the one they have enjoyed for billions of years.

While businesspeople may expound the virtues of diversity within a portfolio, they are actually focused on a singularity: profit. And the only acceptable form that profit may take is money. Satisfaction is not considered a profit, but it should be. Capitalism needs a component of social fairness and justice to function properly. In the form of sensible regulation, social fairness is the grease that makes the markets run smoothly. But many companies are so far off the rails right now that they often practice the opposite. Greed is seen as "good," and in 2018, in America, a Presidential candidate remarked: "You can't be great if you're not wealthy.[59]" This attitude illuminates a blind spot that has been fostered by our success at living in denial of the essential facts of life. Life results from an ecosystem. Ecosystems are complex, diverse, and interdependent. There is no top or bottom; removing any part can take down the entire system. We've been cruising on the resilience of natural systems for some time now. Our actions now seem sure to topple them, along with ourselves.

Monoculture still has an ace up its sleeve, and that card is genetic engineering. Genetically modified organisms (GMOs) are all the rage. Plants that create their own pesticides, super-sweet peppers, corn with twice as many kernels, and so on are the fruits of this endeavor. Proponents will say that genetic engineering is no different from selective breeding, that it's just faster and more direct. But that's not precisely true. While genetic engineering might produce beneficial adaptations that can help increase productivity or drought resistance, baby steps would be safer than some of the massive leaps now being designed. When plants are bred selectively, it takes generations to make one minor change. It can take decades before something is widely accepted and proved, and that's important. While time is not our ally at this particular juncture in history, rash actions can, all too often, lead to disastrous unintended consequences.

GMOs are not part of the established ecosystem. There is no web of interaction

---

[59] That would be: Donald Trump at a rally in North Dakota on May, 26th, 2016 But did he really mean it? Who the hell knows?

with the environment, just a novel mutation that humans have the power to promote without the natural selection that should be exerted on them. Again, they are novel mutations humankind can create, but not necessarily control once released into the wild. No one knows what the long-term consequences might be. We can not predict what the interaction with established species will be when they cross-fertilize with our creations. Many of these GMO crops are "gene drives"; fully active organisms that can hybridize with established species. And they are being released into the wilds.

Monsanto (them again) famously sued a group of farmers after their genetically engineered test crops cross-pollinated with those farmers plants or just showed up. This spontaneous appearance is called "volunteering." (Kimbrell, 2005) They have won dozens of suits because of the copyright system in law. The farmers' loss of these claims benefited Monsanto to the tune of $23 million, and the defendants were regrettably forced to destroy their crops. They had to burn their entire years' investment in seed, husbandry, and labor on top of paying out the cash. (Harris, 2013) Perhaps corporations should pay damages to those farmers for defiling their crops and fields in such cases. But that's not the world in which we currently reside.

The entire idea of patenting life is absurd. While there is a narrow area in which this genetic engineering might be helpful[60], generally speaking, unless you've made life from scratch, you are not creating; you are just editing. Considering the amount of genetic data within every life form, this is akin to someone taking Shakespeare's complete works, adding a sentence, re-publishing it as their own, and then suing every publisher that had ever printed those books previously[61]. Every species on this planet, us included, is here because of a long, hard-fought process. The human race are here because we deserve to be here. I will repeat this so that you'll never again forget it. This privilege is not transferable to our creations.

Using GMOs has a social cost as well. In the early 21st century, Monsanto took their "Roundup Ready" soybeans to Central and South America. There they found a political climate that accepted their untested (in the wild, down

---

[60] As in the case of genetically engineered bacteria that produce lifesaving drugs, or the fungi that produce Heme in a closed laboratory; because those organisms are never intended for release into the wild.

[61] I don't want to put any ideas in someone's head.

lawyers!) plants and began restructuring how agriculture was performed in multiple countries. Demand for soybeans was high in China, so they aided the creation of new mega-farms to capitalize on that demand. In Argentina, 60 thousand family farms were lost to this consolidation. In Brazil, ten farmworkers lost their livelihoods for each one who kept their job. In Bolivia, the use of this intensive cultivation of these crops led to the soil's exhaustion in only two years, on over 100,000 ha (247,000 acres) of land. Much of it had been claimed by slashing and burning the Amazon Rainforest. They converted this land to cattle grazing and it all may soon end up as barren soil or just a new desert. (Altieri, 2004) Corporations love less developed nations because they get to pull off so much crap uncontested there. Weak, authoritarian, "conservative" governments are easy to bribe and roll over by multinational corporations. Given a chance, they'd do the same here, as well. We will soon see examples where they've done just that.

All of these disadvantages are not saying there's no place for genetic engineering. There is, but not as the sole savior of our way of life. And the anti-GMO movement has gone a bit off the rails as well. There are currently very few genetically altered crops in the market, but every package in the health food store has that "No GMO's" label plastered across it. In many cases, on products that couldn't contain them in the first place. Much like putting: "Gluten-Free" on a bag of cornmeal. A vegetarian might scarf down an Impossible Burger, unaware that its taste results from genetic engineering, yet boycott that bag of cornmeal for its modifications. There seems little rhyme or reason, currently. And it's challenging to keep up with the rules and their exploitation by commercial interests. That is why it would make more sense to just label GMOs than rely on unregulated anti-GMO labeling. These labels could then place information on what has been altered so that the purchaser can make an informed decision. Simple and accurate.

The public is currently neither included in a good deal of important food safety decisions nor are they wanted there, to be honest. Balance is so critically important, as nature continually points out. Genetically modified organisms are not: "the One" solution, especially if they work only in concert with petrochemical farming practices. Practices that will soon go the way of the Dodo regardless of whether industry chooses to stop them.

We need to understand our supply chains; we need to question whether even products marketed as "healthy and wholesome" are or are not. GMOs are part of the solution, but not our only solution. They are not even our most desirable solution, as we shall learn. And maybe we'll achieve a safer use of them by walling them off from the rest of the world.

The greenhouse effect may have the effect of effecting an explosion in the use of greenhouses. Greenhouses can protect their wards from the vagaries of weather and drought. They can protect the ecosystem from experimental or highly genetically altered crops. The control possible within them leads to decreases in water use per fruit and a greater productivity per square foot. Industry standards for hothouse tomato production are around 18 kilograms per square meter (4lbs. per square foot) per harvest. Field-grown tomatoes yield about 42,100 k per hectare (37,500 lbs/acre, ~0.9lbs/ft²). This over fourfold increase in tomatoes' productivity grown indoors can be extrapolated to many other crops and hothouses. And while a field can usually only produce one crop a year, greenhouses can have several. The ability to control the environment and the pest load makes this method highly desirable considering future uncertainties.

But greenhouses are expensive. It can cost between $100,000 to $250,000 a hectare ($40,000 to $100,000 an acre) to enclose cropland. What will make this more workable, beyond productivity, will be the unpredictability of the weather that has been initiated by our constant fiddling with the atmosphere. Today people grow high-value fruit crops[62] in greenhouses and space demanding crops such as corn and wheat in fields. Again, GMO crops would be a lot safer if reared indoors for the decades it will take to prove them safe for release into the wild. The bounds of all our assumptions will soon be tested as worldwide crop failures will rapidly force our hands. It would be nice if we could anticipate these effects and speed up the shift to sustainable agriculture now.

Science is of primary importance to our survival as a species. The human race must use this tool wisely and stop the demonization of it. Sure, there is good science and poor science. Science is a grind, and that paper you can read in an hour took months or even years to produce. Science is based on replication:

---

[62] Tomatoes, cucumbers and marijuana

does the same result occur every time? Discovery, to be safe and effective, takes many studies, many trials before understanding is genuinely acknowledged. Pharmaceutical companies are notable culprits in the degradation of these practices. The speed at which their profit wants a task completed is entirely insane. And almost every other industry that engages in science now uses similar strategies. Why not? If big pharma can get away with it, why not us? Fudged studies, regulators that are de facto employees, acceptable risks all add up to potential disasters that societies can't afford any longer. We can't risk destroying even one more piece of the ecosystem; we need it too badly.

Corporate science has given us: thalidomide; DDT; Bhopal, India; Oxy-Contin, and the list goes on. Just stay up past your bedtime, and you can view the endless commercials of lawyers fishing for clients to sue pharmaceutical companies for products that either didn't work or perhaps killed their patients. The fact of the matter is that a pure capitalistic motive will not produce good science most of the time. Profit wants an outcome, and starting with your desired result and working to support it is the opposite of good science. If you aren't prepared to be wrong, if you aren't ready to replicate your studies, you cannot study anything honestly. The people must support pure science and all the dedicated folks out there working on a greater understanding of the planet, its ecosystems, and our interactions with it.

Currently, grant funding for study replication is increasingly challenging to get. When dollars are scarce, why spend it on something someone's already done? Well, because that's how it needs to work. The replication of studies is an essential endeavor. Even the most cautious researchers make mistakes. The only way to confirm an investigation is to repeat it. Supporting science must be more than the hollow patriotism common in America today. Supporting science means just that: fund the work, respect the results. Don't just wear the T-shirt.

 Knowledge is not the curse of humanity. On the contrary, it is our one true salvation.

Which brings us to our last point on food and agriculture, one that hits close to home: What you eat is just as important as how it is produced. In the USA, our citizens overeat meat. Meat tastes fabulous, and it's a status symbol to consume beef every day, or even at every meal. But beef production is incredibly wasteful. It can take 10kg of feed to produce 1 kg (10 lbs of feed to 1 lb) of beef. It also took 3,670 l (970 gals/lb) of water to make that kilo of meat, as well. This practice represents a lot of resources poured into a feebly small output. While I might never advocate for the complete elimination of meat from a human diet, there must be a balance.

The carbon dioxide released by agriculture exceeds that of all transportation and energy use on the planet. A recent UN report has noted that a modest reduction in meat consumption could, alone, reduce the emissions of CO_2 on a planetary scale by up to 8 billion tonnes/year by 2050. (Schiermeier, 2019) This simple action could reduce greenhouse gas emissions more than converting every gas-powered car to electric energy. And it can be happen right now. You don't have to become a vegetarian or a vegan. Just eat less meat. Or change the meat you're eating. There is a red meat animal that is perfectly adapted to the US rangeland. It's the one that pioneering Americans tried to exterminate in our wild west days. It's called the Bison, and the farming of this animal could become a lower impact, higher production industry than our current ranching of an invasive bovine species.

You can reduce your meat consumption without becoming a hemp sock and sandal-wearing Bohemian. You can have a falafel today instead of a burger; you might find out you really like them, have that burger tomorrow. It can be that easy, and if a billion others do it too, we're already on our way to healing the planet. In developed countries like the USA, reducing your meat consumption might just slowly make you healthier as well. You may never exercise those love handles away, but changing your diet might just do the trick. Like many other transformations you will encounter here, this change will save us as we work together to preserve this planet. Eating green often just means eating more greens.

We live on an astonishing planet that surrounds us with life in an astounding array of diversity. You could say it was all given to us, for the march of life has led to our province. But as societies, we seem to have focused on ourelves primarily, ignoring the impressive system that made us. We can now see the allegory of the Garden of Eden from this perspective. Everything we need is here. But instead of stewarding life and celebrating its many forms, humanity has subdued it, polluted it, and exploited it for sort-sighted profit. So now the garden withers, the trees burn around us, and many remain enraptured by themselves and their inculcated fantasies. But our attachments do not fate us to stagnation and the over-consumption of the fruit of ignorance. The previously non-forbidden fruit that allows us to take, and degrade, and discard while feeling justified in doing so. It is a continued enthrallment with our myths and ancient tales that make such denials possible. But we can easily cast aside these attitudes once we find a better template, a complete creation tale, the authentic story of life, and its unfolding.

We combat ignorance with continued learning. There is a need not only to expand our education but to up the ante and begin living the wisdom that it gifts us. We can change. We could do it in a heartbeat if we so chose. Because no one is really satisfied with our current state of affairs, even though many may pretend they are. The change humanity seeks is within reach and is more than a revolution. It is an evolution we aspire to, and we shall have it.

The Green Evolution is the sustainable replacement of the Green Revolution.

CHAPTER FIVE: WATER

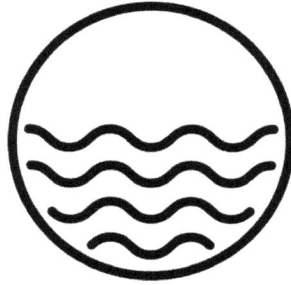

"Whiskey's for drinkin', Water's for fightin'. "
— The Code of the West

The movie "Chinatown" paints a fictitious yet reasonably accurate picture of Los Angles' water struggles in the early 20th century. Superintendent of Water, William Mulholland's wild west water-plundering shenanigans were so legendary that John Houston immortalized them in cinema. In this dark film noir tale, an imperfect and fictional man fights against corruption for a noble (and lost) cause. In the story, as in reality, Mulholland used back-channel deals, deception, and legal trickery to get most of the water from the fertile Owens valley for his growing metropolis. In the end, Jack Nicholson's character, Jake Gittes, fails to foil the scheme. An accurate reflection of the facts, for not only did the "perps" get away with this caper, no one really even stood in their way, unlike this imaginary recounting. This tale remains a testament to the potential wickedness of those seeking water and the power that comes with it. Just as the code of the west tells us, there is no greater treasure than water, and fight for it, people will.

In 1913, the Owens Valley was lush with family farms and small enclaves of pioneer spirited folk. Southern California was a dry region, where pockets of water were sparse, as it remains to this day. But the Owens valley was a place where the water flowed. This area's agriculture could've been a vital breadbasket for the entire nation until Mulholland decided he wanted its water to continue L.A.'s rampant growth. Having the power to do so, he proceeded in what would seem, in retrospect, to be the cruelest of manners possible. Mulholland and L.A. mayor Fred Eaton decided the best way to grab this resource would be by destroying the valley's farming economy. Each new bankruptcy and every farmer sent to the poorhouse was a win and a discount on the final tally in their wicked ledger book. They trashed lives, and burned futures in an authentic American scorched earth tale. The thought of honestly trading for the water appears to have never even crossed their minds.

First, they forged a coalition of the greedy. Using sympathetic politicians and judges, they blocked the US Bureau of Reclamation from building water infrastructure in the valley. A proposed local pipeline would've delivered a clean water supply for the residents and their growing towns. But Mulholland and Eaton didn't want clean water for landowners. Instead, they wished to sweep the table of every drop, like cardsharps in a crooked game. Next, they stoked internal animosity between the locals, playing one off against the other. Their tactics were aimed at ensuring that no one would remain standing to fight back against them, and it worked. Their scheme was: "divide and conquer," still a popular strategy in our war against the environment, environmentalists, and out-groups in general.

In the end, Los Angeles got their water, and the once lush Owens valley shriveled into a desert. Mulholland and his cronies became rich beyond belief. The residents scattered into the winds. And the basin was relinquished to the rattlesnakes and Gila monsters. Sadly, there is no moral to this story.

The human race's pattern of transforming beautiful landscapes into deserts is something to keep firmly in mind as all consider how to negotiate our next few decades. The tactics Mulholland used are essential to remember as well, lest we fail to recognize them when they are inevitably used against our best interests. We must hold this vigilance lest the next few decades be our last. Our civilizations' ability to provide clean water will become a great crisis if something isn't done about it now, while we still have the chance. Nearly 800 million people already do not have access to clean drinking water, and millions die every year because of this (UN, 2013). How our access to water will be affected by global climate change is unclear, but if the people keep putting our faith in this status quo, it will be dire indeed. Water is the penultimate scarce world resource, and wars have, and probably will be, fought over it.

Water is the blood of the land. It covers 71 percent of the earth's surface. The oceans account for 97 percent of all water that flows over this planet. Of the remaining 3%, only about 1/3rd of it, or less than 1% of the total, is available for drinking, agriculture, and other human usages. The rest is locked in the icecaps, glaciers and exists as vapor in the air or moisture in the soil. "Water, water everywhere, but not a drop to drink," goes the old saying, and it's true. Not only is there just 1% of the planet's water available for our needs, but in theory, it should also be shared. That water is there for the fish and the birds, the gazelles, and the foxes, as well.

Our bodies are composed of 60% water by weight; some creatures can be 90% H_2O. None will survive with much lower than 50% water volume[63]. Water is life, and all life forms require it.

Water is a polar molecule made up of one hydrogen and two oxygen atoms. It is highly stable and is often called: the universal solvent. The molecule is shaped like a little "Mickey Mouse" head, with the oxygen being the face and the two hydrogens being the ears. This physical configuration engenders its famous polarity; the hydrogen side has a negative charge, and the oxygen's charge is positive. This structure allows the molecules to link up like a microscopic line of magnets and is responsible for water's characteristic flow. This polarity explains the siphoning effect as well. Water's molecular polarity also allows it to join with other compounds (like common salt: NaCl), effectively dissolving solids into a solution. Changes in the water's physical qualities, such as Ph or temperature, can cause dissolved solids to precipitate[64] out of this solution, giving us limestone as well as rock candy.

Evaporation and re-condensation of water will always render it back into its original form. When rain falls from the sky, it is distilled water that hits the ground, unless pollutants in the air can combine with the freshly recycled water and form acid rain (Sutake, 2001). Unfortunately, with widespread human industrialization, landscapes are now also being drenched in alkaline rain. (Kulshrestha, 2001)

Water is the source of life. Water is the blood of Gaia. It is the ultimate recyclable resource, and the science of hydrology is the study of its cycle. That

---

[63] Some, like lungfish and Tardigrades can go into asuspended animation with very low moisture contents, and while that may be a survival tactic, it sure ain't living
[64] "You're either part of the Solution, or part of the Precipitate." -As the old chemistry slogan goes.

glass of water you drank today once passed through the urinary tract of a dinosaur, as a colleague of mine likes to tease school children. But it's true. Our entire ecosystem acts as an elaborate series of storages and filters that make old water fresh again, over and over. Balance once existed in this system, for life is the great balancer. This recycling was as nearly flawless as imaginable. This system provided clean water for millions of years, interrupted only by massive catastrophes like meteor hits. But some conditions that didn't exist in past geological eras are the pollution, neglect, over-use, and under-treatment that humanity now inflicts on our precious water. Left to its own devices, nature can purify and recycle its water in near perpetuity. But humans don't like leaving well enough alone.

The oceans evaporate massive amounts of freshwater into the atmosphere every day. Wind currents move that vapor over land and sea, where it condenses when it reaches its saturation level or its dew point. Similar to how solids can precipitate from water, water precipitates from the air as rain. It falls onto landscapes called watersheds. Watersheds are simply: an area of land that drains to a particular body of water. The Mississippi River watershed, for example. There are sub-watersheds[65], macro-watersheds[66], and river basins. Sadly, none of them are unspoiled by our activities.

The illusion that water is cheap and just as close as the tap in the developed world blinds us to this situation's severity. As our population expands and becomes ever more confined to cities, the need for clean, fresh water will also grow, as will the competition for its control. Water districts and public utilities are increasingly under political, not expert, management. And this is a concern because politics doesn't need to be rational to be popular. That's not to say that there aren't millions of professionals who take their stewardship of this resource seriously because they do. But the code of the west is in full force everywhere, and it's not always in expert hands that the final decisions on water's fate are decided. Let's look at the example of Flint, Michigan, to understand what is really at stake.

---

[65] 200-400 ha (~500 - 1000 acres)
[66] 400- 1000 ha (~ 1000 – 2,500 acres)

The Flint, Michigan, water crisis began in April 2014, although it would take over two years for it to come to a head. In order to save some money, the local government switched its water supply from pre-treated water delivered from the city of Detroit (sourced from Lake Huron and the Detroit River) to directly drawing it from the Flint River. They undertook this action purely for economic reasons, not technical ones. Because politicians are bound only by oath, and not necessarily by knowledge, experience or empathy, they were able to make what will go down in US history as one of the worst public health decisions ever.

Flint's water supply mains were constructed between 1901 and 1920. These mains were made of cast-iron pipe, and the feeders off the mains were made of cheap and flexible lead (Pb). Lead is not soluble in water, but the addition of specific contaminants or dissolved ions can make it so. And the dissolution of lead by ions was precisely what occurred in Flint's many feeder pipes.

The previously delivered water from Detroit was treated to a higher degree than the newer source. It was clean enough that it had no solvent effect on the lead. The Flint River water, however, was high in chloride ions, that dissolved the lead and delivered it directly into the citizens' mouths. It further poisoned them through their skin when bathing. To make matters worse, when a public meeting was called in mid-2015 after a physician had detected elevated levels of lead in children's bloodstreams, the city pushed back hard. They filed papers with the state falsely declaring their water "safe." They effectively pointed the blame for the lead to some other, unknown source. (Goodnough, 2016) It was not until a year later that the truth was discovered and the water system addressed. During that time, they left the residents to drink and soak in a poisonous brew. This fraud's exposure in 2016 led to a chaotic series of events where blame flowed back and forth faster than the suddenly necessary plastic wrapped cases of bottled water.

Flint's population was around 60% black and other "minorities" at the time of this atrocity. The city's administration was almost entirely Caucasian. Here was not a case of perceived racism; this was racism in both word and deed. The politicians who made the decisions, and stonewalled change, were almost all self-described: "conservatives." This scandal is an example of purely political decisions affecting drinking water's purity and the public's safety, with a bit of racial animus thrown, cherry-like, on top.

Criminal charges were filed against over half a dozen individuals, and that number may go higher. Whether any of them will ever go to prison is anyone's guess[67]. But if this "conservative" worldview is further embraced and even expanded, there will be no guesswork needed to predict what they will do next. The rich will be elevated, and the poor will be indentured or marginalized. No one will be left standing between them. The water, air, and land will continue to be treated as contractually bound objects and polluted or extracted until nothing remains.

Flint, Michigan, is not an isolated incident, as hundreds, if not thousands, of municipalities across America are now finding themselves in this same lead-lined boat. (Pell, 2016) And many others may just not be aware of it yet.

I was working on mitigating erosion issues in one micro-watershed in Colorado after the Waldo Canyon Fire in 2013 when a self-declared "Libertarian" heavy equipment operator informed me at lunchtime that: "The Constitution guarantees free water for everyone." It does not. But our laissez-faire brother had a point, many are now calling for access to clean water to be declared a human right. This action would require a sea change in how societies view water as a resource.

As it stands today, water is property, and its "rights" always belong to someone. Just perhaps not the owner of the property over or under, which it flows. Much like mineral rights, these potentially lucrative holdings are often reserved for those in the monied or powered classes. We, humans, have conferred almost god-like property rights onto the few who are owners. For centuries, these rights have protected their holders from responsibility for what they may, or may not, do on (or with) their "property."

One can find in the literature (at least) three different views on property:

1) Property is Liberty: the ability to own land and chattel makes one free and guarantees individual sovereignty. This view is based on Roman Law and is the dominant perspective today.

2) Property is Theft: the control of vital natural resources by a few is a crime against the many. This idea is the thesis of French anarchist Pierre-Joseph Proudhon's 1840 book: "What is Property?"

---

[67] In 2021, Nine individuals were indicted for this crime, including former Michigan Governor, Richard Snyder.

3) Property is an Illusion: the belief that one can "own" something, in any absolute sense, that has been here for billions of years and will continue to be so long after you are gone is absurd. This last idea is one that is growing among many forward-thinking philosophers and economists. It is simply an awareness that individuals are more transitory than the things we might wish to claim title to, and therefore much of our covetousness is foolish[68]. Humans are not omnipotent beings, despite our best efforts to emulate them. Our lifespans limit our ability as individuals to protect or preserve that which we call our property. Together, we shall have to figure out which one, or combination of these viewpoints, makes the most sense if we wish to see another millennia.

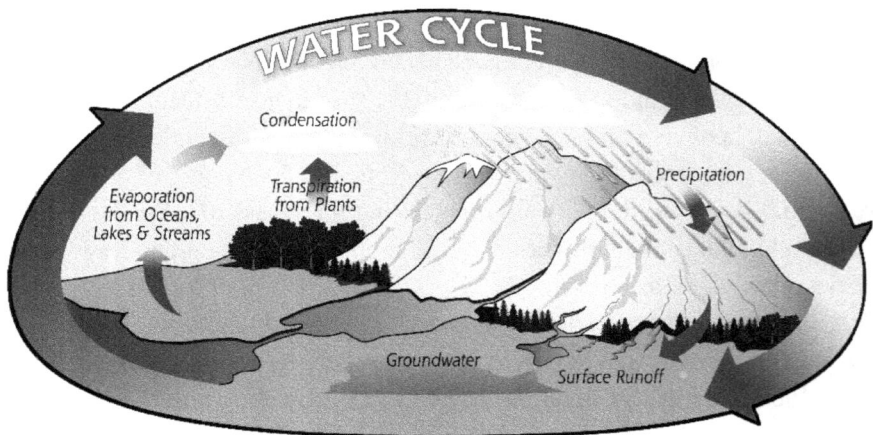

The Hydrological Cycle: Image Courtesy of NASA

The hydrological cycle: evaporation, transportation, condensation, and precipitation is essential for all of us, from billionaire to salamander alike. People have spent centuries damming, diverting, and storing water founded on a particular set of (often selfish) assumptions. Water pacts have expectations of how much water there will be, where it will fall, and where it will flow as the basis for these ownership rights and duties. Assumptions as simple as the rain will fall are no longer safe bets in many parts of the world.

---

[68] Hats off to American Author: Robert Anton Wilson who's absurdist trilogy, Illuminatus!, was a war betwween these three ideas and the sects who held them. He was way ahead of the curve with his use of conspiracy theories to satire our entire culture.

CHAPTER FIVE: WATER

The Colorado River flows 2,333 km (1,450 mi) and is used by as much as 1/5th of the country to drink, grow crops, and generate electricity. It was also the first source of water that Mulholland sought for his thirsty city. When that access was blocked, the resources of the Owens Valley became his fallback plan. Water rights are treated much like mineral rights, as mentioned before. Except that there are no virgin sources of water left to "claim." Dams line the Colorado River's course, and diversions send its flow this way and that to users whose very lives depend upon it whether they know it or not. In the late 20th and early 21st century, drought has depleted significant reservoirs such as Lake Powell and Lake Mead. And no one knows if they will ever be completely re-filled. Societies make water demands that are more or less constant in the form of laws and pacts, but the fountainhead itself is becoming increasingly sporadic.

Water Law is where a particularly tricky problem now lies. For generations, various groups have traded ownership and rights among each other: Farmers/Ranchers, Cities, States, and utilities. These groups all have to work among themselves when nature hands out more or less of that precious fluid. The Colorado River has an annual commitment of about 2.6 million mega-liters (5.4 trillion gallons) to seven western States[69]. The actual average flow is only 1.5 million mega-liters (4 trillion gallons). (Lustgarten, 2013) Even if "normal" conditions were to return, their calculations are still in a deficit. Lake Mead is nearly empty and may take decades to re-fill. If it reaches much lower, it will no longer be able to operate its hydroelectric generators. This failure might plunge large portions of the American West into darkness[70].

Population growth and our lossy agriculture are asking for more, not less, water. As demand grows, even if supplies were to remain constant, there will come the point when the entire system collapses, despite our best legislative efforts. Climate change may present a scenario where our social contracts will be certified null and void by reality itself.

In 2021 the Colorado River supply system collapsed, but there seems to be a continuing denial of that fact. Or perhaps it's more accurate to say there is a continued refusal to face those facts. The usual parties are still working

---

[69] Arizona, California, Colorado, New Mexico, Nevada, Utah and Wyoming.
[70] Arizona, Nevada, Los Angeles: to name a few affected areas.

together as if their compacts and needs have anything to do with the future of precipitation over the Colorado River watershed.

Enter the commodity traders. The current situation where water rights have outstripped actual water existence has paved the way for some people to figure out how to make more money from your misery. Commodity brokers buy and sell "futures." They set a price on some items like coal or cows and sell them beforehand to generate capital to buy this future asset now, with the anticipation that they will then set the price they want on some future date. This enterprise is a very clever type of reverse debt. But make no mistake, it is debt, just like grampa's reverse mortgage. These brokers have been buying future water rights with the explicit intent of cornering the market on all water in the future when it is even scarcer. And this type of chicanery doesn't sit very well with those who want water declared a human right and freely available. So when the management of this one substance, the basis of all life, is threatened by the profit motive, all of us need to look closely at our true values. Water may fall (or not) from the sky for free, but it is already owned in the eyes of the law.

Until recently, it was illegal for anyone in Colorado to collect or use rainwater in any way other than letting it fall upon the ground, the garden, or the roof. The water belonged to the State, and it was not to be tampered with by anyone. In Article 20 of the state Constitution, under the title: "Weather Modification," Colorado claimed all the water that fell from the sky in 1972. Article 20 posits that future advances in weather modification might allow some savvy folks the ability to make it rain over their crops and their crops alone. And that just wasn't right. So, the State stepped in, claimed it then, and in perpetuity, and made rain barrels a criminal device.

"When rain barrels are criminalized, only criminals will have rain barrels."

There was already pressure from downstream rights holders who had Colorado over that same proverbial barrel for committed water that wasn't forthcoming, to be fair. So, in the end, this symbolic act helped in court when Colorado was sued for the water that, while required by interstate compact, just didn't exist. Such convolutions of logic have been and continue to be the state of "Water

Law" — where humankind tries to control natural systems with text written on paper.

The one thing often overlooked by all these grim realities is conservation. We may choose to criminalize behaviors, but perhaps we should just convince people to act responsibly and conserve. Americans generally hate the idea of conserving, well, anything, just as much as they like to be seen as conservatives. I know that's very strange and perhaps a bit daft, but our rules often tend to encourage over-consumption rather than conservation. Think of mining rights for one example. Our crops are needy; many bathe daily, if not twice a day. Recently, feigned outrage arose after several celebrities announced they didn't bathe daily. How did this come up? You might well ask. Let's just say the business of being a personality can cause people to do strange things, just to stay current. Other stars quickly rejoined that they bathed or showered multiple times a day, to calm the nervous masses. This brouhaha allowed late-night comics to have a heyday with society's collective neuroses as the punchline.

Millions of people water lawns consisting of non-native grasses (in most places); they bathe their cars. And why do people wash cars? Probably because they consider them their babies, but they're not. Cars are tools, a means to an end; personification and identification with your car is silly, at best. While this is very human behavior, please knock it off, at least until we can recycle every single drop we wish to slather on ol' Betsy.

Groups ration water when necessary, but most still don't really believe in self-control. All or nothing is how most treat water. And just like the good commodity traders, they may confuse money with actual value. When water rationing began in California after five consecutive years[71] of drought, many affluent folks believed it would be okay to continue to overuse water if they could afford the fines. Local nightly news broadcasts were filled with footage of arrogant people watering their lawns in defiance of the ruling to conserve. It's a free country if you can pay for it! Attempts at shaming seemed not to affect them; because they were shameless.

---

[71] 2009-2014, although it now appears, in retrospect, that Californias drought pattern is far greater than originally believed. They may be in a 1,000 year (or greater) drought.

The population of the United States has doubled since 1970, but our water consumption has tripled. (E.P.A., 2020) Aging infrastructure is a big issue, with leaks responsible for large amounts of waste. According to the EPA, leaks within homes amount to around 378 trillion liters (1 trillion gallons) in loss a year. Burst water mains and outdated treatment plants only add to that deficit. Vast amounts of this scarce world resource could be conserved by replacing worn washers or updating old faucets. But where we could really save massive quantities is by just changing our behaviors. Showers use considerably less water than baths. Switching even some bath sessions to showers could save considerable amounts of water across our nation's population or the world at large. And it's also true that daily bathing can be bad for the skin, stripping vital oils and bacteria. Hair often benefits from less harsh shampooing as well. It seems that while better hygiene is responsible for better health and longer life in our more advanced societies, there may be a balance that has been overrun here as well.

The EPA stopped watering their lawns, and combined with conservation within their buildings, saved over 37% of their water use from 2007 to 2018. Everyone could all do that as well. Irrigating invasive grass lawns is unnecessary and wasteful.

The bottom line is: people waste a massive amount of water for social status, and if we can't exert some self-control, we might just have it stopped for us when the entire system collapses. If social pressure makes us waste, then why not apply this same pressure in the opposite manner. The public can Xeriscape the Whitehouse lawn and enforce restrictions on all public buildings. They are, after all, public property. Then we must then put the same constraints on ourselves. Going about our lives oblivious to the very waste that we feel is within our rights, is a very American default mode. Very protective of our rights and less enthusiastic about our responsibilities, is seen as proper conduct by many in polite society. Waste is not a good look for anyone anymore.

The amount of water that falls from the sky is totally out of our control, weather modification notwithstanding, but what is done with it is not. This issue will have to be faced, and the sooner, the better. Worldwide, civilization has to think about, must begin, living within their means. And we need to hold our politicians and representatives accountable to promote policies that reflect this reality. We have to be big boys and girls ourselves, too. We can't demand unachievable things, or we will find ourselves entangled with characters who will have no problem promising them to us. Politicians who promise results they can never deliver do not intend to do so in the first place. They certainly have some intention, and it's rarely a good one. This water scarcity issue will test us as a species, but there are many water issues we can directly affect today.

People used to drink water from streams, pools, and springs. Folks used to drink water from wells, and many people around the world still do. Now most drink from taps in their homes, and increasingly they drink it from plastic bottles. Water by the liter is the future if the futures traders have their way. If we're not careful, all water will be processed, bottled, and sold to us, 500 ml (~16 oz) at a time. In economic terms, bottled "spring water" costs about 5,000 times more than tap water from a municipal utility.

While waiting in line at the health food store, I recently saw a bin filled with bottled designer water. It was $2.20 for a 0.5 l (~16 oz) bottle. Nearby, there was a bulk filtered water dispenser. It had a price of $0.25/3.7 l (1 gal). The bottled water would clock in at $4.40/l ($16.28/gallon) versus that 25 cents. This inequity is the desired choice for the manufacturing sector. And that fancy water was marked down for quick sale! The irony of all this is that most of those "special" waters come from a tap in a city. Someone then had it filtered, bottled, and labeled as natural. Some of it has essentially been stolen from sources on public lands by multinational corporations.[72] Very little water marketed as "natural" actually comes from a natural source such as a spring or well. And while it's hard to argue that clean water isn't pure and natural (and that's what's keeps these dirtbags safe in court), the profiteering going on right now is anything but. The geniuses who came up with the idea of canning water are planning for a time when they own all the water and sell it for 10,000 times, 100,000 times what it costs today. This market is better than drugs. Because you can actually live without drugs, but without water, you will die.

Please stop buying and drinking water in plastic bottles! Never! Period!

---

[72] Check out: https://storyofstuff.org/

Plastic bottles can release toxins into the water they contain (Cincinnati, 2008). This effect is temperature-dependent but independent of whether the US Food and Drug Association (FDA) deems that container safe. Every year, I see volunteer groups come up to the mountains with those plastic-wrapped cases of bottled water, ostensibly for their health and safety. Blissfully unaware of the game of Russian roulette they are playing with BPAs. BisPhenol A is a synthetic organic compound found in some plastics. BPA is an endocrine disrupter that affects your body's hormonal system. I won't go into some of the sensational claims of what they do to your body. But why would anyone think it's a good idea to introduce hormone-disrupting organic compounds into the bodies of adolescents? Even plastics that are "BPA-free" can affect your hormones. (Yang, 2011)

These same water bottles often end up in the trash. Rarely do they actually end up in a recycling stream. Stern words await anyone who tosses an empty bottle in the forest from our crews, but it still happens. For too long, we've programmed ourselves not to even think about it. Presently, around 26 billion plastic water bottles are produced every year worldwide. Twenty-two billion are thrown away; a scant 16% are recycled. In the United States of America, only ten states have bottle deposit requirements for plastic containers, hoping to encourage recycling. So, these plastic bottles are either thrown away to be deposited into a landfill or just tossed out into the wild. Plastic bottles disposed of as litter that finds their way to a waterway can, and often do, end up in the oceans; because plastic floats.

The Great Pacific Garbage Patch is a floating plastic trash dump twice the size of Texas that exists between Hawaii and California in the Pacific Ocean. in the next chapter I will go into some detail about it and other islands of garbage. But for now, just ponder how littering on land can and does become littering into the sea.

"One planet, under trash."

If you were to view the system of waterways, streams, rivers, and lakes across the planet's surface, you would see something that looks like our circulatory system. This similarity to arteries and veins, capillaries, and organs is why I call water "the blood of the land." This system distributes life's vital and rare essence across the countryside and to its final destination: the oceans. They carry nutrients, energy and life along with them. In our wisdom, humanity just could not leave this order alone. Thousands of years ago, after agriculture began, humanity invented irrigation. And with irrigation, water could be moved from where it occurred naturally to where it was wanted: our towns and crops. The Romans created great aqueducts to deliver water to their municipalities. This public work greatly improved health and hygiene and promoted population growth and stability. (Headrick, 2019) This ability to move water was quite an accomplishment, way back then. And once they had that one under control, why not just stockpile it somewhere nearby? The perfect example of a technology that attempts to control nature, rather than living in concert with nature, is the Dam.

Dams create what are technically called impoundments. Water is collected for our use in places where it is convenient, more or less. The laws of physics place some limits on where dams can be built, based on size. But dam construction allows for stable reservoirs of water for drinking, irrigation, and finally, energy production. Dams can be as simple as a millpond, with an earthen berm designed to make a small pool that could provide the power needed to turn a water wheel and mill grain. Or, dams could be as complex as the Hoover Dam, which constrains the Colorado River and stands 221 m (721 ft) tall and creates Lake Mead, which measures 180 km (112 miles) in length (when, and if, full). Mega-Dams are the largest structures yet built by humankind, dwarfing the Great Pyramids. The weight of impounded water on the planet's surface is now so great that it has a measurable effect on the planet's tilt on her axis. Our reservoirs alter the length of a day by an infinitesimal but measurable amount as well.

Modern dams are technical marvels that collect gargantuan amounts of water behind them. This water, which had once flowed free, is now captured and held by our reservoirs. So now there is a bunch of water in the place of our choice, and it can released it uniformly, over a turbine, if desired. The weight of that water and the force of gravity can do the work of generating power. Nifty, somewhat simple, although building a dam of any size is a pretty complex and expensive undertaking. And of course, dams need maintenance since they continually have millions of tonnes of pressure pushing against them, day in and day out. Dam building cultures are not always doing such a great job with that task.

Dam construction boomed with the "New Deal" during the 1930s in the United States. This job creation program led to the construction of a tremendous amount of infrastructure and helped pull the United States out of the Great Depression. The Tennessee Valley Authority (TVA) installed 29 hydroelectric dams across the South. The idea was to create jobs, produce cheap energy that could be used over a wide area, and "increase navigability" of the Southern US waterways. And it did provide power to many southern states that had never had electricity before. It certainly changed the landscape. But the proposed economic gain of these projects did not uniformly benefit the residents of the Tennessee valley, as promised. They continue to exist at a generally lower socio-economic level than their electrified and irrigated city-dwelling neighbors. As is often the case, these dams displaced many and even more had their livelihoods impacted by their homelands' flooding. (Chandler, 1984) Displacement of populations is a worldwide dam phenomenon. Not only is the ecology of the dammed areas altered, the local social and economic conditions are routinely depreciated as well.

And then there are also a slew of problems few if anyone took seriously enough when they were busy conquering our land's wild and scenic rivers' Problems such as flow variability. Water doesn't run uniformly all the time: season in and season out. Instead, water flows through wild rivers in varying amounts depending on the time of the year and the latitude. In the spring or fall, rains create floods that wash the river banks clean, stir up nutrients, and redistribute them. This high flow/low flow cycle triggers other rhythms in all riparian plants and animals. It is a pulse that nature has been in sync with ever since water flowed over rock and soil.

Flood control was a big selling point when the New Deal began its spate of Dam construction, and in one sense, it worked. Dams can buffer the high-flow/low-flow cycle, which is the natural state of an untamed river, effectively. But what happens to all that sediment that should be washed downstream by the spring floods? Well, it ends up behind the Dam. In general, large Dams can lose up to 1% of their holding capacity[73] every year to sediment in-fill. (McCully, 1996)

There is no simple solution to the problem of reservoir sediment in-filling. Dredging works, sort of, but it is outlandishly expensive and not entirely effective. You can lower your water level before flood season and then try to let it rush out as fast as possible to move the sediment with it. Unfortunately, that doesn't always work well, and it will decrease power generation while you're flushing.

Events such as wildfires will inevitably lead to flooding afterward, and the sediment flows created by a denuded landscape are considerable. After the Buffalo Creek Fire in Colorado, in 1996, landslides nearly filled the Strontia Springs Reservoir with silt and mud. Denver Water, which operates this impoundment, spent $30 million on dredging to remove around 477,000 cubic meters (625,000 cubic yards) of sediment. That's enough material to cover an acre 200 feet deep in mud. (Bruce, 2017). The result was less than satisfactory, and this reservoir will never again reach its designed capacity. The sediment in-filling will continue at Strontia Springs. Eventually, every dam could become just a massive dirt field, with whatever river is left flowing over the top. That is unless (or until) it collapses. The American Society of Civil Engineers graded America's infrastructure at a D+ in 2018, pointing out that as many as 2,000 dams desperately need repair or replacement in the United States. (ASCE, 2018) Strangely, outright removal was not listed as an option in that report.

All of this is in addition to the ecological disaster of dams in general. That flooding cycle is essential for river life. Fish spawning, invertebrate life cycles, and plant health rely on these cleansing and nutrient re-distribution pulses. Salmon need to swim upstream to spawn. Then their young, called smolts, need to swim back down to the sea. They have a limited amount of time to do so, a limit created by the timing of their downstream swimming instinct. The typical trip down the Columbia River used to take a smolt three days; now, it can take 39 days, primarily thanks to the Grand Coulee Dam. These rivers have existed, in many cases, for millions of years in a wild state. The very

---

[73] This is highly variable by dam, with U.S. dams losing as little as 0.5%/yr, and some dams in other nations losing up to 2.5%/yr in holding capacity.

thought that the natural cycles of the river could be so disrupted and everything would be fine was foolish. The Yellowstone is the only river over 1000 km (620 mi) in the United States that remains undammed along its course.

A river exists in a dynamic, fluid diversity that is a fundamental reflection of life itself. A reservoir is a static body of enforced stagnation. Most fish species in these artificial lakes must be imported because river fishes don't thrive in reservoirs. And with these farmed fishes come constrained genetics and sometimes disease.

In his excellent book: "Silenced Rivers: The Ecology and Politics of Large Dams," Patrick McCully lists a plethora of evils involved in our obsession with Dam construction. In China, the Three Gorges Dam displaced over 1.5 million people, polluted the drinking water of tens of thousands more, and caused landslides and heavy environmental damage. After WW II, Stalin set about his "Transformation of Nature" policy and began building great Dams using laborers from the Gulag internment camps. Over 100,000 Prisoners aided in the flooding of 120,000 km^2 (46,330 mi^2) of land. Many have joked that these projects only succeeded in converting the Volga River into a series of stagnant ponds. An example of the Russian's dry humor. You can find an account of this massive enslavement of both men and the land itself in Alexander Solzhenitsyn's famous novel: "The Gulag Archipelago." These projects' social costs dwarf the ecological and technological problems of dams. These include the theft of land from poor and indigenous people to the rape and murder of thousands of innocents worldwide.

The Chixoy Dam in Guatemala was constructed on rainforest land and required the Maya Achi people's relocation. The Instituto Nacional De Electrificacion (INDE) constructed this dam beginning in the late 1970s. They knew they would have to move the tribe, so they tried to trick them into simply signing over their lands. However, when representatives of the people were shown the cramped quarters on desolate terrain, they were to be given in exchange; they declined the offer. And so began a campaign of pressure and murder by the dam proponents, nefarious acts culminating in The Rio Negro massacre.

"On March 13, 1982, 10 soldiers and 25 patrollers arrived in Rio Negro to look for 'guerillas.' Infuriated to not find any men [they had hidden in the forest hoping it would protect the women and children], they rounded up the remaining women and children and marched them to a hill above the village. There, the patrollers began to rape the women, and then kill them. Some were garroted, some were beaten to death with clubs and rifle butts, others had their throats slit or were decapitated. Children were murdered by tying ropes around their ankles and smashing their heads against rocks or trees. Seventy women and 1107 children were massacred. Only two women managed to escape. Eighteen children were taken back to Xococ as slaves for the patrollers." (McCully, 1996)

This slaughter was not some atrocity from the mists of ancient history, it occurred less than 50 years ago, and this work was overseen by a government entity and paid for by the World Bank[74]. People may take water for granted, but those in power, or those seeking power, will see it as a potent mechanism for control. And it will bring out their authentic selves as they pursue it.

We face a terrible conundrum because no one can argue that we don't need water for our growing populations. We do need water for agriculture and to feed our billions of mouths. And the power that dams generate is clean, if not perpetual. As we fix our problems with agriculture and energy, we still have the issue of water for human life, and it's something we absolutely need for our survival. Four days without water will lead to death. We can survive without food for three weeks or more.

Humans are not the only creatures that build dams. The members of the genus Castor are natural dam builders. There are two Castor species: *Castor fiber*, native to Eurasia, and *Castor canadensis*, the North American beaver. The Beaver is not only famous for its lush, waterproof fur, massive front teeth, and

---

[74] The World Bank is an international consortium that has done much good around the world by financing development in the poorer nations of the world. But as we see, they have also financed terror, if inadvertantly.

its broad, flat tail. But also for its industriousness and tenacity. Beavers cut down trees with their specialized incisor teeth, live in family groups, and are keystone species in wild waterways. And did I mention they are a busy, busy lot?

Beavers place their dams in streams to create pools to live out their aquatic lives. They shelter in lodges constructed of interwoven branches and smaller logs and access them through underwater entrances. This arrangement keeps them safe from most predators, except, of course, humans. There were billions of beavers across the North American continent when European settlers arrived in the early 16th century. Over the next few centuries, trappers decimated their populations, harvesting their fur to make coats, and notably hats. Their fur's water-repellent characteristics were essential for constructing the Top Hat, which was all the rage back on the continent. Abraham Lincoln would later become recognized for his affinity for that same Top Hat. And like many a 19th-century gentleman would actively sport a dead rodent on his head, suggesting that fashion has always been a bit daft.

The current US beaver population is estimated to be around 6-13 million, a precipitous decline (Baker B. W., 2003) from their heyday. This is perfectly in line with our treatment of the American continent and its inhabitants.

The beaver pond was an essential part of the native ecosystem. It provided a habitat for fish and aquatic insects, plants, and birds. These diverse communities' serial nature worked to slow water flow in the high mountain stream reaches, controlling erosion and enriching the environment. The water from these natural impoundments would percolate slowly into the ground, charging the local water tables and creating a buffer for times of drought. The sediment build-up that is so catastrophic for large-scale human-made dams worked to improve the local soils and nurture exotic plant life that, in many areas, are no longer present. Many rare temperate orchids that once flourished around beaver ponds are now nowhere to be found. Being small in capacity but plentiful in number, flood events could easily top these pools and roll down the hillsides with minimal damage to the dams or the local ecology. Beaver pond complexes were essential for migratory fowl and provided clean water for deer, elk, and countless species of mammals, reptiles, and amphibians.

This seemingly harmonious situation became an issue as pioneering farmers and ranchers came to covet the rich dirt collected this way and sought to destroy the dams so that they could claim the fertile soil. Often, they then faced flooding when the displaced beaver re-built a dam elsewhere. Many of us in the natural resources fields are familiar with contemporary instances where a rancher might hire a crew to blow up a dam that was flooding a dirt road, or a field, at the cost of thousands of dollars. The sun would set, and by the time it rose again, the dam would be back, and the pond re-filled because that's what beavers do. Busy, busy, busy. While this may make dedicated conservationists smile, it doesn't seem funny at all to the ranchers who tried to take out the dam. This outcome is predictable and is why the shrewd rancher will trap and re-locate or just destroy the animals. The trapping of beavers is heavily regulated nowadays because of their rarity. But this is the one situation that will allow a person to get a permit with minimal hassle. Those occasions when they can be designated "nuisance animals."

Beavers have become a topic of discussion in ecological restoration circles recently. Several studies have suggested that perhaps the best and least expensive stream restoration technique in the American West is as simple as putting the beavers back where they belong. (Burchstead, 2010) (Baker B. W., 2005) If it seems like an easy decision to return a foundational species to its habitat as an excellent way of restoring that habitat to its proper function, then you aren't aware of how land policy is presently negotiated. Land management is a result of the consideration of many differing points of view. Stakeholders are those people and entities with "skin in the game." Federal agencies, State agencies, landowners, and corporations go through a process to create rules and laws. Rarely is nature at the table, and beavers, even less so. Indeed, at times, the Federal and State managers act as advocates for nature, to the extent their politically controlled organizations will allow. But the decisions will almost always favor private and corporate interests. Ecological decisions, all too frequently, are made based solely on economic factors. Often, they will be made by men behind closed doors and dictated to those below them, the ones who actually know how ecosystems function but whose hands are tired. These choices also may only have a very short-term view of value. Nature's value must be measured in millennia and not fiscal quarters or even decades. That is, if we're interested in the long-term survival of humanity and not just making a buck on the quick.

A rational approach would be to return beaver to their place in the environment and let them rebuild the ponds and recharge water tables. This tactic would allow the plant communities to rebound and the birds, frogs, and salamanders to return. Beaver restoration is one of the most straightforward and easily understandable methods of using the ecology to restore itself with our help. And many still resist it at nearly every level. Groups and individuals who have powerful motivations to return beavers to their properties have had a miserable success rate in Colorado and many other Western States[75]. The fragmentation of our remaining wildlands and political pressure to keep "growing, growing, growing" has made even such a seemingly simple action complicated. The number of permits and meetings required to move a beaver from a place they might be consider a pest to where they are wanted is nearly impossible for most to endure. And they can rarely achieve it fast enough to save that animal's pelt. Contrast this with the ease with which one may get a nuisance animal permit, and the inequality is clear. The rights we have granted ourselves can come crashing into reality itself as we bind our ability to act in our own best interests. This failure is well illustrated in how we deal with water, native wildlife, and the habitats that they have created for millions of years.

The wetlands created by our beaver friends are often called swamps, a demeaning label indeed. "demeaning" is a linguistic method of devaluing an idea without actually having to prove, well, anything. It uses trigger words to elicit a rapid and inflexible response to a concept or idea. "Drain that swamp!" is something we want to do, if for no other reason than it must be a good idea, right? Here are multiple examples of what a bad idea that really is.

Sure, swamps are scary places, and they harbor dangerous creatures. One of the most dangerous of the swamp dwellers is the mosquito. Dangerous to humankind, that is. Bear in mind; mosquitoes can indeed spread blood-borne diseases by virtue of their womenfolk's blood-sucking natures. And battling diseases like malaria is a critical human endeavor, but how we choose to do it

---

[75] Just ask Sheri Tippie, Licensed Beaver relocator and Colorado's "Beaver lady".

is an example of the imbalance in our understanding of natural cause and effect. For some time, public health officials have looked to eliminate mosquitoes as a desirable tactic in the war against malaria, West Nile Virus, and other pathogens. They've sprayed DDT on ponds and Malathion from the skies, attempting to eliminate these insect pests. Governments have happily poisoned their towns and cities in the process.

Now there's a new idea: Genetically engineering the genocide of mosquitos. The results of a study in Burkina Faso have demonstrated the effectiveness of killing mosquitos using the fungus *Metarhizium pingshaense*. This cousin of the *Cordyceps* zombie fungus was genetically infused with spider venom. Their Franken-fungus eliminated all mosquitos in their test site in 2 generations (Lovett, 2019). The annihilation of a local species in 2 generations is an impressive feat — for a psychopath. So here is another triumph of "science" that needs to be filed in the "we can do it, but should we?" card cabinet.

As annoying as they are, as dangerous as they may be, mosquitos are also a keystone species. Mosquito larvae form the base of the food pyramid in still aquatic environments along with other "freshwater plankton." They lay millions of eggs just for just a few thousand of the flying vampires to make it to maturity. They feed the small fish and other aquatic insects. In turn, those consumers are eaten by creatures on the next rung of the food chain, and so on. The collapse of the food chain by pulling out a principal nutrient source has happened repeatedly and is technically called a: "bottom-up trophic cascade" (Matsuzaki, 2018).

We need to get past this idea that we can just wipe out sections of an ecosystem for our whims, and everything will be okay. It doesn't work like that, not in the real world or human culture either. Every piece has its place. The elimination of malaria is a serious issue, and I'm not suggesting nothing should be done. But killing the messenger has never been an effective strategy, and that's the essence of this technique. If you just take out a vector of disease, the disease can still effectively exist.

"Life, uh, finds a way," as the now mythical character, Ian Malcolm from the film "Jurassic Park," was want to say. And despite his swagger (and the fact he was a fictitious character), he was right.

In 2021, a potential vaccine effective against malaria began field trials. (Datoo,

2021) So it would seem there might be relief for the thousands who contract this disease annually without wiping out a foundational member of the environment. That would be an outstanding achievement, once people get back on board with getting vaccinated.

If you destroy a piece of the ecosystem for any reason, you will feel the effects down the road and possibly magnified in severity. The collapsing ecosystem around us results from just such actions. Many of our behaviors seem minor or inconsequential from a personal perspective, but the sheer number of these micro-actions can snowball into catastrophe. We alone choose which behaviors to exhibit and when. Only with an expansion of our awareness can we see better ways to act and interact with our fellows and the biosphere. Nature is built on cycles interacting with cycles; we can choose to enhance some for our benefit. But to destroy one will always increase our jeopardy. We have enough peril as it is.

Water is the ultimate recyclable compound. People recycle water all the time. It's called wastewater treatment. In developed countries, used water is colleced in sewers and transported to treatment facilities, where it is filtered and decontaminated before returning to the natural water cycle. There is a good possibility that water treatment will be much more of a closed process in the future. Water will be treated and returned directly to the tap. This is a closed loop system and represents a proper way of thinking about our needs and how they might be satisfied in a more balanced manner. Right now, this isn't workable because our treated water, in many cases, is not quite fit to drink. Our standard practice is to dump somewhat under treated water back onto nature for a final "polish."

As a culture, many, if not most of us, treat our drains as some sort of magical portal that we can use to dispose of anything at any time. And most believe it will just disappear. They dispose of solvents (oils and paint thinners etc.), toxic chemicals (insecticides and many cleaning compounds) down the same drain used for low contaminant wastewater. The water from our showers and dish sinks, which we call gray water, is generally pretty clean, containing only organic contaminants, materials easily broken down by bacteria. Our treatment systems are designed to deal with skin cells, tiny food scraps, and soap. Even the water from our toilets is far cleaner than a drain where industrial chemicals are disposed of. However, standard practice is to mix it all together, making the separation and decontamination process far more difficult. We dump everything down our drains, and we need to stop doing that.

Wet wipes and "flushable" products are horrible, as they can physically clog pipes, and even if they don't, they will need to be mechanically removed by someone downstream. A "crappy" job if there ever was one. There is virtually no regulation on what people can flush. And commerce really doesn't seem to give two hoots and a holler about what they sell to a generally under-informed populace, under-informed in terms of where all this stuff goes and what has to be done to it at the treatment facility. Tampons are not flushable, yet many folks still do. In 2019, American women purchased over 5.8 billion tampons (Borunda, 2019). In the UK, up to 50% of women flush them, even though they know they shouldn't. (Blincoe, 2016)

Many of us who live in the country get our water from the natural water table through wells. Our wastewater is disposed of in a septic tank. Septic tanks are simple water treatment devices that separate solids from liquids in a large tank where bacteria can break everything down. The upper layers of bacterially pre-digested liquid flow to a leech field, slowly percolating back into the ground. Natural processes, plants, and soil bacteria complete the recycling process, and this water will eventually rejoin the water table. When you live with a septic tank, you need to be very careful about what you dump down the drain. Because if it's not organic and recyclable, you're going to end up drinking and soaking in it, eventually. If it's a harsh chemical or a bottle of antibiotics, you risk exterminating the septic's essential biological processes altogether. People who live in these areas obsessively look at the labels on all soaps and cleaning products they purchase. "Septic Safe" is a serious deal, and it should be for those who live on municipal systems as well. If our urban sewage systems were treated like septic systems, our waste management costs would dramatically decrease. And treated water quality would increase concurrently and almost overnight.

Unawareness of these facts can become an issue in rural communities when people move from the city and do not know what is happening with their water cycle. A real estate agent recently told me about a new mountain home buyer who asked her: "What does 'on well and septic' mean?". It means you might not want to move in if you aren't already aware. Because in many rural areas, you are your own water treatment district. You are also your own road plow and pizza delivery person. If anything goes wrong, you need to either be able fix it, or wait for someone else to do so. And so when they discovered the messy truth and tore the contract to shreds.

In the mountains, one uninformed person can contaminate the water shared by many, and depending on what they spoil it with; there might be no fix. Usually, this is not done out of malice. It's just ignorance; and a continuation of learned behaviors based on a consumer mentality. The indiscriminate trashing of "stuff" through the commode is what millions of people do every day in their cities and towns. It's what many manufacturers tell them to do. So this lack of awareness surprises no one in wastewater treatment. Exasperates them is more like it. People imagine that mysteries below the ground will wondrously render everything they drain or flush out of existence. And if it doesn't, it should, and that's someone else's problem. But the problems created by our lifestyles and our waste products are ours, and ours alone.

Black water is the term used to refer to toilet water. It's a different animal than gray water, to be sure. But it's still is not as wild and dangerous as it may seem on the surface. Many have a cultural aversion to excrement, for some good reasons, and some foolish ones as well. Let's get the gross part out of the way. You could eat poop. Bears do it all the time (so does Fido, check his breath). I don't recommend this, but it's just a fact. There is no inherent deadly toxicity if the stool is not infected with a communicable disease or parasites. You can drink pee as well. It's also just not recommended and will require your body to remove all those metabolic wastes once again and re-excrete them. As you've probably seen on some survival television show, if you have no other choice and need water, you can and will drink urine.

The actual issue with Black water is infectious diseases. It's human waste, so any disease contained within it is directly contagious to other humans, with some caveats. The human digestive system is powerful, and most bacteria will be killed by stomach acids quite effectively. But contaminated water that enters the bloodstream via a cut, scratch, or tear is a genuine concern.

The larger the population mixing their blackwater, the greater the chance that some of it will have a dangerous pathogen mixed in. *Escherichia coli*[76] is an especially harmful bacterium that can easily cause extreme gastric distress and potentially death if ingested while still viable. Even in the developed world, parasites, such as tapeworms and roundworms, can occasionally occur. But they are manageable and easier to deal with than some of the other mechanical toxins in our wastewater stream.

Poop is organic and can be recycled to great benefit. Often, the screened solids from its treatment are processed and then composted. This composting process will kill any remaining pathogens and their spores/cysts. The material can then become a pretty high-grade fertilizer. However, when pesticides or organic solvents contaminate it, it might become unusable. A pure Black water stream would make waste recycling easier, less hazardous, and more economical. It is also a replacement for petroleum-based fertilizers and is never in short supply.

"Back to Nature" types and even some survivalists already practice separating gray water and black water to beneficial effect. They can use gray water directly in irrigation as it contains many valuable nutrients. Black water can then go solely through the septic, reducing the load on the household system and diminishing the frequency of expensive pumping later on. Some people, and some friends of mine, use composting toilets, essentially making their own plant food from feces on the spot.

Municipalities have not taken up this dual stream practice for the most part. It is exorbitantly expensive to put in an entirely separate water return system, even though treatment costs for either black or gray water would be exponentially lower than the "mixed stream" system in current use.

The Bill and Melinda Gates Foundation has been working for years to spur the development of improved sanitation systems for the underdeveloped world. Their Foundation seeks to make progress in many humanitarian fields, sanitation being an important one. Their "Reinvent The Toilet Challenge" has offered hundreds of millions of dollars in the effort to create systems that are:
1) Suitable for remote installation

---

[76] E. coli, a gram negative rod shaped bacteria that is ever present in the lower digestive tract.

2) Do not require industrial water or power supply, and

3) Operate inexpensively

Such systems could dramatically improve life in much of the world and reduce child mortality. It could also be a boon to the developed world by effectively removing blackwater from the treatment equation, making the current single-stream system all gray water. Once the masses stop pouring chemicals and "flushables" down our drains and toilets, that is.

There are yet more types of contamination to consider, unfortunately. A big one is pharmaceuticals. In the developed world, folks take a lot of drugs; most of them are potent compounds that are not entirely metabolized before excretion. The most problematic of them are the anti-biotics and the endocrine disruptors (synthetic hormones, Them again). Our over medication has had some pretty devastating side effects so far, and there isn't much hope in sight. Antibiotics are expelled right out of our tubes and can kill the bacteria needed to break down organic compounds and effectively treat our wastewater. Endocrine disruptors such as PCBs[77], PDEs[78], BPA[79], and even the metabolites of birth control pills enter the water system and are not efficiently removed by current water treatment methods. Many pesticides and herbicides fall into this category as well. When these compounds pass through the water treatment system and are released into the environment, they can wreak havoc on the natural world. (Mai, 1996) And then there are chemicals that are still dumped directly into the wild water system as the by-products of industrial processes such as C8.

C8 is a secondary product of synthesizing one of the "greatest" chemical discoveries of the modern world: Teflon. Teflon is a fluoridated organic compound[80] that repels water, oils, stains, etc. It was introduced in the late 1950s as a miracle non-stick coating for pots and pans. They marketed Teflon to a naïve public as a time-saver for our over-tasked wives enslaved by

---

[77] Polychlorinated biphenyls- contained in synthetic lubricants and refrigerants.
[78] Polybrominated diphenyl ethers- contained in flame retardants and some plastics
[79] Bisphenol A- contained in plastic bottles, even many that are listed as "food safe"
[80] Fluoridated organic compounds have another broad use: nerve gasses, and poisons such as Novichok. Yummers!

suburbia. C8, ammonium perfluorooctanoate, has been released directly into waterways, with virtually no treatment for over 50 years. All because the manufacturers claimed it was harmless, and the regulators replied, "Okay, we'll take your word on that." Now, the foaming horror continues to be pumped directly into our rivers and streams. It is linked to cancers, congenital disabilities, and increased human and animal mortality, which the manufacturers have known for decades. This same sad song keeps repeating itself in our world, and it generally leads to increased regulation, but maddeningly, not an increased will to do better. Or so it would seem. C8 is now in your bloodstream and everyone else's. Its transmission vectors include the air we breathe, the water we drink, and those crappy non-stick pans we still use[81]. You'll need a scorecard to keep track of the pollutants modern civilization has injected into your blood. Ask your doctor and make sure you're sitting down first.

Natural systems have a hard time dealing with these various compounds that are dumped and excreted into their finite waterways. Industry has created chemicals that never existed in nature and often treat the whole situation with an offhand attitude. Fortunately, there is hope within the natural world that could help us. Again, we look to the funga. The enzymes used by saprophytic mushrooms to break down tough natural compounds like cellulose and chitin work by severing carbon/hydrogen bonds at the molecular level. And for the decomposition of exotic organic compounds, this is precisely the chemistry set needed.

Myco-filtration is a very infant subject of study that has shown considerable promise so far. The fungal enzymes and peroxides excreted by mushroom mycelium have been shown to reduce bacterial contaminants such as *E. coli*, a potentially dangerous water contaminant. (Taylor & Stamets, 2014) The discipline of mycoremediation has shown efficacy with employing fungi to break down PCBs, PCEs, and even crude oil and hydraulic fluids. (Singh, 2006) Perhaps the even more exciting news is success in isolating and

---

[81] I anticipate pushback on this, as the pans are still listed as "non-toxic", the exception being if the pan is heated above 260° C. (>500° F.). And what are the chances of that when pans are heated on the stovetop? If one of these pans scorches after being left accidentally unattended- dispose of it. Better yet, don't wait-just dispose of your Teflon pans now and avoid the rush. Check out TerraCycle and the Calphalon program. www.terracycle.com

stabilizing these enzymes for municipal scale filtration installations. These methods of using natural fungal compounds have shown promise in the destruction of those annoying endocrine disruptors and other organic contaminants. (Falade, 2018) When the fungal enzymes are molecularly stabilized, they can become capable of thousands of destructive interactions with the pollutants before breaking down themselves. Whether these methods could work on C8 is so far unknown, but is worth continued investigation. Mycorestoration is the discipline that investigates and implements many facets of inquiry into the use of fungi to degrade and detoxify everything from wastewater to oil spills. It is a powerful tool to remake and restore materials at a molecular level without needing more industrial chemical processes or laboratories. And it is precisely the kind of ecological solution needed to deal with the mess we've made of this world.

So, can we consider living within the same protocols under a municipal water treatment system as many do when living with a well/septic system? If the wastewater we generate is basically "clean," the treated result will be first quality water. This is just the starting point we need to reach before achieving a true closed loop system of water recycling.[82] Just cleaning up the wastewater stream by stopping our incessant flushing of toxins and wipes will instantly improve every current treatment method. Again, a modicum of self-control can quickly reduce the stresses placed on all of our systems. If a product claims to be flushable, it probably isn't. Toilet paper is about it[83]. Flush nothing else. It's that simple. Conservation of this most precious resource is desperately needed and will come naturally when we finally realize what we are doing and what it's doing to us.

---

[82] Recently, an extreme "conservative" Congresswoman publicly denounced what she called: "Poop water" In an attack on Bill Gates and his philanthropic enterprises. This demonstrates such an ignorance of the hydrological cycle that one may be tempted bang their heads against the wall. Please, don't anyone tell her that a dinosaur once peed out her water while she's drinking, and record the response.

[83] I know of some folks that don't flush TP either. They collect it in a separate container and add it to their compost. They claim the "human scat" smell deters bears and raccoons.

Water is the Blood of the Land. And its contamination is a reflection of the pollution our ways of living has injected into our own bloodstreams.

Water is life, and it's limited in both quantity and quality. Still water reflected light and was our first mirror. Unfortunately, it seems we mostly saw ourselves in that reflection. We are over 60% water, just slightly less than the surface of Gaia herself, echoing the world that created us in every sense. Our interactions with the hydrological cycle are symptomatic of our disconnections from our proper place in this world. The human race may be the greatest species in the singular, but we are not the greatest in terms of the totality of life. We are still a part of a system. A system we have been damaging for far too long.

Water should be a human right.
And an animal right.
And a plant right.

Because without it, everything dies.

CHAPTER SIX: THE OCEANS

"Water which is allowed to enter the sea is wasted."
— **Joseph Stalin**

The Deepwater Horizon was a state-of-the-art offshore drilling rig that held the world's record for the deepest drill of its time. It towered 97.4 m (~320 ft) above the waves and was positioned 66 km (40 mi.) off the Louisiana coast on April 20th, 2010, when it exploded. At 8:45 PM, local time, the well "kicked," disrupting the carefully balanced pressures of drilling mud pushing down and oil and gas pushing back up the slender bore pipe. Activation of the automatic Blowout Prevention Device (BOP) failed, allowing, instead of avoiding, a blowout. And out this rig did blow! Eleven crew members died in the initial fireball, and 17 more were injured. Some personnel leaped from the deck, 34 meters (111 ft.), into the dark waters below in panicked escape. Specialized firefighting helicopters battled the blaze but were helpless against the towering inferno. Most of the 126-member crew were safely evacuated by airlift, and two days later, on April 22nd, the entire rig crumpled and sank 1500m (5,000ft) to the gulf seafloor below. Many wiped away the brow sweat beneath their hardhats, thankful that this mishap was over. But the carnage of the Deepwater Horizon was only getting warmed up. As the rig rushed towards Davy Jones' locker, the drilling pipes twisted like pretzels and tore loose from their deep moorings. The BOP had activated when the original blowout occurred, but had failed to shut off the oil flow. Redundant systems, the AMF Deadman Blue and Yellow Pods, were designed to seal the well in the case of catastrophic failure and with no need of human intervention. Both, however, failed due to miswiring. In a strange irony, each system was miswired in completely different fashions. The Blue pod was cross-circuited such that it had drained the battery that should have actuated a solenoid. That solenoid would've then triggered a shearing ram to seal the pipe. It failed; the battery was dead. The Yellow pod was misconnected such that only one side of its solenoid coil became energized. Thankfully, this failure still allowed the weak activation of that single solenoid. Since the solenoids should have actuated in pairs, this triggered only one shear/ram device. And because the pipe had buckled during the collapse, the blade missed its mark, severing instead of sealing the pipe. (CSB, 2014) It's impossible to say if this made the problem worse because it's difficult to imagine that this disaster could've been worse by any imaginable measure. As the technological marvel continued its downward plunge into the briny depths, it became apparent that the real Deepwater Horizon disaster was just getting started. It also became clear that the growing oil slick on the surface was not the result of the rig's destruction. Instead, the black pool was still expanding because of a leak nearly a mile beneath the ocean's surface. Without the rig intact, and a functioning BOP or Deadman, there was no way to stop it. At first, it wasn't even apparent where the leak was coming from or how much oil was leaking. It would take over a week to answer those two vital questions. And thus the worst oil spill in United States history was underway. Out came the hordes of television crews and well-meaning oil clean-up volunteers. Our screens were plastered with images of disaster and interviews with the fishermen whose livelihoods had just been flushed down the drain. Talking heads chattered, but no interviews with the company were forthcoming. No crew or direct witnesses were available to give their recounting of the accident. But the images were telling. Catastrophe just didn't quite seem to describe this scene. Here was the carnage of war, humanity's war against nature revealed. It took the CEO of British Petroleum over a week to set foot on the beaches his company had defiled. And this man, Tony Hayward, exuded privilege and arrogance from every pore when he finally did so. He dodged responsibility, stonewalled blame, and famously lamented: "I'd like my life back" on camera. Yeah, you, the fisherman, and that sea turtle as well, Tony.

In private, he had expressed dismay that he should have to deal with this problem at all. He was supposed to be making more money for the shareholders from the safe confines of his plush office. Hayward wanted the blame to be placed squarely on the shoulders of BP's sub-contractors. (Shogren, 2011) So, while waves of black slime covered the beaches and birds, and celebrities tried to grab the spotlight with their pet oil-spill solutions, the majority of the leak drifted silently out to sea. British Petroleum spent millions addressing the situation. And in a fine status quo manner, they probably made it even worse. By the time they capped the well, in July of that year, it had released an estimated 5 million barrels of oil, 780,000 m³ (210,000,000 gals.).

So, what do you do when you've released millions and millions of gallons of crude oil into a shrimp and shellfish nursery?

BP decided that "sweeping it under the rug" would be a fitting solution. Their initial response was to use chemical dispersants on the massive body of floating oil to make it break down and (hopefully) drown away into the salty brine. A dispersant is basically some sort of soap. The idea was to break the chemical bonds and allow the oil to dissolve into the seawater or perhaps just sink. When you're a master of the universe, faced with an embarrassing problem, out of sight is out of mind (and often, out of liability). The chemical dispersant of choice was Corexit, a product initially developed by the Standard Oil Company and manufactured by Ecolab. Corexit is composed of 2-butoxyethanol, propylene glycol, and a proprietary Sulfonate compound (protected by trade secret, and why would they want to tell us what they were pouring into our ocean?). While it definitely came from a lab, one must wonder what part of it was "eco"(logical)?

BP spread an undisclosed amount of this arguably toxic stew onto their inarguably poisonous oil spill. While they weren't very forthcoming with figures, later examination of government documents estimated they used at least 3,179,000 liters (1.8 million gals) of Corexit. They asserted that this somewhat mysterious substance would allow naturally occurring bacteria to consume the oil, rendering it into nothingness rapidly, and then everything would be peachy. Such was not the case, of course. They made the oil slick appear to go away, but not disappear. No, no, no, it did not disappear in any way, shape, or form. That oil is still there, to this day, on the deep ocean floor. The shrimp industry has not recovered, and an uptick of mutated shrimp covered with ghastly abscesses has continued to be documented.

Recent studies continue to remind us that Deepwater Horizon's effects are far from over. The decaying organic mush they created with their petroleum/Corexit stew is attracting crabs, millions of them. They arrive to breed, possibly attracted by newly formed exotic organic compounds that resemble sex hormones. They smell it, and then they converge on its source, like a convention of traveling pot and pan salesmen. And then they fail to reproduce as they mutate on contact with this undulating black Blob. It has decimated the local diversity of seafloor fauna. The collapse of the crab life cycle portends the future doom of all species that feed upon them. Here is yet another bottom-up trophic cascade in action. At best, the toxins that now infiltrate their bodies will be passed ever upwards in the food chain. (McClain, 2019)

Ultimately, if humanity succeeds in killing our oceans, that death will be passed on to us as well, paying these human-created horrors forward in equal measure.

British Petroleum paid enormous fines. Sub-contractor Haliburton paid fines for their role in the wellhead construction. Every company involved, including the BOP manufacturer, paid hefty fines (okay, not the Ecolab folks). But it will never be enough for the fishermen, who lost their careers, and the fragile ocean ecosystem at whose losses we can't even begin to guess.
So how can it be, when economic forces adversely impact an industry, like coal mining, are the environmentalists blamed and demonized? But when an oil company blatantly obliterates another enterprise, the gulf fishing industry, for example, is not one peep to be heard?
The answer is simple: The human race, our societies and governments, are addicted to oil. And therefore, we protect our dealers and their cartels.

Climate is the interaction of the Earth's surface and atmosphere on a mechanical level over an expanse of time. The Oceans, which comprise 71% of the planet's surface, drive our climate. Water has roughly 3,000 times the heat-storing capacity as air. This ability to hold heat is due to water possessing 784 times the mass of an equal volume of air. Water heated at the equator flows towards the poles, distributing this heat as it moves. These conveyor currents are not only crucial in the distribution of heat over the planet's surface, but they also circulate nutrients throughout the seas. The winds are the drivers of the surface currents, and those winds are, in turn, created by differences in air temperature and humidity. These two factors combined yields the atmosphere's density. Denser air will always flow into areas of less dense air in search of equilibrium. Everything tries to reach balance, mechanically and ecologically speaking. That they cannot is what powers the planet and drives life forward.

The ocean currents are like the wind, instead manifested in water. Temperature, and salinity, and therefore density drive the sub-surface currents. The sought-after equilibrium of these mechanical systems would be stasis, everything equal and unmoving. These differentials are then the operators of dynamic change. And they are chaotic, constant, and undeniable. Solar radiation, cloud cover, evaporation from the oceans all work together to ensure that there can never be some perfect state. And that's a key driver of this living planet. In that, the atmosphere will always keep moving. The water will be "recycled" and re-deposited, the nutrients are redistributed, and energy is shared. Equality of temperature over the entire globe, absolute harmony, would lead to stagnation and death. Thankfully, the winds blow, the rains fall, and our weather is birthed from the greater mechanisms of climate, compelled by the interactions of the sun, the atmosphere and the oceans.

All rivulets, streams, and rivers flow to the oceans. They hold the lion's share of Earth's water and clearly illustrate that this is an aquatic planet. Over three billion years ago, there was barely a spit of land rising above a worldwide ocean in the Archean epoch. Without crustal uplift and the polar ice caps, most of the world might still be underwater. And soon, much of it will be again, at the cost of many of our settled coastal areas. Humans have always flocked to the shores, not just to maximize beach time, but because areas of environmental transition stimulate diversity and productivity. Such a zone is where major rivers pour into the seas.

River sediment is welcome in the great deltas, where this deposition creates new, fertile lands. The mechanisms of erosion untamed have given us those beaches of which people are so fond. Sediment transport also gives the world the wetlands, cradles of living diversity. Many of our favorite table fishes live near or have life cycles intertwined with the system of river deltas.

Along with the soil and organic matter flushed from the surface water system, come mineral salts that will join with over a billion years of washing to impart saltiness to the seas. As water evaporates from the oceans, it will leave them just a tiny bit saltier than before its departure. So freshwater flowing in from the wild rivers is required to replenish this loss. Salinity levels exist in a moving balance, like a dance, as everything rushes to and fro, seeking a final reckoning that will be forever denied. Like all Gaia's systems, the oceans thrive on variability and shun standardness and uniformity, justly called stagnation.

The warmest part of the planet, the equator, is that area that stays pointed towards the sun as Gaia wobbles back and forth on her axis. This oscillation is the mechanism of the seasons.

North of the equator lies the tropic of Cancer; south of it is the tropic of Capricorn. They demarcate the limits of the planet's tilt, as they represent the bracketed confines of the Sun's overhead travel. Between them, the sun can be directly overhead. This curiosity can lead to eerie situations where no shadows are cast in the full sun of noon. An optical illusion that may trigger anxiety in a mind subconsciously trained by life in the temperate zones. Beyond the tropical parallels, the sun will always be at somewhat of an angle, hugging either the northern or southern horizon. This central bracket is the equatorial zone, the tropics, and it is the area of the most significant and constant solar heating.

The poles, receiving the least direct radiation, stay the coldest. Water precipitated on them tends to stay put as massive glaciers, the ice caps. They represent an immense store of freshwater. They also act as giant mirrors, reflecting light and heat out of the atmosphere. The average reflectance of the earth's surface is called its albedo. Reflection is a primary mechanism for sustaining the moderate climate of the planet. The ice caps help regulate our surface temperatures in concert with the greenhouse effect. The levels of the greenhouse gasses act like a thermostat. They modulate how much heat is re-radiated and how much stays put. So, greenhouse gasses have been here for a very long time and used to be one of life's great allies. When our industriousness overdosed the atmosphere with them, it essentially broke the thermostat. We need to recognize that a very dynamic system exists in a balance that allows our Mother, Gaia, to maintain life as we know it. When we tamper with one thing, we affect all the other parts of the structure.

This system is complex and dynamic, but it tends to function more or less predictably, with drastic changes taking many hundreds, if not thousands, of years to manifest themselves completely. Our introduction of gargantuan amounts of greenhouse gasses over the past couple of centuries has altered that balance. The planet are warming up with increasing rapidity, and since many of us don't seem very alarmed by this, the damage will continue, and predictably, worsen. Our lead time to get in front of this problem is already gone; our potential actions get fewer with each passing year (month, day). Fortunately, the solutions to our self-made problems lie in a systems approach, just as that system's disruption led to our troubles in the first place.

As the planet warms, the evaporation from the oceans increases, and this amplifies the ocean's salinity and density. The seas' currents and nutrient cycles are very dependent on the exact value of the water's salinity. As the warm seawater flows towards the poles, ice formation extracts pure water through ice formation the farther north it goes. This action further increases the salinity, making the water even heavier. This denser water sinks. Where the depths cool it, making it denser still. This cold, dense water then turns back and flows south along the ocean's floor, distributing life in an endless conveyor. The gigatons of sinking water is a driver of the Gulf Stream in the northern hemisphere of the Atlantic Ocean. This flow is the famous Atlantic Meridional Overturning Circulation (AMOC).

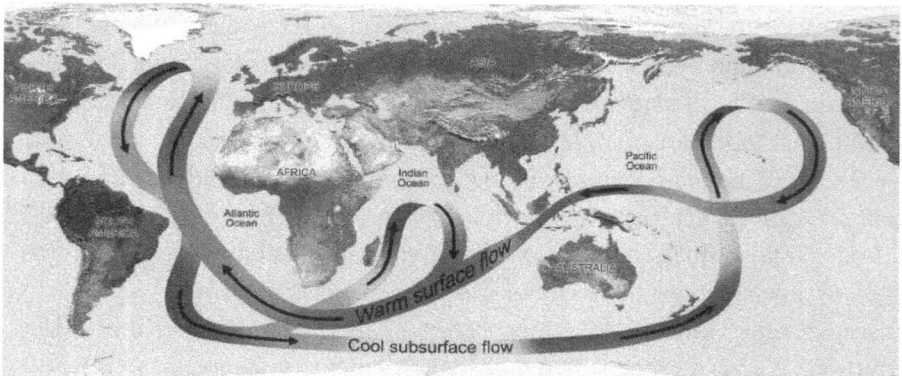

Figure 1The Atlantic Meridional Overturning Circulation. Courtesy of NASA/JPL.

And it is currently slowing. Climate scientist Stefan Rahmstorf's foundational paper and the work of others show the likelihood of significant disruptions in ocean currents much sooner than previously believed. This disruption suggests a potential future breakdown in this system, perhaps its complete failure. (Rahmstorf, 2015)

Another fly in our ointment is the melting of the icecaps, decreasing the salinity of the water locally and diminishing its density so that it no longer sinks as readily. This melt-off will also decrease the totality of the planet's albedo, reducing reflection and leading to more heat retention in this fragile area where most radiation should reflect into space. The system is so complex that its disruption's likely outcomes are too many to put a pin into what this portends. But if the currents were to stop and the winds calm, it doesn't take a scientist to understand that everyone is in for an actual world of hurt.

And changes in temperature and salinity are not all that is going on, unfortunately. The increase in CO_2 within the atmosphere has also led to a decrease in Ocean Ph. When CO_2 dissolves into water, it forms carbonic acid (H_2CO_3), and estimates are that 30% or more of the excess CO_2 created by humankind ends up dissolved in the seas. (Millero, 1995) Megatons of CO_2 are dissolving into our oceans, lowering ocean Ph planet-wide; this is called "ocean acidification." Increased ocean acidity will alter the rate and types of chemical reactions that can occur. It affects coral reefs that require immaculate and very basic (alkaline) water to survive. It is predicted that Ph changes will lead to massive phytoplankton die-offs and migrations. Phytoplankton forms the base of the ocean's food pyramid. Phytoplankton also generates 50% of the planet's available oxygen.

As the surface temperatures of the planet rise, the ice caps continue melting. This fact is widely reported and makes for dramatic news footage, but it is difficult to understand what can be done directly to halt it. There are these harrowing images of polar bears stranded on shrinking rafts of ice. But that is literally half a world away for most of us. Even though we want to help, there's seemingly no way we can. And secretly many thrill when they see city sized chunks of ice calf from the glaciers and crash into the ocean.

An ice sheet is defined as any glacial-like ice mass greater than 50,000 km^2

(20,000 mi²). While it may seem incredible that any ice fields of that size exist, there are several. The Antarctic and Greenland Ice sheets contain 90% of the planet's freshwater. The presence of so much water may seem like an answer to some of our problems right there. But as an old rancher friend of mine used to say: "You 'caint get there from here." Using this ice for drinking water is a problem of location. The thirsty masses are too far away from the supply for it to do them any good. And water is heavy; transporting it would require an engineering feat, and energy expenditure, of a nearly unimaginable scale. It requires the power of a functioning planetary system to move the waters of life. And as certain many are that we master such energy, we do not.

They're rapidly melting anyway, so there's no time for new technologies to be devised to overcome these obstacles. So far, all estimates of how long it will take for the icecaps to disappear have been sadly incorrect, as change continues to advance faster than our efforts to estimate it[84]. Melting is proceeding with increasing swiftness. The ocean's tipping point has passed. Their triggers pulled, the icecaps are now predicted to be extinct this century (DeConto, 2016). The sea rise that will occur could be in excess of 6 m (20 ft). And then, the concurrent decrease of planetary albedo will lead to more warming still. This process is called a positive feedback loop, but its effects will be anything but positive.

In the past 100 years, the sea level has risen about 15-20 cm (6-8 in); this rising will rapidly accelerate. The expected amount of sea level increase could nearly obliterate the State of Florida, just as I asserted to my young friends so many years ago. And since most of the world's population lives in these coastal areas, the human displacement will be staggering. Global warming might evict two billion people from their homes by the year 2100. That's over two-tenths of the human race. (Giesler, 2017) The refugee problem created by these phenomena could lead to societal de-stabilization on a global scale. It will probably lead to massive riots, famines, and the total breakdown of order in many places. It would be wise to plan for these contingencies before there's

---

[84] It should be pointed out that a popular tactic of the climate change denier is to point out that the predictions made by scientist keep proving themselves wrong. The fact that the error is because the scietists' calculations have been too conservative, and therefore the effects are worse than predicted,is always left out of their arguments.

no longer a chance to do so.

The loss of sediment from the Missouri River's damming and the channelization of the Mississippi River, combined with the subsidence of the delta lands due to oil and gas extraction, has led to a reversal in the function of its massive estuary. The Mississippi delta, which once created new lands, beaches, and vital fisheries, is now shrinking by around 10,000 ha/year (~25,000 ac/year) when it should be building new shores. (McCully, 1996) Rising sea levels will cause saltwater intrusion into coastal and delta agricultural areas and threaten food production by 2050. (Norwegian Inst, 2017) Considering this situation's seriousness and the reluctance of many in politics and policy to address the underlying causes, one must conclude that global warming denial is insane. It's not just borderline crazy; this climate change denial is insane[85]. Carbon pollution will create the greatest catastrophe humanity has ever encountered. And it was so preventable; now it seems merely predictable.

You can see the dramatic and concerning disruption of these mechanical ocean functions, but this is just the tip of an unfortunately massive, and mostly hidden, iceberg. Ocean currents and evaporation drive the weather over land as well, the lands where everyone lives. What happens in and over the seas will quickly be experienced in our own backyards. We need to understand the extent of the energy involved because this is the power available to either decimate or transform our world. And the amount of energy this system contains is staggering. Around 1000 watts per square meter (1.1 yd²) of solar energy continually hits the planet's surface. When the sun is shining, of course. A square kilometer is 1,000,000 square meters; that's 1 billion watts per square kilometer. The planet's surface is 510 million square kilometers (196.9 million square miles). While only one face of Gaia points sunward at any given time, that's still a staggering 265 million million (265 X 10^{12}) watts of power continually cast upon us. (NASA, 2009) Over 70% of this energy falls on the oceans, powering the currents that distribute nutrients worldwide. This stimulates evaporation and the air movement that circulates water vapor over

[85] Just as insane as an American president predicting Covid 19 will: "Just disappear". EXACTLY that insane.

land and sea. These currents move in great masses, carrying both heat and moisture wherever they may travel. Where cooler air and warm, moist air meet, we get storms.

Hurricanes, cyclones, and typhoons are ocean-driven storms that differ only by location and, therefore, title. Hurricanes form in the Atlantic Ocean and the Northeastern Pacific. Typhoons are the same phenomena occurring in the Northwest Pacific. A cyclone forms in the South Pacific and the Indian Ocean. Meteorologists use the generic term: "tropical cyclone" for any group of thunderstorms that form into an organized system that begins rotating. Once such a system establishes and reaches a sustained wind speed of over 119 kph (74 mph), it becomes a hurricane, typhoon, or cyclone. It also gets its name. People who live in coastal areas fear these storms, and rightly so. Property destruction and loss of life are routine, especially when these storms grow larger.

Meteorologists use the Saffir-Simpson Hurricane Wind Scale to rate cyclonic storms. A Category One storm is one that reaches the minimum wind speed of 119 kph (74 mph). It becomes a Category Two at 154 kph (96 mph) sustained wind speed. A Category Three clocks in at 178 kph (110 mph). Category Fours designated as "Catastrophic damage will occur" begin at 209 kph (130 mph). And then there are Category Five's, massive systems with a sustained punch of 250 kph (157 mph). There is no Category Six — yet.

In 2005, Hurricane Katrina made landfall in Florida, Mississippi, and Louisiana as a Category Three storm. Its wind speed had peaked at 281 kph (175 mph) only hours before landfall, a strong Category Five. It was the third-largest hurricane ever recorded on that day, August 26, 2005. But by the end of that year, it had fallen in the rankings to the fifth largest storm because Hurricanes Rita and Wilma ultimately snatched away its crown. Over one million people had evacuated the area before the storm, but miscommunication and incomplete paperwork left even that effort wanting. Then there's the "can do" American spirit, that is, sometimes a bit too optimistic in the face of danger. People stayed for no other reason than because they insisted it was their right to do so. But both the populace and the authorities were woefully unprepared for what was about to occur. Not only unprepared but lacking the tools to comprehend what only a day before had seemed inconceivable. The image of President #43, George W. Bush, continuing his vacation instead of rising to the challenge of this disaster was indelibly etched into the minds of the residents of New Orleans that day.

Gulfport, Mississippi, received the first hit and was nearly obliterated. When Katrina soon arrived in New Orleans, hundreds of thousands had already evacuated, leaving over 250,000 to shelter in place. The Louisiana Superdome events center became a safe harbor for 26,000 of those folks. Then, on August 29th, a storm surge hit and submerged around 80% of the city, areas built on land below sea level. The floodwalls surrounding New Orleans gave way to the powerful surge, and it inundated entire Parishes. Those sheltered in the Superdome watched as two sections of the roof peeled away over their heads as the storm rushed inside, like a slavering wolf.

Seven hundred people died in New Orleans, thanks to Katrina. Across the entire storm footprint, they estimated that 1,700 lives were lost. There was a great controversy over the entire emergency response to this storm. That is beyond our scope here: except to point out that this may be the shape of things to come as superstorms become more frequent and manifold disasters (hurricanes, fires, floods) coincide. If this one event could stress the United States' emergency preparedness, what will happen when multiple overlapping tragedies become common? The disturbances to come may leave no stone unturned, including the rocks under which many climate-change deniers currently hide.

As the planet warms, the oceans will receive the majority of that energy, the same power that creates these storms. Current research suggests that not only will this lead to more extensive and destructive hurricanes, but that it already has (Patricola, 2018). While the actual dynamics of cyclonic storm formation are quite complex, the basis for their power is simple: more energy into the system equals more energy out. We can't build our way out of this; we must reduce the warming to affect cyclonic storm formation. And to minimize warming, we have to change how we conduct our business as humans in many fundamental ways.

The oceans are the proverbial bottom of all hydrological flows and the principal region where evaporation lifts water back up to the top. When water evaporates, it doesn't just make the oceans slightly saltier; it now also concentrates our pollution. Turning our gaze west from Louisiana, towards the Mississippi River's confluence and the Gulf of Mexico, brings us to another human-driven phenomenon: Dead Zones[86]. The 2019 gulf dead zone was the second most massive on record, following only the 2017 event's severity. Dead zones are caused by excess nutrient loads dumped through our waterways into the ocean, through the riverine estuaries.

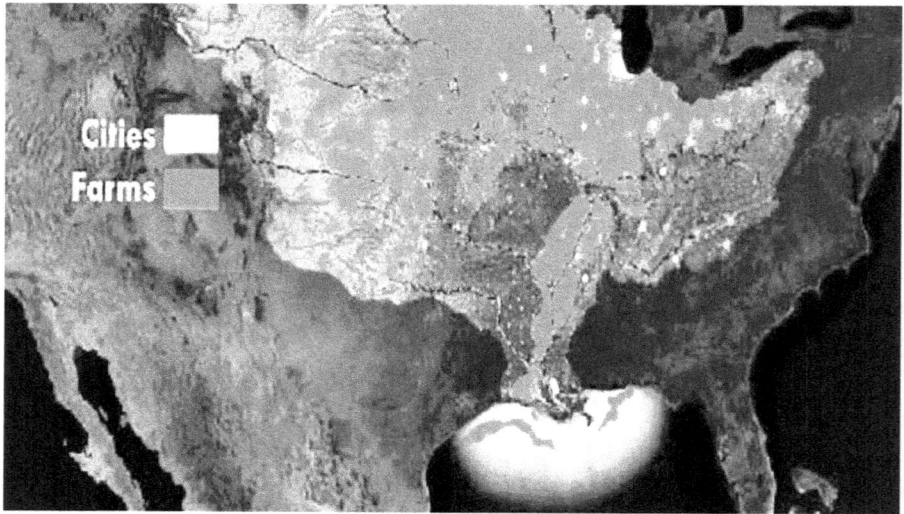

The 2019 Dead Zone at the areas that fed it. Image courtesy of NOAA.

This toxic transport is the result not only of agriculture but the previously discussed water treatment issues. Excessive nitrogen and phosphorus-based compounds both washed and flushed led to rapid spikes in ocean algal growth. After the growth peaked, the micro-plants died, and their rotting corpses fell towards the bottom of the sea, consuming all available oxygen with their decomposition. Once this area, an estimated 20,200 km² (7,800 mi²), lost its oxygen stores, everything else died of suffocation. The rancid death of sea creatures who can't swim away, such as marine tube worms, urchins, and crabs, further compounded this effect into a bacchanal of death and decay. (NOAA, 2019) Again, most of the world's fish catch is of species that are

---

[86] Can't get away from Stephen King, does it make you wonder if we're in one of his books?

linked, in one way or another, to the estuaries and delta systems worldwide. The actions that are the root cause of this parading circus of death are that our wastewater and agricultural run-offs are highly polluted. Many of these pollutants are not even considered as such by the EPA. Because they are not technically poisons, they are nutrients. Nutrients are just that: food. And there is no good way to stop them from washing out of the soil of farms across the country. Except by massively reducing or stopping their use in their current forms. Megatons of these compounds come from Papa's turf builders and brand-name laundry detergents and dish soaps. But the primary sources are petroleum-based fertilizers that, as you have seen, are applied widely and highly concentrated[87].

In addition, the very wet spring of 2019 increased soil runoff as the planet's increased warming sped up deluge-type precipitation events. Too many of our activities form a cycle that directly opposes those of nature. And while humankind will never overcome nature's mechanisms, they can undoubtedly overcome us. If we don't fix our problems. This side effect of modern living is yet another illustration of how our current way of life leads to one final conclusion: our extinction. Humans have been playing Jenga with our oceans, and it is now time to begin planning our moves carefully, and not just reclining on the beach with an umbrella festooned cocktail, watching it all go to hell. We can, and must, stop this.

The oceans are a vital link in the human food chain. A billion people feed on the bounty of the oceans every day. Up to 15% of our high-quality protein comes from seafood worldwide. People love their Salmon, halibut, lobsters, and Crab. Folks eat Orange Roughie and Red Snapper with gusto. Heck, we'll even snarf some eel, raw squid, and slurp down sea urchin eggs. Seafood is considered healthy meat, with high levels of omega-3 oils that are good for heart health and smooth skin. We love it so much we have seriously over-fished the oceans. Since 1970, oceanic fish stocks have been reduced by almost 50% because of overfishing. Some, such as the Tuna family, have seen stock reductions of 74% in the same half-century (Tanzer, 2015). Many economically important species are on the verge of extinction by our over-

---

[87] The elimination of chemical fertilizers and their replacement with compost and soil regeneration could solve this particular problem in a jiffy.

consuming ways. The fishing industry now operates in a manner that doesn't leave much stock unmolested, and even when they do, the next guy might not because he is likely a pirate (parrot optional these days).

The mechanization of industrial fishing is a far cry from the old days of family-owned trawlers setting out at dawn to try their luck. Now fleets are dispatched with satellite data, equipped with sonar trackers, running seine nets between multiple boats to drag entire schools into their holds. This approach sounds familiar by now, or at least it should.

Deep fishing is accomplished with purse nets that drag the ocean floor and trap everything in their path. When they catch something they don't want, they throw it back, often dead or dying. This "bycatch" can be dolphins, sharks, rays, and even sea turtles. Oh, they'd take those turtles if it was legal, but for most species, it is not, so they just toss their corpses back into the ocean after the submerged net has drowned them. Turtles and dolphins breathe air, and there is none when you are trapped in a net, far below the surface. Up to 20% of this wasted bycatch can belong to endangered and threatened species. (Tanzer, 2015) Corporate fishers will call the final feeding of these mammals (dolphins, dugongs, small whales): depreciation. As they wait to die or be released, these sea predators will feed on the captive fish school, reducing the catch's profit. This loss often leads to angered fishermen, who might now be more inclined to choose to kill instead of release. (Read, 2008) Out there in the middle of the ocean, no one can hear them scream.

Sportfishing is where you take out your pole and give it a shot, casting into the water, hoping you have the right bait and a bit of serendipity. There is no sport involved in commercial fishing. It is a ruthless harvest conducted with the highest level of technology and the lowest level of humanity. If 50% of our forests had disappeared in the last generation, or half the deer and elk vanished, you'd think people would be up in arms. But the residents of our seas swim below the surface, never seen by most of us until served up with lemon slices and an excellent Chenin blanc. Again, out of sight is out of mind for many. If our ocean stocks collapse, it's going to be a lot worse than no more fish sticks.

This crime is occurring in the 71% of the planet that has already lost half of its population! There is a moral and a pattern here we have to recognize because this will soon be our fate as well; if we don't fix this world. If the human race succeeds in eliminating ourselves, no one else will care, certainly not the dolphins and whales. Human survival is a human problem.

Perhaps the best remedy to the problem of overfishing is just to step away and let nature heal itself, but that's never going to happen. Too many people need ocean protein to survive, and this is an international issue and dilemma, so it needs a global solution. That's a tough ask when so much anti-globalism propaganda is floating around out there. Tribalistic politicians and their supporters are already outraged by the prospect of world-based rules and the fear of losing even one Fillet-O-Fish sandwich to mutual cooperation. The United Nations (UN) operates under a cloud of not so thinly veiled racism and animosity in the United States of America and other developed nations. But they have a solution to this problem. The Food and Agriculture Organization of the UN (FAO) has been studying this problem for some time and has issued plans for the sustainable harvest of our oceans. (FAO, 2009)

> "These Guidelines have been developed in response to requests for further information on the practical adoption and application of the ecosystem approach to fisheries (EAF), with a special focus on its human dimension." -F.A.O.

The ecosystem approach to sustainable harvest considers what extraction levels would be reasonable to stabilize and improve any given sea stock. It places substantial weight on the economic and cultural values of the people involved in the trade. Since these plans recognize that people are a part of the ecosystem, they are designed to improve both the human communities and the dwindling ocean fisheries. And it works!

In the early 1990s, the tuna fisheries around Fiji collapsed precipitously. Demand for fresh tuna for the sushi and sashimi industries encouraged a mega-harvest based on these waters. Fishing was so successful that the tuna were on the verge of eradication in this vital nursery. The FAO stepped in and developed guidelines for the sustainable harvest, and less than a decade later, the fish population rebounded. The industry did not collapse either; it merely slowed as new licensing structures were instituted and catch limits enforced. In the end, the entire fishing enterprise was saved by these rules, and a way of life was preserved. (FAO, 2009)

In the end, the entire fishing enterprise was saved by these rules, and a way of life was preserved. Because of proper regulation.

Had there been no regulation, and the tuna had gone extinct, that would've been the end of that story for both the fish and the fishermen. Regulation, properly applied, will increase, not impede progress.

So the feared outcome that so many continually raise when complaining about regulations or the institution of any form of self-control proved to be empty. Again, it needs to be noted that environmentalism is true conservation and should be considered the actual conservative value. And if self-imposed rules can achieve ecosystem renewal faster, then regulation should be labeled a conservative value too. When did "everything in moderation" become a radical idea?

It should also be noted that the United Nations has been at the forefront of climate adaptation and human population protection for quite some time. Their opinions and programs do not always square with profit expansion, hence their demonization by the other multinational concern: corporations. The UN's solutions often favor the smaller, family, or group-owned businesses over the kinds of corporate entities that prefer endless expansion and profit above all other concerns; since they are interested in sustainability and social fairness. The UN comprises a culturally diverse group of people. They are deeply concerned with the survival of the planet, the human race, and the bio-diversity that thrives in concert with both.

We need to get out of our own way, as much as we need to get out of the way of nature, to implement the all-or-nothing salvation that is our last hope. Our disrespect of nature is a perfect reflection of our disrespect towards each other. And over-fishing is also a disrespect for the web of life of which all are a part. Out there, in the ocean, there are no borders nor nationalities. There's just life clinging on by a flipper, hoping humans will pull our heads out of our blowholes before it's too late.

New York Harbor was once a pristine ecosystem; over 400 years ago, that is. The Hudson River's flow poured into the Atlantic Ocean there, and this mixture was filled with natural nutrients and life. An oyster reef filled over 81,000 ha (200,000 ac) of it in 1609 and was the backbone of the Lenape peoples' way of life. This reef fed the Lenape and protected them from storm surges since time before memory. Like every reef, it was a living underwater mountain. It was a treasure that had built itself over thousands of years and held the promise that it would stand for thousands more. So here is another parable, a sad repetition of humanity's impact on nature all rolled into one. And it is also a tale of hope for our future.

The City of New York grew up around that harbor. At peak production, New York Harbor sourced 50% of the world's oyster production. It was a hectic time of harvest and consumption. But by the early 20th century, the reef had virtually disappeared, eaten by the population, and buried in sediment by our commerce. For over half of that century, the harbor was lifeless, rendered toxic, and vile. They dumped sewage directly into the harbor, storm drains spewed into it, and industry regurgitated their waste seaward with wanton abandon. The oyster reef was once where the people ate, and then New Yorkers took a dump all over it, as was (is) the way of unregulated commerce and unbridled growth.

When Congress passed the Federal Water Pollution Control Act Amendments of 1972, these waters were finally protected from their previous onslaught. In 1977, these rules were amended and rolled into the Clean Water Act. That Act offered the hope that the harbor might be healed. But it took another 30 years before the toxicity had declined enough to re-introduce the Oysters.

This pause in pollution corresponded with a period of heightened ecological awareness in the US. But, although several dozen cases for environmental pollution were brought before the bench in New York State during that time, none resulted in a single corporate officer's conviction. The administration of Governor George Pataki, a self-described "conservative," was proud of this record of business-friendly decisions, policies, and actions. So while many types of environmental damage were criminalized, the "perps" still always got away with it in New York. The idea that pollution can somehow be good for business is a very nearsighted view indeed. Dead customers make for poor commerce.

Then, in 2014, a non-profit called the "Billion Oyster Project" began work to restore the reef and return a billion oysters to the Harbor, just as its name suggests. Murray Fisher had founded the New York Harbor School as an environmentally focused high school. He began cultivating and releasing oysters (*Crassostrea virginica*) back into their historic range with his students. Oysters are filter feeders, and they can percolate 3.8 L (1 gallon) of seawater per hour per oyster. Calculating upwards, they concluded that approximately one billion oysters would be the goal for re-establishing a self-sustaining colony capable of effectively cleansing the harbor. This organization now works with its students and volunteers from across the area to raise and place seed colonies in an extensive and well-coordinated citizen science program. They enlist local restaurants to donate oyster shells from their waste streams and use them as bases to grow the larval oysters upon. Finally, they take these plates of new life and return them to the harbor. Little actions such as these, performed by many people, will eventually have an enormous impact on their local environments. Just as other little actions had multiplied over the years to disrupt them. By centuries' end this project will have affected the re-establishment of the reef, along with an accompanying ecosystem renewal and a natural breakwater that will protect the area from storm surges, just as it had done before the blind pursuit of profit consumed it.

Meanwhile: some engineers and politicians are advancing the idea of building massive mechanical walls to protect the harbor from the storm surges[88] that are expected because of sea rise and climate change. These types of massive structures are what many seem to assume to be the solutions to our problems. However, they are merely a continuation of the mindsets that created our problems in the first place. Futuristic mega-structures that emulate mechanisms from science fiction, and reinforce our supposed mastery over nature, will not solve our problems.

The Billion Oyster Project is yet another example of the use of nature and her systems to restore and re-balance what has been broken. This project is an example of "Ecology as Technology." Undoubtedly, engineering firms would happily profit from the proposed massive sea walls of concrete and steel. And

---

[88] See: www.riverkeepers.org

these structures might serve some of the same functions as a reef, but at a princely cost. The cost of energy to make the concrete and steel, the cost of designers and workers to install the structures, and the cost of maintenance as the sea's assault relentlessly wears it down.

On the other hand, oysters can do it for free, and they heal themselves after storm damage and improve the water quality as they grow. So, if we would only change our mindset from pretending to be the masters of this universe and accept our role as stewards of a small piece of it, not only could we work with nature to save ourselves, we would be happier in our hearts and more settled in our minds. So, which solution makes the most sense: more massive energy-intensive machines or ecological restoration?

Machines can't heal themselves; reefs can.

The other arguably more famous type of reef is just as afflicted as the temperate oyster reef. Tropical coral reef systems are well-known to everyone, from schoolchild to corporate banker. The dazzling colors and otherworldly forms of this, the rainforest of the oceans, create images that never leave one's imagination, even if you've only seen them in pictures.

There are over 1,000 species of coral, and like oysters, they are filter feeders. Coral belongs to the Phylum *Cnidaria,* and rather than creating shells like the mollusk oysters, they build elaborate, often branching structures of calcium carbonate ($CaCO_3$): coral reefs. The animal itself is a tiny polyp, a very primitive life form that has demonstrated its fitness for this world for millions of years. Over the millennia, these reefs have become large enough to be seen from space. This vibrant ecosystem is home to over 50,000 unique plant and animal species. Marine scientists believe the actual number may be closer to a million. This diversity arises because the reef acts as a protection from the sea's currents and storms. It is an oasis of nutrients and a safe harbor for reproduction. Trillions of tiny polyps continuously filter this water, so the water quality is pristine, almost beyond belief.

In Australia, the Great Barrier Reef is 2,300 km (1,400 mi) long and occupies an area of 216,617 km² (134,600 mi²). Located in the aptly named Coral Sea, it gains its moniker from the simple fact that it protects Northeastern Australia from storm surges: The Great Barrier.

These and other tropical coral reefs are disappearing rapidly, the most notable lethal phenomenon being "coral bleaching." The coral's vibrant colors come from *zooxanthellae*, photosynthesizing dinoflagellates[89] associated with the corals that live inside the polyps and paint their white calcium backbones with a rainbow of colors. This mutualism is similar to the associations seen in lichens. When the coral animal is exposed to water temperatures above their standard range, the *zooxanthellae* are expelled, and the structural framework is exposed, allowing the ghostly white skeleton to shine through. This vision is a ghastly and genuinely terrifying thing to encounter. The coral is not quite dead yet, but they have lost an essential supplementary food source, their photosynthesizing pals. The corals can recover if the heatwave passes quickly enough. But only if the heatwave passes.

As the oceans continue to warm and bleaching events become more common, actual reef death is accelerating. Everywhere you look, in every domain, death is advancing, and time is running shorter.

Other threats come in the form of that aforementioned ocean acidification. The entire coral system relies on pristine water conditions. Even slight increases in Ph can alter their delicate metabolic functions irreparably. In addition, highly alkaline waters support the calcium carbonate structures that are the backbones of the reef. Acidic waters will dissolve them. These monoliths represent a massive sequestration of carbon. If they dissipate and die, then there will be even more carbon dioxide released into our oceans and atmosphere.

Coral are also the victims of too much human attention. Tourists traveling for that once-in-a-lifetime vacation swim amongst the coral. Unbeknownst to them, the suntan lotion they wear can damage the reefs at the genetic level. The sunscreen agents oxybenzone and octinoxate are the primary culprits, and they are toxic to coral at levels of one drop in 6 Olympic-sized swimming pools (62 ppt[90]). (Downs, 2015) Other less damaging but still concerning human actions include physical damage to the reef by divers and coral harvest for the aquarium trade.

---

[89] Dinoflagelates are single celled crtitters with little hairs that lash about and allow them to swim: flagella. They can also turn the sun's energy into sugars just like a plant. Isn't life marvelous? You want to stay a part of it, right?

[90] Parts Per Trillion

I need to point out that virtually all aquarium coral in the US and Europe is grown by specialists and hobbyists, and none of them would ever buy wild-harvested specimens. Indonesia has cracked down on de-coralization and outlawed exports not strictly cultured in farms. Regretably, that is not true everywhere, and the paradox here is that those living collections, in the hands of private individuals, may one day become a vital gene bank for the future. The point here is: the elimination of wild coral harvest would not affect the legitimate aquarium trade, but the elimination of that hobby might well harm the future of coral[91]. Some of our well-meaning advocacy groups currently want to stop activities such as the aquarium hobby or the exotic pet trade[92]. This prohibition would be a band-aid that could lead to yet more unintended consequences if posterity loses the species diversity that now lives in peoples' homes. Homes that may be the most stable environments for the next few decades.

The wild coral harvest should be stopped, and indeed almost all collection is currently illegal under the Convention for Trade in Endangered Species (CITES). Rare species, in private hands, should now be protected as well. If the coral reefs are lost, we potentially lose a million other species of unique and beautiful life. If nothing is done, up to 90% of the world's tropical reefs may be gone by 2050. (Burke L. , 2011)

Stoney Coral Tissue loss Disease (SCTLD) was first noticed in Florida in 2014. It causes rapid tissue loss as abscesses grow to envelope entire colonies. It affects over 30 different species of coral and has continued to spread in the intervening years. At this point, over 50% of the Florida reef tract has been affected. The Florida Reef Tract Rescue Project (FRTRP) has enlisted Zoos and accredited facilities across the national to take coral samples removed from danger, far ahead of the spread. Their partners will give these creatures safe harbor for as long as is necessary, many using hobbyist grade reef aquarium technology. The reef hobby has made lower-cost massproduced equipment for just such projects easily available. And that's good news for the

---

[91] Sorry, PETA, but this genie is out of the bottle.

[92] This is a very sticky wicket, because the exotic pet trade includes tiny dart frogs as well as tigers in the suburbs. Tigers in the back bedroom are a bad idea. Captive bred frogs and lizards, not so much. There is a need for some regulation here, but it must be specific and rational.

coral, who are now even spawning in captivity. These colonies produce millions of eggs and sperm in chorus. They do it once a year, on exactly the same schedule as the wild corals, with one main difference being: In the wild over 90% of the larvae will die at some point, in captivity over 90% survival is possible. In the aquarium your only limit on harvesting the larvae while they are still free-swimming is how efficiently you can capture them.

There is currently no word on what causes SCTLD. So far, there is only hope for the wild Caribbean reefs. Finger crossing levels of hope, but still no technique. Scientists will keep looking until they find the answer, or no longer have any subjects to apply that solution to.

Dr. Ruth Gates, Director of the Hawai'i Institute of Marine Biology, is a proponent of "Assisted Evolution." This discipline is the practice of leveraging genetic knowledge within organisms to aid in rapid adaptation to abruptly shifting environments. This group has identified hardy coral strains, individuals who show resistance to increased temperatures, and then cultured them in the lab. A single coral group is clonal. They are genetically identical, as they reproduce asexually. While coral also reproduces sexually in dramatic mass copulation events that occur during full moons, their primary method of reproduction is budding. Operating similarly to the Billion Oyster Project, the Hawai'i Institute of Marine Biology grows their charges for release back into their ocean habitats. Other scientists are reproducing heat-resistant coral and also returning them to the reef. Using the techniques of selective breeding and generational conditioning, just as humankind has done for thousands of years with crops and animals, there is hope that Gaia can recover her reefs using nature's inherent ingenuity and our assistance. (J.H. van Oppen, 2015) There is more than hope. These are factual demonstrations that this is a path through this thicket. We just need to make sure they are funded and encouraged to expand.

Sailors have characterized the oceans as a vast wasteland, a dangerous expanse with a never-changing horizon. Sea-farers have always been a stalwart breed. Their life expectancy was once but a fraction of that of the landlubbers. And yet, the allure of adventure and sites unseen would always produce more people ready to meet that challenge. The world seems smaller these days, as the once seemingly endless currents are all mapped, and nary an undiscovered tropical island exists. There is, however, a new class of islands touched on earlier that now bears closer scrutiny.

The Great Pacific Garbage Patch (GPGP) swirls and drifts between Hawaii and California and is currently twice the size of the State of Texas. This floating catastrophe comprises over a trillion pieces of plastic refuse, and its weight estimates are somewhere around 80,000 tons (88,149 tons). It is just one of many. These swirling testaments to human indifference, formed by ocean currents called gyres, continue to grow because our civilization lets them. The GPGP is estimated to reach over 30 meters (98 ft) deep in places and is fueled by the millions of tons of plastic produced globally/annually. Greenpeace estimates that 10% of all plastic produced may end up in the sea. Groups have, and will continue to, form with the intent of harvesting and cleaning up this mess. The company Parley has been collecting small amounts of this trash and recycling it into a plastic fabric. This plastic fabric is used to make designer tennis shoes, and companies: Adidas, Nike, and others offer them for steep prices. As "feel good" as this might be, it is far less than a drop in the bucket and not the solution to this problem. It is mostly a trivial action intended to focus on the issue, not solve it. So far, no technique has had a measurable effect on this massive and remote synthetic atoll. That's because people just keep making disposable plastic containers, wraps, and pointless do-dads. And then everyone throws them away willy-nilly. And out there: they all float. There are at least two "garbage patches" in the Pacific and others forming worldwide; this is a global issue for which, while there is plenty of press coverage, there is little workable technique to address. These rafts of trash act in some ways as floating artificial reefs. But they are toxic reefs and adversely affect the carbon, nitrogen, and phosphorus cycles within the truly massive areas they cover. Acting as floating lids, they reduce evaporation and dilute the waters' salinity under them. They can kill fish by literally sucking the salt from their bodies as they attempt the swim beneath them.[93] Very few creatures can survive a passage under two Lone Star States fast enough to avoid this fate.

---

[93] This is because of that simple principle from high school science class: Osmosis. When the fish are acclimated to a certain salinity, all is good. But if the water suddenly loses its salt, the salt in the fishes bodies can be sucked out as the solutions seek equilibrium. It can work in the opposite manner, reversed, if the water becomes too salty.

The light blocked by such rafts is also decimating the phytoplankton population, which should be photosynthesizing and releasing more oxygen under the open waves. Larger sea creatures, such as dolphins and whales, accidentally ingest plastic bits and chunks with fatal results. The gyres' churning action grinds some of the waste into floating dust, which washes up on Hawaiian beaches, and everywhere else. This dust drifts from pole to pole. A million sea birds and their chicks are now dying every year from plastic ingestion. (Greenpeace, 2015) Everyone should be able to agree that this needs to stop.

Our industrial culture, based as it is on the misconceptions that nature is both infinite, and made for our exploitation, has led to the dumping of everything imaginable into our oceans. This frontier is where all our drains end up. And the same lack of care exercised at home is exponentiated when waste disposal hits the industrial scale. Toxic effluent from chemical plants poured into rivers will inevitably end up in the seas. Since a significant percentage of the human race lives near the oceans, they may dump many toxic cocktails directly into the oceans.

Psychologists will tell us that babies typically develop the concept of "object permanence" around 4-7 months of age. It is around then that they realize things don't cease to exist just because they can't see them. Before then, objects that are out of sight are assumed to be absent from existence. And yet, this is precisely the way grown adults deal with waste in our modern world. They bury it in the ground; they throw it into the sea. Who cares it's gone now. Adults hold adult positions and make adult decisions, yet as societies, people often act like children. And not just children- but babies. All the toxic chemicals discussed, and the human waste in our blackwater, is often released directly into the sea in many places worldwide. Barges carry tons of trash out past the breakwaters and dump it with impunity. From the middle to the end of the last century, several countries just dumped nuclear wastes into the deep oceans as well. (IAEA, 1999) You can probably guess that the United States was one of them[94].

---

[94] But the number one oceanic radioactive waste dumper was: Russia. Followed by: the UK, then Switzerland (which is really interesting since the country is landlocked) and finally, the good old U.S. of A.

CHAPTER SIX: THE OCEANS

Life arose in the seas. It took almost 3 billion years for nature to crawl onto land. The oceans are the cradle of life. The human womb is like a tiny ocean, a salty liquid vessel that gives a safe harbor to life at its most fragile. An egg, similarly, protects its nascent being within a liquid embrace. All life arose from the watery domains, and if humanity succeeds in destroying them, what should our fate then be?

Just as once all roads led to Rome, all rivers and streams lead to the oceans. All the crap and contamination exuded into our waterways will eventually concentrate in the seas as well.

You've seen that the insults inflicted on our oceans are chemical (toxins), mechanical (warming), and extractive (overfishing and extinction). This trifecta of carnage is largely out of sight but not without consequence, as this frontier is 71% of the planet.

Our cultures have uniformly adopted customs of throwing things away, and the ultimate garbage dump is the ocean. We need to stop! And then we need to do everything possible to fortify natural processes and reverse the damage we've inflicted. We can't do this too quickly.

Just as the little things people do add up to massive negative impacts, much of it can be reversed by taking small positive actions in concert. We must replace inaction with action. Societies must recycle, regardless of the startup costs. All people must stop using toxic chemicals where less toxic or completely non-toxic alternatives exist. And primarily: folks need to stop acting like spoiled children who inflict hurt out of pure spite.

They used to call the oceans the last frontier before humans stepped into space and immediately began filling it with trash as well. The seas act as a giant buffer for the planet's heat and mediate its intrinsic chemistry. The oceans are the forges of terrestrial life. Our combined assault on them has been so unwise that it raises the question whether the human race possesses much usable wisdom at this point. The ocean is a metaphor for our collective unconsciousness as well. And just as in dreams, the crap we try to hide from returns to haunt us there. Our seas have a staggering capacity to absorb insult. But even they have limits.

All it will take is a modicum of restraint and effort to help our oceans return to their proper trajectory, and nature will do the rest. But first, each of us needs to decide that life is really what we want. And that will require an evolution in our ways of thinking and interacting.

We must evolve how we interact not only with our oceans but with all of nature. This is the Green Evolution.

CHAPTER SEVEN: AIR

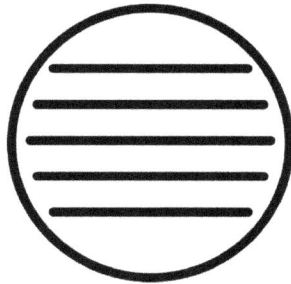

"We've got to pause and ask ourselves: How much clean air do we need?"
— Lee Iacocca

In 2015, multiple forest fires occurred in the vicinity of Palembang, Indonesia. These wildfires developed across many "sustainable" tree plantations, creating a smog cloud that choked Sumatra, Singapore, and Malaysia. Smog had long been a serious issue in this part of the world. Slash and burn agriculture, heavy industries with little stomach for regulation, and homes still burning primitive fuels all contributed to this problem. However, this event was different, and it became known as the "Southeast Asian Haze of 2015." But the name trivializes this event's severity. Over 28 million people were wrapped in a toxic cloud for months; 9 people died, and up to 140,000 cases of respiratory illness were reported. (Koplitz, 2016) This event led to the declaration of multiple national emergencies. They estimated in this same cited study that up to 100,000 deaths might eventually be attributed to this disaster. So what happened this time?

In order to increase their profits in the paper pulp and palm oil businesses, many producers bought swampland tracts and converted them into tree plantations. Across multiple islands and countries, it seemed peat bogs would be the ideal places to develop farms, or nearly ideal as the case may be. Peat is a highly fertile substance. But when in-bogged, it is soft, too weak to support tall, fast-growing trees. The trees of choice were the Acacia and the oil palm, both of which would grow to a certain height and fall over in the mushy ground. This result was less than optimal from a farming perspective. So, the businesses drained the swamps to "firma the terra." That solution sort of worked, at first. Good news: the trees stood up just fine. Bad news: entire plantation complexes caught fire and burned down, spreading a toxic pall across Oceania.

Dried peat is an excellent combustible. Civilizations have used it for centuries around the globe as stove fuel. It has a lot of stored energy and is easy to light. That's great when you need to warm up the yurt rapidly after a long day tending the sheep. But in this case, it lit off too easily. Drought caused by global warming had already primed the entire system, with kindling-dry conditions and low relative humidity. This situation became the definition of a tinderbox. The moisture that had been in the peat was pumped into the sea, drained and discarded, just to pollute the waves even more with a tannic acid tea.

Once an ignition sparked, it spread with wanton abandon. Massive ember storms arose and spread the fire from one fuel bed to another, sometimes from one island to the next. As the cloud rose, the authorities found they had no way of fighting an inferno on this scale. They had no choice but to stand by and try not to choke on it. The billowing clouds of pollution rose into the skies, blocking the sun and poisoning the land—just another day at the office for the profit-driven human race.

The fires were finally extinguished when the rains came months later. In the end, it was natural forces that put an end to this catastrophe. Again, often only nature has the power to undo what has been done. And she works on her own schedule. A schedule that is often unaligned with our timetables and not concerned with humanity's fate one whit. These can be expansive time scales far longer than any single human life or even multiple generations, but at least one can comprehend them. You can understand the time span of a million years, even though you will never experience it. Human civilization could last even longer, but will it[95]?

Not if we don't get our act together.

When you walk through the countryside and breathe in the pure, fresh air, you immediately feel alive and more in touch with your natural being. With each breath, the stifled scents of indoor living melt away, and all your senses

---

[95] Ask your local bookie about that one.

sharpen. Lungs open, and your chest expands beyond its previous capacity as you drink in the sweetness of pristine oxygen. You can smell the perfumes of the flowers and the dusky musk of the moist ground. Above you, the clouds drift against a deep blue sky, which may often bring the shapes of animals or human faces to mind. Walking with a partner, you may feel a heightened sense of camaraderie or even sometimes, love. The seemingly endless canopy that arches overhead is a curtain between our terrestrial life and the heavens.
You don't usually think that this air, this atmosphere, is akin to a rarified ocean. But it is, and you are walking across its floor. The birds swim past you as do fishes in a sea, and the shrubbery is a terrestrial reef of green, nearly as dense with life as the oceanic type. This body of air suspends aloft the dust that rises with your footsteps, just like mud swirls around your boot when you step into a puddle. It transmits sound waves, allowing us to speak with one another, and delivers our pheromones to communicate messages of an entirely different nature. This vaporous body holds bacteria and spores, viruses, and a host of microbial life aloft. For life dwells in every estate on Gaia.

You may think of the air as oblivion if you think about it at all. One can move through it with such ease that it might almost be a vacuum. There are land and sea, but the air sometimes seems as if it were but an expansive volume of nothingness. You can fly through it in dreams, over the land like a bird or butterfly. You may surf our pillows over fields of emerald in deepest REM sleep, but you know that experience isn't real. And yet, somehow, the atmosphere can support a jetliner filled with people traveling back and forth at 1,000 kph (600+ mph). Wait, a bit for the wind to rise, and you'll know that the air is indeed something. At 112 Kph (70 mph), that ether will knock you on your ass. Because what is call "Air" is a mixture of gasses, and gas is one of the four primary states of matter[96]. These gases are nitrogen, 78%; oxygen, 21%; and argon, ~1%, with trace fumes, including carbon dioxide, making up the rest. Air possesses mass, as undoubtedly as any ledge of rock or wave of water.

Our atmosphere is indeed voluminous but finite, nonetheless. It has a measured volume of 1.7 billion km³ (470 million cubic miles). This dry ocean is 100 km

---

[96] Solid, Liquid, Gas and Plasma

(62 mi) deep. And air, unlike water, can be compressed under its own weight. So, the atmosphere at our sea level is denser than the air at the mountain tops. Where I live, at about 2,800 km (8,500 ft.) of elevation, the air is quite a bit thinner than down by the beach. And as many a tourist will tell you, that means the oxygen levels are noticeably lower. There is still the same proportion of oxygen as the other gasses; there are just fewer molecules of each for any given volume of air. In our case, fewer per lungful.

Mount Everest climbers are well aware of the "Death Zone." At the elevation of 8,000 m (26,247 ft.), there is no longer enough oxygen to keep a human alive without a breathing apparatus. Recently, in 2019, a climber died on their Everest ascent simply because he waited too long in a line. The overcrowding and elevation proved fatal. Others, eager to show their superiority, have done the last pitches without breathing assistance. Such a feat only demonstrates that the levels of damage a body can sustain at the behest of a driven mind are extreme, indeed. So many climbers have died on Everest, chasing that dream of conquest that it has become a significant problem. That same year, 2019, volunteers pulled 11,000 Kg (24,000 lbs.) of trash off the mountain, along with four additional but unidentified corpses. The ego drive to "conquer" this inhospitable spit of land was so great for some that it even overrode their will to live. If denial were a river, perhaps this is its headwaters.

Respiration is a metabolic exchange that all terrestrial life engages in, and is manifested as the intake of certain gasses and the output of different gasses. This swap is essential for all life. Animals inhale and exchange oxygen, which they need for energy production, for CO_2, which is toxic to us. Plants intake that CO_2 and exhale oxygen, a very nice balance. We complement each other. This transaction is a form of gaseous recycling, and it is fundamental to our systems of life. Plants strip the carbon atom from CO_2 to use as organic building blocks for tissues and sugars. They do this with the energy of the sun in the process of photosynthesis. This chemical reaction leaves the oxygen molecule (O_2) free for re-use by their more active, distant cousins. Animals, like people, use that oxygen for their energy production. They "burn" carbon with oxygen to produce energy and, most importantly, heat.

Respiration/digestion, the drivers of our metabolism, can be referred to as "Cold Fire." For even though they produce the heat that keeps us warm, it is a controlled reaction that won't run away like a burning forest (spontaneous human combustion aside).

Interestingly, fungi inhale oxygen and use it just like animals do. O_2 in, CO_2 out. CO_2 is one of the primary currencies mycorrhizal fungi trades with plant roots. In nature, everything is a cycle. There is never a final product, and there is never a waste product. Life is chemistry, and all living organisms are chemists. And Gaia is the tapestry of life embroidered upon the canvas of the inorganic world.

Our current troubles hinge upon the atmosphere. Humans have become external chemists, no longer just intrinsic ones. And we have been manipulating matter in many useful, but also precarious, manners. While the chemistry that all life engages in is pretty safe and internal, we humans have really uncorked the genie. Our industrial revolutions have increased the productivity of each of us with an astounding release of energy. This energy comes primarily from combustion, and it releases genuinely staggering amounts of gasses into our atmosphere, almost all of which are troublesome. In the 1970s, a wave of environmentalism was focused heavily on air pollution. Air quality was indeed suffering because industry had poured smoke and chemical by-products out of billions of chimneys with little to no concern for over a century. Society became acquainted with the term "acid rain" and were justifiably upset to learn of the release of mercury and other toxins into our air and water. (Laden, 2006) At the beginning of this chapter, Lee Iacocca's quote says it all about how the status quo at the time reacted to this drive to stop polluting. His great argument: "stop whining, you wimps, and suck it up!" was neither great nor much of an argument. But it was a typical "conservative" line of reasoning at the time. The bully's tactic is to make your perception of a problem a problem with your perception. Their blame is always somehow your fault for noticing it. (He who smelled it, dealt it?)

So, air pollution continues, even though we've done some good at reducing many of the more toxic substances, such as mercury, released into the atmosphere thanks to the EPA. But, the real problem was almost totally overlooked at that time. Our problem was not just the toxic chemicals released from the millions of smokestacks belching skyward. It was "Hot Fire," combustion itself. When a carbon fuel burns, atomic bonds are ripped apart, and atmospheric oxygen joins to the free carbon atoms. This simple reaction creates heat for our furnaces and power for our cars, just as it releases energy within our bodies. It also produces, as its main by-product: CO_2.

Carbon dioxide is not inherently bad, as we know the plants love it. "Bring it on!" they sing. But there is always a balance that needs to be maintained. Atmospheric inputs had equalized with outputs throughout millions of years. But after having pumped so much CO_2 into our finite atmosphere, there was just no mechanism to use it all. As a result, the balance has not been lost so much as obliterated. Rampant CO_2 levels have built up, and they continue to build. The result of this ever-increasing CO_2 load above our world is global warming. There are no ifs, ands, or buts about it. Nor should anyone temper our language by using more acceptable terms to describe it.

Suffice it to say; this is a hot topic of conversation for all but the most contrary and stubborn among us. And even the deniers talk about it. If only to pooh-pooh the science and logic, hopefully insulating themselves from the consequences of their own actions. The polluters don't talk about it much, though, as they seek to deflect both cause and blame.

Carbon dioxide emissions and atmospheric concentration (1750-2020)

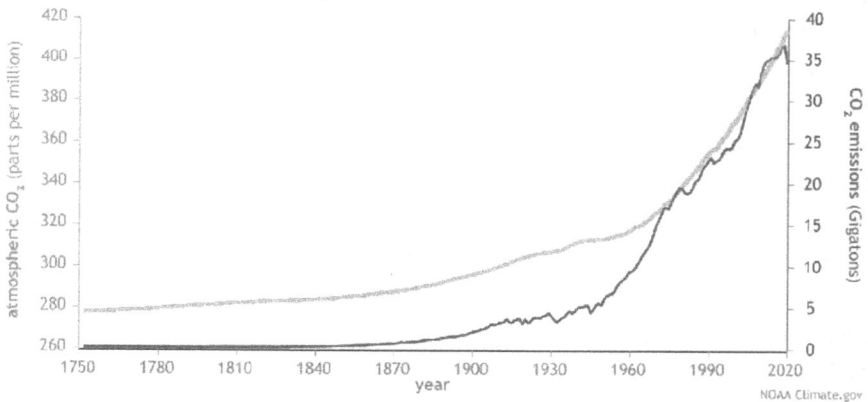

CO_2 emissions corelated with atmospheric CO_2 levels. This graph is self-explanatory. Image Courtesy of NOAA.

The polluters knew in the 1970s that carbon dioxide exhaust was just as harmful as mercury or sulfides. But they must have felt blessed by the cover these more toxic substances gave for this particular type of pollution. Because you cannot burn anything without releasing CO_2, it is the basis of the process itself. The only mitigation available would be to stop (or drastically reduce) fossil fuel burning and replace it with something clean. That just wasn't going to happen on their watch.

I cannot overemphasize that this is everyone's problem and everyone is at least partially at fault. Humanity did this, and pointing fingers should take a back seat to how we can fix this problem. If you lived with a remote tribe somewhere, with little to no contact with the "civilized" world, then you might be nearly blameless. If you live in one of the many impoverished areas of the world, your guilt is slight compared to that of an American or a Chinese citizen, but it still exists. And otherwise: if you are reading (or writing) this, you are almost undoubtedly somewhat culpable, and the bill that is about to come due is yours. Sure, oil companies lied about this. Politicians made hay off of denying these realities. But folks let the patriarchs of the status quo get away with this by continuing to elect them and frequenting their businesses (or via our developed nation's widespread apathy).

These corporations and petroleum conglomerates could still become valuable partners in our solutions. We're going to need all the technical and financial resources we can muster. Only after they refuse to help voluntarily, should we consider compelling them. But we could compel them; there is no magical sovereignty owed to the profit-takers. There is nothing special about those who have more (most). In our civilizations, the difference between the haves and have nots is often just the toss of a coin. Whether such a thing a "luck" exists or not, humans certainly reward it.

Be wary of the anger, and mistrust that the profiteers have fostered. This anger is palpable, dangerous, and sadly misdirected. Many corporations have intentionally misled the public and fomented distrust, and even hate, for those who want to save our world. They've used propaganda machines, and their misinformation flows in torrents across the media and the internet. Politicians use this anger and fear to distract the populace from the genuine dangers we all face together. Some citizens in the United States so violently reject science that they show their disdain not only in social media rants but in open violence. They also show their contempt for nature by creating vehicles intended to pollute the air even more than normal ones.

"Coal Burners" are diesel trucks that have been modified to produce immense clouds of black smoke as they noisily rip around our cities and towns, emulating a coal-fired vehicle. "Eat this Libtards![97]" is their message, but this messenger is sorely misinformed. The black smoke released by Coal Burners is heavy in pure carbon particles produced by an incomplete combustion cycle. These carbon particulates will fall to the ground eventually as black soot. The carbon that makes up this soot did not become CO_2, and it is relatively inert in this form, for the short run at least. So, the Coal Burner community has made their vehicle's combustion cycles less efficient (at an actual monetary loss to themselves) and modified their trucks to convert unburnt fuel into pure carbon that is not readily available for life forms to incorporate, and is resistant to chemical reactivity. You should now see this showy exhibit for what it is: ignorance in motion. This display is not stupidity, for there is a logic behind it, just incorrect logic. These folks extol a naïve reasoning that lacks basic information about the carbon cycle. A Coal Burner actually pollutes less for each liter (0.26 gallon) of fuel because of its incomplete combustion cycle[98]. Now you can temper your shock if you see this audacious display and instead react appropriately, and laugh. This joke is on them.

A little education can easily cure this type of ignorance. But you don't educate someone by arguing with them or yelling in their face (which is what they are doing to you symbolically with their trucks). Huge clouds of black smoke are not a political statement; they are red flags. The only way to get through to someone successfully is to talk to them reasonably and avoid insulting them. The Coal Burners' very existence is intended to insult the planet and a particular type of person, the kind who cares. Still, please show them some little empathy because these altered vehicles will soon be economically impossible to maintain and worthless for future sales or trade. Useless, just like nearly all petroleum-powered, internal combustion vehicles will quickly become. Only a very few will have value as collector items or museum pieces.

---

[97] "Libtard" is a derogatory nickname for those with a liberal world-view, smushed together with the word retard. A violent Portmanteau, that despite it's lofty title, is really just a schoolyard taunt.

[98] And it does reduce the miles per liter/gallon, because less fuel completely combusts.

As for the politicians whose fortunes have risen on the tide of this hate and anger, don't debate them in an effort to change their minds, just vote them out of office. The newly informed and ecologically aligned economy will most likely find them void of any resale value as well.

In order to repair our air, we first have to understand how it functions. As we know, the air is a combination of multiple gasses that co-mingle in a homogenous mixture. In most instances, they just flow around each other and rarely combine. Combining gasses with one another or with other elements requires energy. This energy could be the heat of the sun or a flash of lightning. The mere jostling of the wind is rarely sufficient to induce atomic reactions in these gasses.

When air is at room temperature and pressure, is acts as a dry, transparent, ethereal substance. It can be compressed, and doing so will increase its temperature and store energy. Releasing this pressure rapidly through a tiny orifice will lead to cooling. This phenomenon is the basic principle used in refrigerators, although the preferred refrigerant gases are often fluorocarbons, not air or any of its primary constituents. Extreme cooling of gasses renders them liquid; some can be further cooled to a solid. Liquid nitrogen and dry ice are good examples. These are the standard states of gases, and each has its own rules. An important rule to remember is that air can easily accept many different gases into its amalgam.

An influential gas that occurs in the air is water vapor. In its liquid state, polar attraction holds water together, as you have seen. But when heat is applied to water, the molecules will begin to shiver and shake. This movement is an expression of kinetic energy. And as the water warms more, this movement can become violent enough to break the polar attraction, and single H_2O molecules will escape into the breeze. Each freed molecule takes a tiny bit of that heat energy with it as it leaves, cooling the mass below by an equally small amount. But when you have billions of these escaping molecules, each taking their wee bits of heat, it leads to a measurable cooling effect. Here is the process of evaporative cooling.

Once the molecule is free, it finds itself lifted into the air; the water molecule being just a bit lighter than the mixture of gasses that make up the atmosphere. So they float upwards. After the air has absorbed some amount of water vapor, its mass will rise as well because a volume of dry air is heavier than the same volume of moist air. You can see this manifested as the air mass displacements that create weather fronts. And when those air masses have different temperatures, condensation can occur at the interface, and you might get a nice spot of rain.

The amount of water that air can hold before it falls out is called its saturation point. Relative humidity (RH) is a measure of how much water vapor is contained in the air versus how much it could hold at full saturation. 100% humidity means the air has all the water it can take- and "it can't take no more." The amount of water that any air mass can hold is directly related to its temperature; that's the relative part of RH. Warm air can hold more water than cool air. When warm air is cooled by another air mass (or its altitude), it might pass its "dew point" and release that moisture. Again, you have those lovely little raindrops, but this precipitation could also arrive as snow, depending on air temperature. And in violent storms, when the air currents keep sucking the precipitation back upwards and freezing it in layers until it becomes too heavy to hold aloft, you get hail.

This rising effect of moist air also leads to one of our favorite atmospheric phenomena, clouds. As the air rises and the atmosphere gets cooler, approaching the deep freeze of space, some of the water will condense out of the air, and when it can coalesce around a nucleating particle, big puffy clouds form. Those clouds will often sit at altitudes designated by layers within the atmosphere. Just as temperature and salinity create layers in the oceans, humidity and density create invisible stratums in the sky. This phenomenon is why the bottoms of clouds often seem flat. And why, eventually, giant cumulus clouds will have their tops sheared off. Wind currents will sculpt our clouds, creating seemingly endless forms. Including the saucer-shaped lenticular clouds that conspiracy theory folks love to obsess over (they do not hide UFOs, at least not effectively).

The compression effect of jet aircraft wings slicing through the skies can cause an evaporative cooling effect and create contrails: string-like cloud forms. This is because compressed air heats as it's forced over the top of the wing, and when it completes its trip, the pressure drops rapidly, causing it to cool. This cooling causes moisture to condense, making the contrail clouds. Air follows the rules of gases. This is simple physics and is a fascinating subject for further study if you are so inclined.

There is a massive amount of water vapor in the air globally. It can vary from almost nothing to around 4% of the total volume in any given air mass. When calculated, it is found that the entire atmosphere contains about 142 million billion liters (37.5 quadrillion gallons; 37.5×10^{15} gal.) of water, on average. That is an enormous amount of not only fresh but freshly distilled water. Most evaporation occurs over the oceans, but a significant amount of water evaporates from the land as it is exhaled by the trees, squirrels, and the hikers beneath them. In ecology, they measure the respiration of life forms and land masses as evapotranspiration. This is the breath of Gaia.

If only there were some sci-fi-like method of harvesting some of that water for drinking?

There is! Hygroscopy is the phenomenon of condensation, and it occurs on your lawn every cool morning. As you now know, warm air can hold more moisture than a cooler volume can. A localized drop in temperature alters the air's water holding capacity and causes it to deposit that excess water as dew. Again, the temperature at which this occurs is called the dew point. It will condense on almost any surface, a glass of iced tea or your spectacles when you step inside from a chilly morning walk. It is known that one could place refrigerated surfaces in the open and harvest water from them, but the energy cost is outlandishly high. If you think water is expensive now, the sticker shock of water gathered in this manner might give you a heart attack. It would be nice to have a method to extract the water vapor without the need to spend tons of energy to do it.

Enter: *Stenocara gracilipes*, a humble beetle that lives in the Namib desert. It has developed a unique adaptation that allows it to condense water from the ultra-dry desert air with only the shape of some bumps under its wing covers. Did someone say it condenses water with shape? Yes, you heard that right! Researchers at Harvard University have developed a material that simulates those wings' function in a field of study called Biomimicry. Combining that shape with specialized material properties, they are developing a passive water harvesting material. (Park, 2016) Biomimicry is a discipline that looks directly to nature to understand complex physical phenomena, such as how a gecko can climb up a glass sheet or how a bird can maneuver so gracefully through the air. This line of inquiry suggests that many of the problems we have created may already have solutions in nature. Many solutions exist to problems we have yet to encounter as well. We'll just have to find them. If this is not a solid argument that we should endeavor to survive and continue as a technologically advanced race, I don't know what is.

An unfamiliar way of looking at atmospheric interactions is the "Airshed"; a concept that first appeared in the EPA Clean Air and Clean Water Acts in the late 1960s and early 1970s. It received further definition in the amendments to those acts in 1990. An Airshed is:

> "The geographic area that produces a significant amount of the emissions that contribute to atmospheric deposition in a watershed."[99]

This designation was made so that scientists could begin pinning down the localized effects of air pollution with greater accuracy. Power plants, cities, feedlots, and even large open farming operations produce tremendous amounts of emissions. While some pollution may travel around the globe, most of it will come to roost in a relatively confined region. There it will end up in the local water, on the soil, and in your lungs. Airsheds are still somewhat controversial with industry, as they see them as yet another basis for more regulation, hampering their free enterprise. The sad reality of regulation is that if organizations or industries could police themselves, there wouldn't be a need for so much of it. Unfortunately, since they continually demonstrate that they can't, or won't, self-regulate, statutory rules become necessary.

The Airshed concept also begins a necessary investigation into previously overlooked environmental actions. Mapping airsheds and analyzing their functions will give us valuable data for a future time when someone might figure out how to filter the air directly. Or perhaps there will be a pivotal function to help us speed up natural processes to absorb or transform CO_2. Maybe it will help us find better places to situate facilities to lessen their damage; as we move towards a genuinely carbon-neutral future. Remember, carbon dioxide belongs in the atmosphere. Without the greenhouse effect, the

---

[99] The EPA definition

planet might be uninhabitable (for us anyway). The problem is that our activities are adding too much of it to the air. We must seek carbon balance, as we should seek balance in all things.

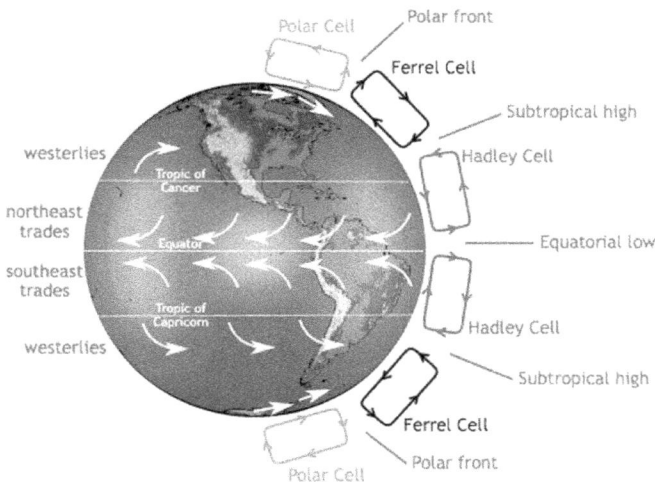

Atmospheric currents. Courtesy of NASA (by the Center for Multiscale modeling of Atmospheric Processes)

Sands lifted aloft over the Sahara Desert predictably come to land over the Amazon basin. Likewise, spores from Europe often seed Antarctica. Mapping these patterns more accurately will lead to greater understandings of this ocean that flows above our heads, within which all swim. The thing about pure science is you don't need to know exactly where your line of investigation will take you. In fact, it is always better not to second guess your results. One big reason we are behind the 8-Ball on climate science is that we have let organizations and corporations control the agendas of our inquiries into nature. Airsheds are a novel way of looking at atmospheric systems and will soon become better known. One way they will come to prominence is through carbon offsetting.

There are a lot of discussions these days about carbon offsets and credits. Currently, the public in the USA has a poor understanding of them. Cap and Trade systems, carbon markets, and heaven forbid, carbon taxes will come into play soon. So it's a good idea to get your briefing now. Our societies pour gigatonnes (gigatonne = 1 billion metric tonnes, 1.1 billion tons) of carbon into the atmosphere annually globally. All are already feeling some adverse effects, with much, much more to come. And, boy, is it is going to get costly. When climate scientists and economists look to quantify this carbon release's monetary costs, they find it's staggering. Recent studies have set the environmental, social, and economic costs at over: \$200/tonne (\$198/ ton) (Moore, 2015) for CO_2 release[100]. This number sounds crazy, right? So, how can that possibly be?

In the next chapter, I'll go into more detail about the carbon/energy cycle, but let's look at this quickly. 1 ton of coal releases almost 2 tons of CO_2 on combustion. One ton of coal costs about \$50 (it's a commodity, so that price fluctuates wildly). Under this analysis, the damage of the CO_2 emissions of 1 ton is just under \$400. That's almost eight times the cost of the fuel itself! Let's let that sink in. There must be something seriously wrong with our methods of calculating worth if there is such an imbalance between these values. (Hint: there is)

1 kWh (1000 watts of power for one hour) is roughly equal to the output of 5 laborers for the same time, in pure energetic terms. Five men hauling some load would cost at least \$50/hour. 1 kWh of electricity costs less than \$0.30. There it is again. There is a serious disconnect between what these things cost and their actual value. That difference is the "ecosystem service" of the atmosphere absorbing the emissions needed to generate that power. That is something people pay nothing for, but that's just because we've been in denial of the actual economy that supports us: the thermodynamic economy. First, let's give those laborers a raise because they can't feed their families for \$10/hr. And next, let's talk about real value.

The reason the carbon cost is so high are the damaging effects on humanity and the environment that global warming is poised to deliver, not only on our cities but our entire economies. Crop loss, infrastructure destruction, the carnage of natural disasters are already being felt, and they can be directly linked to those carbon emissions. And it is understood that this will worsen.

---

[100] In 2021, the Biden Administration put a preliminary price on carbon- from \$14 up to \$76/tonne. This is a ten-fold increase in the valuatution bandied about (but never codified) by the previous administration.

When coastal cities disappear under the tides when trillions of dollars in infrastructure and property disappear, these effects will finally come into focus for everyone. But why wait? We know what is happening and why.
The damage caused by climate change is cumulative as well. We've been pumping out a lot of CO_2 for a long time with no care or restraint. A tonne spit into the wind today is worse than one released a century ago. Simply because we have passed the tipping point, and we keep adding more straws to the poor camel's back. So, this carbon cost reflects what the actual loss to National GDP[101] is expected to be. This $200/tonne is an accurate measure of the damage we are doing, and that should be pondered seriously by everyone.

It is well known to those of us in the Resource Management and Environmental Restoration fields that the cost of mitigation is a logarithmic function[102] of the cost of the damages that are now occurring and will occur. For example, it may cost 1/10th the amount to thin a forest to control wildfires than to pay for the fire suppression and clean up after the wildfire. People lose homes. Families lose lives. And as ghoulish as it may seem, economists will place a monetary value on the loss of life. That is not because you really can, but because not doing so is akin to lying about actual impacts. It costs ten times more to fix our messes after the fact than to get ahead of them; that's the mitigation advantage. And even though this is common knowledge, budgets include fire suppression and aftermath costs (in a narrow sense), but there is very little money for mitigation. And you'll find this same lack of foresight across all spectrums of human endeavor. So where does this mitigation money, investment in ecological insurance, come from now?
Today, the pittance available for ecological mitigation comes from government grants and contributions from industry and those few people who will donate their time or money to a good cause. It does not come from the insurance industry, or generally, from ordinary people or property owners. Where it could come from is a carbon tax.

---

[101] Gross Domestic Product
[102] A logarithm is an inverse exponential value. Instead of values advancing rapidly, in a logarithmic progression the values recede slowly.

Cap and trade is one form a carbon market might take. The cap is a limit on the quantity of greenhouse gasses that may be emitted annually. And this would be based on projections of what is needed to slow the release of CO_2 towards zero. The trade is the carbon market, and it's similar to any commodity market. Let's say I'm Mr. Burns, and my power plants release an ungodly amount of CO_2. I can't just reduce this release easily. And my brother-in-law owns the coal mine, so there are some family issues there as well. I know I will exceed my cap, so I look for Hector. He's a botanist in Brazil. And he's been re-foresting thousands of hectares of land with millions of trees thirsty for CO_2. Hector creates carbon credits for his good works by calculating how much CO_2 his trees drink in and then sells them to Mr. Burns. This is an offset. The plant continues operation because it "bought down" the emissions. Hector now has enough money to keep planting trees and buying more land to conserve. And maybe Hector can now hire a couple more ecologists to improve on what he's already achieved. This arrangement is more than just some financial trickery that lets polluters keep pumping. The intake of CO_2 by Hector's trees has to consume not only the plant's emissions but a bit more on top. How this is quantified is in the protocol.

A carbon offset protocol is a stringent model for assessing and calculating carbon usage and emission quantities. An independent agency oversees the protocol, called a registry. In order to create a single credit, detailed modeling and reporting are required. Yet another independent group of certified monitors inspects the projects on a schedule. A carbon offset has to be honest, and it has to continue to function, or the registry eliminates it. And then Mr. Burns has to find a new project to fund. Despite our unfamiliarity with it, this is a sensible system to generate capital that will be used to help everyone. This framework can support mitigation now, instead of just waiting for our economies to collapse under the financial burden of the catastrophes we've set in motion.

The carbon tax appears when the producer of the CO_2 is not just Mr. Burns, but you and I. Every gallon of gasoline produces three times its weight in carbon dioxide, and a tax will someday occur to offset that carbon cost. Electric cars will bypass this tax, but it might be somewhere in your electric bill if carbon-based fuels produce that power. This change is coming; it is the future. Now would be an excellent time to get rid of your gas guzzlers because their value will plummet the day this happens. There will still be value in their steel and aluminum for scrap to be recycled, so there's that.

No one likes taxes, but this is one tax that can be wiped from existence. Once everyone starts to conserve, this tax will evaporate like so much paper in the wind. Poof! It will vanish once the energy system develops into renewable sourcing and our carbon emissions reach an equilibrium, carbon neutrality. Meanwhile, it gives those fighting against the tides of the status quo to save our world just a bit of cash to continue their good works. In a way, this is putting regulations on ourselves. And it is the proper and mature thing to do. Adults can temper their actions and delay their gratifications; children can not.

The form of oxygen that life relies on is the molecule O_2, two atoms of oxygen locked in an embrace. There are other forms that an oxygen molecule can take, and an important one in our atmosphere is ozone (O_3). Lightning strikes naturally create ozone. Their electrical discharges become the zap of energy that can bind those oxygen atoms into a cozy threesome. Ozone is also created by combustion, and the petroleum fuels burned in automobiles create a fair amount of it. Ozone is considered a pollutant at ground level and is not the best thing to inhale because its high oxidation potential destroys tissue, in your lungs for example. Even worse, it is strongly oxidizing and can damage tissues in the lungs and air pathways. In the stratosphere, however, it is a bonanza. The ozone layer surrounds the blanket of gasses in which life thrives and filters out the UVB[103] rays that cause skin cancers. Once upon a time, human civilization almost destroyed it.

The ozone hole was first described by Joe Farman, Brian Gardiner, and Jonathan Shanklin in a 1985 paper in the journal Nature. This group of British Antarctic researchers noticed a strange phenomenon. While the atmospheric ozone layer was well documented and contiguous around the globe, they saw a large opening in it forming over the South Pole. And it wasn't clear why this anomaly was occurring.

Ozone is also naturally created in the high atmosphere where oxygen molecules collide, and sunlight disassociates the O_2 into single O atoms, which

---

[103] UltraViolet B rays- The type of UV rays that cause sunburn, and contribute to skin cancer.

can then join a paired O_2 molecule. When ultraviolet type B (UVB) radiation hits an O_3 molecule, it has just the right energy to split it back into an O_2 pair and a single O atom. The cost of this reaction is the wave particles' energy. And that's how ozone filters out these damaging rays; by absorbing them and stealing their energy. So where did this hole come from?

It turned out that the culprit was ultra-stable halons, human-made chemicals created as refrigerants and aerosol propellants. Our Frigidaire's and hairsprays were releasing CFCs[104]. And when they drifted upwards, they had their own destructive chemical reactions with the ozone. They broke the ozone apart with no light energy needed. Thus, a vital atmospheric reaction was interrupted almost permanently by AquaNet and leaky Coldspots. And by ultra-stable, I mean that one molecule could last for years, decades even. And it might engage in tens of thousands of destructive chemical reactions that rip apart those oxygenic three-way parties like some ruthless Chicago "Morals Squad"[105] in the sky.

American society had a bit of a debate, similar to the one that is currently raging, about the potential need to eliminate these CFC compounds. First, the status quo attacked the science, and next, the scientists themselves. Then they claimed that changing the way the status quo did things would destroy the economy. Subsequently, the debate came down to the unwillingness to change just for the sake of that unwillingness itself; "conservative" values, stinking hippie environmentalists, researchers on the government dole. Does this sound familiar? It should seem old hat by now because these are the same nonsensical arguments used against changing our carbon polluting ways. However, in the end, governments outlawed these types of compounds for many uses and strictly controlled them for others. It worked. The ozone hole is now closing over 30 years later. The world did not end. The economy did not explode. And "conservative values" still survive to this day. If no one had done anything and the ozone layer was lost, millions might have died of cancer, and

---

[104] Chloroflourocarbons- halogenated paraffin hydrocarbons, produced as derivatives of methane, ethane, and propane with clorine and flourine.

[105] The Chicago Morals Squad, in the early 20th Century, found it worthwhile to violate all kinds of personal rights and liberties in the name of "chastity and modesty". They tore apart as many normal families as "kinky" ones. Thankfully, they were finally hung out to dry. Like the wet blankets they were.

the effects on the ecosystem might have been profound as UVB rays ravaged the landscape. I have often heard it said: "If you find yourself arguing on the side of cancer, it might be time to take stock of your life." And yet, many made that argument, and many still do today. It seems that Lee Iacocca was a pioneer in creating empty, mildly insulting debate strategies. Many carry his banner still.

Some suggest that society use our technology to deal with our carbon-based air pollution differently, in a way that doesn't require changing for the better. Geo-Engineering is a broad label covering many different techniques, not a few only theoretical. In particular, there is the idea of building giant machines to filter CO_2 out of the atmosphere. There isn't any accepted design for what these monolithic mechanisms might be, but they bring to mind images of the giant alien machines from the sci-fi movie "Total Recall." Large multinational corporations are currently funding investigations into the possibility of these types of mechanical carbon sponges. Indeed, it might seem that many industries that have contributed so much to our atmospheric problems are now looking at how they might profit from "fixing" them. Here is the status quo at its finest.

Solar Geo-Engineering is the idea that sulfur compounds, aluminum, or other substances could be injected into the upper atmosphere to reflect sunlight into space before it hits the planet's surface in the first place. It is understood from natural phenomena[106] that this could lead to a global cooling effect. But it completely ignores the root of the problem. This "fix" would do absolutely nothing to address the crisis of the CO_2 already in the atmosphere, and indeed, it may be intended to allow us to continue adding even more. So the oceans would continue acidifying, and atmospheric chemical reactions would become even more unpredictable. Dimming the sun would also have an unknown effect on the rates of photosynthesis worldwide. But if we end up trading mechanical cooling for healthy, natural carbon sequestration by plants and the biosphere, we will make the wrong choice. And the sulfur, if injected into the atmosphere, would eventually render into acid rain, again increasing ocean acidification. Sulfuric acid, carbonic acid, when will we ever learn? Such

---

[106] Volcanos, and their dust clouds.

actions could unleash a new cycle of destruction onto our oceans.

Can we say: Enough!?

These types of over-engineered solutions are just more status quo quackery. More machines consuming more energy, requiring more mining for materials, and generating more profit for the shareholders is the problem, as it is intentionally sidestepping actual solutions. Society needs to stop burning carbon-based fuels for our energy and releasing carbon dioxide and other excess greenhouse gasses. While we can't completely eliminate combustion, it can be reduced to a tiny fraction of what is now our standard practice.

The very idea that the way out of the atmospheric alterations already made by our activities is to alter them even more is ludicrous. It completely ignores the need to consume less and shepherd the ecosystem to do what it is so good at, restoring its balance. The biosphere is our carbon sponge; it is the original solution to this problem and still the only one needed, the only solution that exists on the scale of the problem. As long as we don't completely destroy it, that is.

Our current oil stocks (and what are the chances our fanciful carbon-absorbing machines would run on oil?) and mineral stores are dwindling. And just because companies keep finding more, remote deposits of raw materials give us no reason to use them. Even if there is more oil, iron, copper, and aluminum to be found, what gives us the right to take them and use them all up? We've already thrown away several civilization's worth of these materials into landfills. Don't recycle, don't conserve, it seems some are saying. Is there no concern about the future? Do the generations yet to come deserve none of it? Apparently, they do not, according to the "conservatives" of the status quo.

There is one more substance we have put into our atmosphere with which we are just now reckoning. Our skies are filled with plastic.

Our modern world loves plastic. We wrap our food in it. We make computer cases out of it. We cover our skin in it (nylon and other engineered fabrics). Then we discard it at mind-boggling rates, and like most things, we pretend that when it hits the trash can, it's gone. But hyper stable compounds don't disappear so easily. When they rub together in the garbage truck; when they are driven over on the road by our cars; when those synthetic clothes spin in the drier, they break down into dust. Much like the actions of those oceanic gyres.

From a few millimeters down to microns[107] in size, plastic particles are just as reluctant to decompose completely as discarded water bottles and supermarket bags. They are, however, capable of becoming airborne, and that is what they do. Recently, microplastics have been discovered in the snow of the Arctic and on top of the Swiss Alps. (Bergmann, 2019) Ice sheets give us a great medium on which to measure this pollution. Snow forms around these particles and captures them as it falls, just like rain captures sulfur compounds. A recent study has estimated that an industrialized country's average citizen consumes over 50,000 pieces of microplastic a year. (Cox, 2019) People inhale these microplastics, and they eat them. This free benefit of industrial society may be linked to lung cancers and is just plain disturbing.

In high school science class, we were told that most of that dust dancing in light rays were dead skin cells. But now, face the fact that many of them are powdered plastics. There is no realm on the planet that has not been soiled, would not be fouled even more if left to continue as we are heading.

I will keep repeating this: science can help save us and the planet upon which all depend. But science has no will. It is not a philosophy. So it is up to us to decide to use it wisely so that survival might be won. Survival is its own reward. The ultimate wages of the status quo's unbridled profit-taking are now seen to be extinction. Continued primitive existence after society collapses will be neither desirable nor perhaps even possible. A "Road Warrior" type world will not happen — those guys needed gasoline. And without technology, there won't be any (it has a shelf life, too, you know). The altered atmosphere and its runaway warming will destroy our food crops and kill our wildlife, leaving the remaining humans powerless against the deadly tide their ancestors set in motion. If societies collapse, there will be no means of adapting to the multitudinous changes that will continue, even if all emissions dropped to zero tomorrow.

The clearing of smog that the world saw because of the global COVID-19 pandemic shows how quickly nature will move towards balance. But even such an extreme decrease in greenhouse emissions will not remove the excess CO_2 that is already in our atmosphere. Thus, adverse climate effects will

---

[107] A micron, or micrometer, is one-millionth of a meter: 1×10^{-6} meter. Pretty dang small.

continue until the concentrations of these greenhouse gasses are reduced, either slowly by nature or more rapidly by our ingenuity. Or, hopefully, by a combination of the two.

We need to keep all the knowledge we have and that we are rapidly gaining if we wish to survive. Indeed, all need to do a better job of practicing balance in every aspect of our technology. Still, this can only be achieved if we refuse to let our expanding knowledge be interrupted by tribal politics, war, or targeted ignorance. Lacking the power of high technology, humanity just won't survive the changing world already set in motion. There's zero chance. We must find ways to sustain ourselves in the ecosystem we have, and fortunately, we can.

Everyone will feel the impacts of climate change; in fact, everyone already is. We've now accomplished increasing the amount of CO_2 in the atmosphere to over 420 ppm in just over a century. The question is: will humankind reverse that trend? There is currently no technology, no silver bullet, for just pulling all the greenhouse gasses out or cooling us down. Instead, natural processes are our only levers to control our greenhouse gas concentration, which will take time.

But what we can do now is STOP.

Stop driving everywhere. Stop buying gas-guzzling vehicles when you don't need their utility. Eat lower on the food chain. Don't buy a new smartphone when the one you have is still functional. If you have electronics that have gone bad, recycle them, don't just throw them away even if it costs you money. That is wealth that must be kept and shared.

Use your body and mind for what nature designed them. Exercise by walking to work, the store, or even around the block. Breathe some fresh air. Think of solutions you can implement and support while you take that walk. I can guarantee that when you make your body move, your mind will do the same. Humans evolved in motion, rebelling against that has not worked out so well for us. The population must get out of their chairs and act. Plant trees and urban gardens. Make some compost. It won't turn you into a bleeding-heart liberal or a hippie. It will, however, save your life. Recycling is the way of nature: emulate it in every way possible. Support renewable energy sources.

The reality is that none of this is radical or anti-conservative. On the contrary, conservation is the ultimate conservative value, as the name suggests. Environmental awareness is not a political stance; it is a genuine concern for all of us. We will soon come to recognize this.

The atmosphere is like an ocean with no shore; it moves freely everywhere across the face of the Gaia. Air is the one primal element that all share. It connects us all. Air carries our voices to each other, making human communication possible. It brings us the sweet smells of life and each other. It transports dust and pollens as well as oxygen and contaminants without respect for borders. So treat our atmosphere as a beautiful, pure ocean, shared by everyone equally, and stop its defilement!

We need fresh air as much as we need pure water and unpolluted soil.
We must guide its evolution into a healthier state.

CHAPTER EIGHT: ENERGY

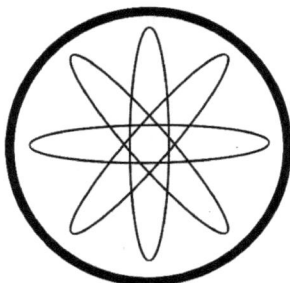

"Energy can neither be created nor destroyed."
— ***First Law of Thermodynamics***

In 2006, George W. Bush (#43) signed into law the Postal Accountability and Enhancement Act (PAEA). This act took the highly respected and profitable US Postal Service (USPS) to bankruptcy's door. Long a "conservative" goal, the destruction of the agency was sought because it gave proof to the lie that private, for profit, companies were always better managed and operated more efficiently than public services. At the time, the USPS drove over a billion miles a year, delivering mail to almost every physical location in the nation. Twenty-six thousand six hundred and eleven post offices delivered over 213 million pieces of mail and had a cash flow of $73 billion dollars per annum. What a prize if only it could be dissolved and taken over by private corporations (at a bargain price, of course). Toppling this giant was second in line only to the privatization of Education started by Ronald Reagan (#40).

Not by coincidence, that very year, the Postal Service had bragged in their annual report they were poised to move away from fossil fuels and electrify their entire fleet of mail vehicles. With 216,000 vehicles, this pledge to spend billions to alter their carbon footprint would've changed the entire United States transportation industry. They could've completed this transformation by 2020, accelerating the technology and proliferating charging stations nationwide.

Enter congressman John McHugh (R-NY), as chairman of the congressional oversight committee's subcommittee on the postal service, he was just the man to put a stop to it. He was also a member of ALEC- the American Legislative Exchange Council. ALEC is a shadow group funded by the Koch Brothers (more about them later) that writes bills for "conservative" legislators to present to their respective chambers. And hopefully, to change laws for the benefit of ultra-large corporations and, specifically, the oil industries. ALEC had a neat little pre-fab law called: Unfunded Pensions Liability Act. It became the backbone of the Postal Accountability and Enhancement Act. The sling that would almost slay the giant. Passing both houses of congress on a voice vote, so that none of the cowards would have to go on record as either voting for or against it, this act required the Postal Service to fund all pensions for 50 years in advance, immediately. It also forbade them from raising postage to cover this new expense, estimated at over $5 billion a year. This requirement made them stash away the full health care and pension costs for future employees not yet born when the act was passed. Also, the Post offices would have to deliver mail 6 days a week now — do more work with less. GW Bush (#43) signed this bill into law with his trademark Texas oilman's smirk and the promise of electrical post office vehicles died right there on the resolute desk. They probably expected the USPS to fold within a year of two so that some investor could walk in and buy it for pennies on the dollar. And then the price of postage could rise with no restraint and services could decrease to whatever level would be required for the profit of the shareholders. But that, thankfully, didn't happen. The mail deliveries kept on, through sleet and snow and the onslaught of "conservative" BS. Now, in 2022, the electrification of the postal fleet is on the chopping block again. The standard rotation of vehicles could finally allow for the switch to emission-free transport. Could.

Enter, Louis DeJoy, Postmaster General appointed by President #45. Prior to his installation at the USPS, DeJoy was the CEO of New Breed Logistics, a freight company and a competitor of the US Postal Service. His job was, again, to destroy the US Postal Service. Which he began by instituting new policies that forbade overtime, yet removed high speed sorting machines, slowing the movement of mail. He also began removing mail drop boxes nationwide just prior to a presidential election where many people would vote by mail because of the COVID-19 pandemic. His primary targets: inner-city neighborhoods where liberals and people of color lived.

Conflict of Interest is never an issue with a certain type of personality, for they will seek their selfish best interests with zero conflict from conscience or ethics. There will be no electrical vehicles on DeJoy's watch[108]. Again, our transition to renewable, sustainable power is delayed by those with no real knowledge of what power actually is.

Energy permeates the universe. It is the essence of what is called motion; intrinsically connected to our experience of space and time. Life is based on energy, chemical energy, kinetic energy, and thermal energy. Living things eat to gather the energy they need to move and reproduce. Plants store energy captured from the sun in chemical bonds by making sugars. A mother takes a substantial amount of her energy and gives it to her forming child so that it can grow within her. Energy can change from form to form in a flash. It seems ethereal, neither structure nor substance but the essence that gives rise to both. Energy is movement and never dwells. What energy a life form takes, it either uses or passes on.

Energy is equivalent to matter, as one is a manifestation of the other. An equivalency that is expressed in Einstein's most famous equation: $E=MC^2$. A substance's intrinsic energy is equal to its mass times the speed of light squared. The elements are power condensed, massive amounts of power. The three laws of thermodynamics[109] broadly define the interactions of matter and energy. Life operates within the thermodynamic economy, energy is conserved, and intakes will always balance outputs; it cannot, in fact, be any other way. Energy can neither be created nor destroyed. Every action has an equal and opposite reaction. Energy is the coin of the thermodynamic economy. Existence is the product.

---

[108] Actually, his hands may be tied and there will be some electric mail trucks, just as few as possible.

[109] The First law: Energy can neither be created or destroyed, Second law: Entropy always increases within closed systems, Third law: Entropy approaches a constant as temperature approaches absolute zero. (But it seems that absolute zero may be an unattainable temperature, although it can be approached by infintessimal amounts. So, in practicality, sub atomic motion never ceases, the energetic root of matter is never silenced.)

Our advanced societies have a slightly different meaning when they talk about Energy, and have economies based upon it. Energy, with a capital "E," is a product that is bought and sold at a profit. This Energy, indeed, changes from form to form and is shipped from place to place in a market-based manner. But it has lost its equality and universal value because one group may pay more and another group less for the Energy they consume. The sun may continue to shine on all of us equally, but perhaps that's only because no one has figured out how to patent or copyright it–yet.

A human has a certain amount of work that they can accomplish in any length of time, based on size, fitness, and age. The world of humanity has survived for millennia with that limiting factor placing constraints on what any group of people could achieve within any given season. The Pyramids are an excellent example of just how effective human power can be. When the Energy era first began, we found we could exponentially increase the work one person could achieve with machines and massive amounts of Energy. This became so ingrained in our thinking that many now find it hard to believe the Pharoah's could've built their monoliths without machinery. Yet, all solid evidence shows they did, and that no extraterrestrials were around to help them.
It is easy to appreciate that one human power is less than one horsepower. One human power is about a seventh of one horsepower, with elite athletes capable of up to a third of horsepower of energy release for short periods. One horsepower is roughly equal to 750 watts/hour of energy use. Your car may produce between 100-200 horsepower, and you probably never even think of that as you use it to get around every day. That's the equivalent of 750-1500 human powers (or 300-600 peak Schwarzenegger's). This calculation is staggering when you think about it in this way. You might imagine teams of dozens of horses or hundreds of people pulling your old Escort down the highway to visualize what power has been harvested for your ease of travel. Our world would just not be possible without massive amounts of Energy. Energy in bottles, at the outlet, and in the gas pump. Our living standards would crash, and progress might roll back centuries if society suddenly didn't have access to these valuable resources.

And the status quo knows this; The human race is addicted, seemingly hopelessly, to cheap and easy Energy. This time we're over the barrel, a shiny oil barrel.

As measured in terms of our Gross National Product, our Energy economy represents nearly 9% of the United States' total productivity. (EIA, 2011) This less than 10% doesn't give us a good grasp of its importance to society, though, just how much it represents in terms of cash flow. It is well understood that access to cheap, plentiful Energy is essential for a healthy technological economy (Kraft, 1978). So, there's a pretty sweet deal going on here, as people have exponentially amplified the work every one of us can achieve day in and day out. This bounty of power is the basis of our elevated living standards and rank as a first-world nation. You can travel around the world in 80 hours (or less). We possess seemingly endless cool gadgets and tools, all of which would be pointless without access to cheap Energy. Nearly every object surrounding us has an Energy price tag, and that price is meager when measured in dollars. Now is the Golden Age of Energy, as much as technology, and everyone is the beneficiary of this bounty. Well, perhaps not everyone.

So whence did all of this energy come? Science hasn't yet learned how to crack open the fabric of space/time and harvest limitless energy, so where and what is it?

When our solar system formed, cold matter in space, remnants of stars from the generations of suns before ours coalesced, in the wake of the Big Bang, into a swirling accretion disk. This galactic gyre was further compacted down into celestial bodies by the force of gravity and their momentum. Thus, the two initial culprits are gravity and the energy released by the Big Bang. Remember that energy is not being created or destroyed; it is just changing from form to form. The energy spent to snap your fingers and tap your toes has been passed down to you over billions of years. Now it's spent, as heat from the friction between thumb and finger, and the sound wave sent forth into the ethers. It's never coming back.[110]

---

[110] And, strangely, it's never going away. You have created an effect on the universes's state it is now forever altered. So, please, snap your fingers on the beat.

As our infant solar system formed, most of the lighter gasses collected near the center as centrifugal forces spun heavier elements outward. This mass of hydrogen gas would eventually outweigh the rest of the solar system by an exponential factor. Gravity gathered this central hydrogen cloud into a focal region of space, where its colossal mass acted as an anchor for the nascent solar system forming around it. Internal attraction slowly coalesced this cloud into the vast gas giant that would eventually become the sun. Similarly, our planet collected, harvested, may be the better word, the debris in its orbital path and underwent similar compaction until it became a shrinking ball of rock.

The compression of any type of matter: gas, liquid, or solid creates heat. For Earth, this led to a complex molten core and violent seismic activity expressed as geologic upheaval. But, as for the sun, there were bigger things in store for our super-sized ball of gas. Heated and compressed, the hydrogen atoms kept colliding together, fighting back at the crush of their mutual attraction, until they reached a super-stimulated state. Slowly, this massive sphere heated to a dull glow, its temperature increasing incrementally, eventually making nuclear fusion possible at its core. Once this chain reaction initiated, it spread in an instant, and the entire ball ignited. The explosion would've been terrible if not for the fact that our Sun had enough mass, just enough, that the force of its gravity held it together. So, our gargantuan miasmic sphere erupted, some of it transforming into plasma, the fourth state of matter. And it became a star, our star: the sun.

As they say, the rest is history, and our solar system evolved, as all things do. The Greek philosopher Heraclitus[111] once said: "change is the essence of the Universe," and he was not wrong. In fact, he was perhaps more insightful than he knew because that little phrase pretty much sums up physics, astronomy, and biology, from the quantum level to the most massive of galaxies. It is also the direct opposite of our previous philosophies that stated everything was perfect and unchanging. But it turns out it wasn't just an idea, it was fact. And, so, while our sun was going about its new life, fomenting a scheme that would last for billions of years, Earth was up to some exciting developments

---

[111] Pre-Socratic Greek Philosopher (circa. 535 –475 BCE) hailing from the city of Ephasus, which was part of the Persian Empire back then.

herself. Life was forming on this planet, and it needed energy, which it got mainly from the sun's rays. This went on for billions of years. Life grew up and spread across our home planet transforming the dead rock into Gaia. And some of it, upon death, collected into massive deposits and became fossilized. This aggregation occurred at the bottom of the ancient oceans. Eventually, accumulations of biomass within the giant forests, perhaps buried by terrible volcanic eruptions became similarly impounded.

The fossils you might commonly think about; that you might even have on your desk or see in a museum display case are mineralization artifacts. The animal that died is no longer there; its tissues replaced by a solid stone that filled their body cavity like a casting within a mold. But there's another type of fossilization that preserves the organic material itself, and those fossils come in three primary forms: petroleum, coal, and natural gas. These are the "fossil fuels," and they have been the backbone of our current Energy revolution.

All fossil fuels are solar energy, concentrated and preserved.

It is essential to understand this fact. There is some energy harvested and stored by living organisms from the thermal energy of the planet's core, called geothermal energy. And some primitive organisms can use compounds such as hydrogen sulfide to produce their energy in a process called chemosynthesis. But the primary method of tissue creation on our planet is and has been photosynthesis. The organic material within the bodies of all these plants and plankton that have become fossil fuels resulted from the sun's energy. It took eons to create the fossil fuels, so the phrase: "They ain't making it anymore" seems uniquely accurate. This process is slow and grossly inefficient. We've been benefiting from the fact that while fossil fuels have been available for millions of years, nothing else has capitalized on their existence. Why has no other life form discovered this energy-rich organic muck and used it for some purpose, as energy or perhaps as food?

Because it is poison.

The complex carbon/hydrogen/oxygen bonds within raw fossil fuels hold a considerable amount of stored energy. These compounds are called hydrocarbons, and there are many impurities in their naturally occurring state. Cyclic hydrocarbons, common in crude oil, may act as a solvent on organic tissue, slowly disassembling flesh if ingested. They are especially damaging to DNA and can cause mutations and cancers. Sulfur, heavy metals, and nitrogen compounds are common bonus contaminants, and each presents their own problems. But even without these toxic hitchhikers, the burning of hydrocarbons is problematic. When you burn a carbon-based fuel, you are essentially releasing the ancient energy stored in that material's molecular constructs. When the bonds break, new compounds are created as the freed atoms rejoin with atmospheric oxygen. This process is an exothermic reaction, one that produces heat. If confined, it will also express kinetic energy as the compounds expand in volume. While it seems like the fuel is disappearing and flames or heat are born, the mass is not consumed. It is merely changing form as the laws of thermodynamics demand.

One primary result of burning carbon is the creation of carbon oxides, carbon monoxide, and carbon dioxide[112]. Once a carbon/hydrogen bond is broken, in the process of pyrolysis, single carbon atoms latch on to two oxygen atoms, forming carbon dioxide. The energy released when forming carbon dioxide is approximately four times the energy required to break the original Oxygen/oxygen (O_2) molecular bonds and the carbon/hydrogen bonds in the fuel source. Oxygen weighs more than carbon, so the resulting CO_2 molecule, containing three atoms, weighs 3.67 times as much as the original carbon atom. And while it may look like the fuel has disappeared, the post-combustion exhaust always weighs more than the initial fuel. The additional oxygens come from the atmosphere; petroleum will not burn in an oxygen-free environment. So just like energy, mass is conserved and is merely changing its form. Since matter and energy are synonymous, the laws of thermodynamics apply to them equally.

The energy released had waited for millions of years in the organic matter of the plankton or plants that made up the fossilized fuel. And the original energy that built those organic bonds came from photosynthesis. Which means the initial power came from the sun. You can now see that petroleum-based fuels are merely toxic, wasteful, third-hand forms of solar energy. Armed with this

---

[112] There is also carbon trioxide, CO_3, a very short lived oxide that is difficult to produce. While chemically similar to the carbonate ion -CO_3 $^{2-}$, it is not a stable carbon store like limestone ($CaCO_3$).

knowledge, one can make a rational argument that the move to renewable energy sources is merely the equivalent of "cutting out the middle man" in our future Energy endeavors.

Coal is, was, and can never be: clean. Coal is the crudest form of fossilized Energy and was laid down during the Carboniferous[113] era, some 350 million years ago. The conversion from plant and animal bodies to the forms society uses as fuel occurred in deep layers of rock and leveraged the heat of the planet's core and the compression from heavy strata above them. These pressures slowly morphed organic detritus into the materials we burn at a manic pace every day.

The coal body is mined by the removal of the rock layers above it. And as you have seen, industries have been stripping away mountains to reach the coal beneath them for decades. This practice destroys natural landscapes, releases toxins, pollutes waterways, and subjects the miners to dust that can (and often will) eventually destroy their lungs. This pulmonary insult ofttimes leads to a torturous death. Destruction, it might seem, can change from form to form too. But at least these pursuits are somewhat fruitful, from an economic standpoint. The Btu's[114] captured by these processes are far greater than the Btu's spent to harvest them. For it always takes energy to get Energy. In this scenario: the mining, drilling, and refining of fossil fuels are all energy-intensive tasks. Burning them just releases more power than is spent collecting them, substantially more energy. Just as the process releases considerably more toxic byproducts into the biosphere.

Coal is approximately 50% carbon by weight. While this may seem like it would be a good thing, in terms of stored energy, it is not. The simplicity of the molecules in the coal store less energy than an equivalent weight of, say, gasoline. If one measured energy potential kilogram for kilogram (pound for pound), you'd find that gasoline stores almost one and a half times more energy than coal. Since coal is 50% carbon, we can go back to our previous reckoning and see why one tonne of coal produces almost 2 tonnes of carbon

---

[113] ~360-299 Million years ago.

[114] BTU: British Thermal Unit, an international standard that equates to the amount of energy required to raise the temperature of a pound of water 1° F. So definitely an Imperial measure, and not metric one. Go figure?

dioxide. For one ton of coal, calculate half of it as carbon. When the carbon becomes CO_2, it takes on two oxygen atoms that each weigh bit more than that carbon by itself. A carbon atom weighs 12 AMU[115], oxygen weighs 16 AMU's so the sum of 2 oxygen and one carbon is 44 AMU. Divide this by carbons weight and you get 44/12, or 3.67. So carbon dioxide weighs 3.67 times as much as carbon. So you can finally see how one ton of coal, at 50% carbon can weigh nearly twice as much once it's been burned (0.5t x 3.67 = 1.835t). This chemical reaction is why coal will always be "dirty." If we consider the production of CO_2 as pollution, as we should, then coal is not a clean-burning fuel, nor can it ever be. Also, coal contains varying amounts of sulfur, which converts into sulfur dioxide on combustion (SO_2). When mixed with water, sulfur dioxide yields sulfuric acid (H_2SO_4), which is where acid rain come from. The next time rain happens in an atmosphere contaminated with sulfur dioxide, diluted sulfuric acid will shower us, our cities, and our crops. Add the previously mentioned heavy metals and other contaminants, and you have one incredibly dirty belch of toxic black smoke. And all of this happens for the privilege of obtaining "cheap" energy from coal.

Coal is torn from the earth with little concern given for environmental safety or workers' rights. And yet, politicians of a particular stripe extoll the coal industry as something virtuous, and worth protecting, a part of our cultural heritage. Among an unfortunately large constituency, this argument seems to work. Humans are enamored with our words. Often to a point where many will not even take a moment to consider what they might actually mean. This rhetorical manipulation is how marketing and behavioral conditioning works. Phrases like: "traditional values" have great power if little definite meaning. Whose tradition and from when? Oh, you must mean all those guys crammed into that mine elevator.

To harvest petroleum, as discussed, crews drill into the planet's crust and extract it. Petroleum drilling gives us gasoline, natural gas, plastics, and pharmaceutical and agricultural chemicals. Oil is used to create asphalt for our roads and the additives to rubber for the tires that glide over them. Petroleum is used to make jet fuel so you can travel to Europe and South America (or visit

---

[115] AMU- Atomic Mass Units. Based on Hydrogen which weighs 1

Granny, who just won't move away from the family home in the rust belt). Oil is genuinely ubiquitous[116]; it makes up the keyboard keys I am using right now to write these words and the lenses of the reading glasses you might use to read them. Here is an addiction humanity can't easily quit, but could massively reduce, and quickly. As we will soon see.

Our oil extraction process is another inefficient practice, with much less than 50% of any reserve ever being harvested. (Tzimas, 2005) Our methods of withdrawal may lock away or contaminate the rest for "eternity." The initial removal of oil from a well rarely exceeds 15%, with the rest extracted by pumping "mud" into the oil body to displace the crude or with steam injections. Injuries and deaths are common in the oil extraction business, with 1,189 deaths documented from 2003 to 2013 in the United States alone. (Mason, 2015) Waste is a hallmark of the oil business, as it is emblematic of our entire way of life. But this is another industry with considerable political support. It is a huge economic driver and is the backbone of the developed world. As its origins show, it is also petrified death in a barrel and will hopefully soon become a novelty of a bygone day.

Except: there's just too damn much money involved!

Gasoline powers our cars, our boats, and our planes. Natural gas heats our homes and generates our electricity. Coal is also a significant producer of Energy, but that is transforming rapidly. Coal just weighs too much, contains too little energy, and even those who would lie about it know that it pollutes too much. Other energy sources are outstripping its transportation, extraction, and processing costs. Contrary to popular myths promoted by some, it is not the "enviro-Nazi's" [117] killing coal; it is simple economics. This is actually rather sad, as the pollution and death caused by coal should be reason enough to discard this industry, but that appears not to be the case. No one seems to care about people dying in the Energy sector or from lung disease downwind of power plants. Much like the narcotics industry, the rush is just too good to

---

[116] Ubiquitous- Everywhere, everything, all of the time.
[117] Another portmanteau from the schoolyard of folks who may not have an argument, but can sure call you names.

spend any time worrying about how you got your "hit." People need power, we need it now, and we need it close by: in a wall outlet, for example.

Like many of us, I was taught in school that Benjamin Franklin discovered electricity when he put a key on a kite string and flew that kite into a thunderstorm. Staining his spats as he sloshed through the mud, his firm grip holding tight the slippery silk string, he stood fast as Saint Elmo's fire raced towards him to illuminate a dangling skeleton key. The sizzling sound, the rising hair, every nerve alive as the potentially fatal charge ran safely to ground — Thus sparing his goose seconds before cooking. There was no lightbulb yet to go off in his head, but his wits saw a connection even without one. For the first time, he had proven that the spears of Thor were indeed powerful physical phenomena. This magic was the force of electricity, and he'd just captured it in a Leyden jar[118]. Taming it would only be a matter of time and study.

Paintings, etchings, and school textbooks have canonized this tale, but it was never explicitly claimed by Mr. Franklin. He sent a letter describing such an experiment to Peter Collinson in France in 1750. It described an experimental design for the capture of electricity from a tall tower using what is now called a lightening rod[119]. When this account was translated into French and published by Collinson, Thomas-Francois Dalibard took it upon himself to actualize the experiment, and indeed, he captured lightening in a jar. One month after Dalibard's proof of Franklin's concept, Ben published a statement in the Philadelphia Gazette that this same experiment had been performed in the city of brotherly love. This time with a kite. He was rather vague about exactly when or who conducted the experiment. It was not until 15 years later that an article by Joseph Priestley laid out the Ben Franklin and his kite story that it became American cannon. Apocryphal stories such as this are littered throughout our fables and popular histories. There is reasonable doubt that he performed this dangerous stunt, but similar doubt exists that no one ever tried it. This tale, for all its drama and uncertainty, is just another great American tale.

---

[118] A Leyden jar was a contraption used to store an electrical charge- a crude capacitor that nonetheless was very useful in the early days of electrical science.

[119] Another invention of Benjamin Franklin.

Ben Franklin did not discover electricity. His design showed that lightning and electricity were the same. And that was good enough, and another laurel to add to Printer; Philosopher; Postmaster; Humorist; Statesman; Playboy, etc. So who cares who actually did it or when? Benjamin Franklin was a serious and creative scientific investigator. And however you view it, Ben's invention of the lightning rod saved countless homes and buildings from what before had seemed the whim of the gods.

Humanity had known about electricity since the time of the ancient Greeks. They just didn't know what to do with it. And neither, to be honest, did Ben. The invisible movements caused by static had made many magicians successful in their careers. A small lodestone, palmed in a skilled hand, made for a serviceable proof of ghosts and the spirit realm. Still, the realization of what electricity actually was, the power of the movement of electrons and the magnetic force radiated outward and perpendicular to this flow, would take significant effort and expense to achieve. It would fall upon others to discover the hows and whys that have led to our modern energy distribution systems. As for Ben, he wouldn't have waited for the dust to settle on his Leyden jar and would already have set off on his next adventure. But discover his contemporaries would, for the tale of the United States of America is also the tale of the energy revolution. A coup that would arguably become more important and lasting than the one we Americans were all taught was the greatest revolution in the world's history[120].

Oh Electricity! The marvelous phenomenon that capitalizes on the flow of electrons from one state to another. We love electricity in all its forms. Wireless gadgets that run on tiny batteries connect us to each other. Electricity powers great atomic colliders that peer into the mysteries of creation. Soon we may wear it and perhaps even bathe in it. But what is it?

As electrons move from an area of overabundance to an area of deficit, they confer power; a *lot* of power. We call this: the electromotive force. Electrical current moves from a positive pole to a negative pole, an anode to a cathode. Electrons, being negatively charged particles, flow from the negative pole to

---

[120] The greatest revolution if you lived in the U.S.A., anyway- 1776 anyone?

the positive. At first, this may seem counterintuitive; the force moves in opposition to the electrons' flow. Yet, that is the physics behind the power. So how does this seemingly intangible force travel through thin metal cables to light our lights and animate our flat screens? A simple way to visualize what is happening is to think of one of those desk toys, a Newton's cradle, with metal balls suspended on a frame so they can whack into each other. When you lift a ball on one end and release it, the energy moves through the other balls, which remain stationary, making the very last ball fly outwards. This toy demonstrates the conservation of momentum[121] and is only an approximation of electricity moving through wires. But if you think of each ball as an atomic core and the energy passing between them as the current, you can visualize how force might move from copper atom to atom without consuming the wires themselves. And in our form of electrical transmission, the electrons move mostly over the outside of the wire in what is called the skin effect.

Now let's scale it all up: the first ball is a power plant, and the last ball is your toaster. The balls in-between are the atoms of the power transmission lines. Where the energy comes from and its use is manifold and far less interesting than the electromotive effect itself. At least to us nerdy folk.

We generate electricity with giant turbines that spin rapidly. The turbines have a magnet that continually revolves inside a stationary coil, which induces the electrical current. This current runs through transmission lines at extremely high voltages to nearby transformer stations that step the current down and deliver it to your home. Standard electrical current, which can be generated by simply sliding a magnet over or inside a coil[122], is called "Direct Current" (DC). And this is the form delivered by a battery and runs your phones and laptops. However, for long-distance transmission, we use a method devised by Nikola Tesla[123] whereby the magnets' polarity rapidly shifts between positive/negative (North/South), creating what he dubbed: Alternating Current

---

[121] p = mv; Momentum(p)= mass(m) times (v) velocity. Another of Newton's rules based off the first law of thermodynamics.

[122] Michael Faraday's coil. For extra credit: It functions according to Faraday's law: E= $d\Phi/dt$

[123] Tesla- a Serbian American engineer and inventor. If you've never heard of him, as soon as you finish this book, go out and get a book about his life. You'll learn a lot about science, and how American industry really works.

(AC). This polarity shifting is the buzz you feel when you stick a fork in that toaster.

And this is an excellent example of how energy changes from form to form. Some initial energy spins the turbine, often coal or gas boiling water to make steam. And the exchanged form is the electricity that flows outward. The two are equal, with some loss because of mechanical inefficiency and line/load resistance. In a perfect system, the input and output would be identical. And in reality, it is there's just loss from the imperfection of our mechanisms. This ideal exchange is called a unity; the input and output match: $1/1=1$. This equivalence is also what that YouTube video promoting over-unity power production doesn't understand. Over unity would mean they produce more energy than they put into the system, $2/1=$ nonsense in energetic terms. You can't violate the rules of thermodynamics. You can, however, claim to have done so. And then you can gain a million likes on social media. Perhaps that is getting something for nothing; it just isn't a helpful something.

Nikola Tesla is considered by many to have been a visionary and has lent his name to Elon Musk's electric car company, albeit unintentionally. His contributions to the field of electrical Energy were second to none. Unfortunately, he died penniless and alone, his works confiscated by government officials in a move that has given birth to countless conspiracy theories. Tesla wanted to create a method of harvesting electricity, free of charge (no pun intended), from the earth's intrinsic electromagnetic field. And suppose you think of the planet as a giant turbine, with a magnetic core spinning out an electro-magnetic field beneath a piezoelectric crust. In that case, you might come to the same conclusion he did: that the planet itself is an electrical generator. And indeed, it is. Earth scientists call it the Geodynamo. Lightning, then, is the charge equalizing as electrons move from a cathode to an anode. Anything that gets in between must either conduct the current or get fried; lightning rod or founding father alike.

We have to assume that Nikola failed to figure out how to construct his worldwide free energy grid, despite disturbing experiments on Tava (Pike's Peak[124]) outside of Colorado Springs, Colorado. You can see these depicted in

---

[124] Pike's Peak is named for Zebulon Pike, an American lieutenant who traveled the west in the early 19th century at the behest of President Thomas Jefferson. He named

several films revolving around his, now nearly legendary character[125]. He was not popular with Thomas Edison and was an enigma to George Westinghouse, both of whom saw electricity as an opportunity for a massive cash grab. History would say they won this battle, and that is why you get an electrical power bill every month.

Most of our electricity is generated using fossil fuels. Currently, natural gas (Methane, CH_4) leads with approximately 38% of US electrical generation. Next is our problem child, coal, at 23% of the total; Nuclear power weighs in at 20%; with renewables coming up from behind with 17%, and finally hydroelectric at 6.6%. (US EIA, 2020) That means 76% of our electrical energy systems use polluting or environmentally damaging sources. And this marks an improvement over a few decades ago when that percentage was 94%. This 18% reduction in polluting fuels is not enough, nor will it be, until society can reach close to 99% clean energy. This is achievable, but will take a hell of a lot of effort and considerable will. The will and effort to move a massive ship before it hits the iceberg. But we can figure that out. First, it will require that we stop entertaining nonsensical arguments promoting fossil fuels from politicians. Folks who are possibly reaping its financial benefits at a very real cost extracted from the rest of us.

Nuclear power currently looms over our whole electrical Energy portfolio. It's a power source that is already in use and is absolutely carbon emission-free. This has led many to tout this much-maligned method of electrical production. Recently, former US President Barack Obama became one of the Nuclear power industry's' cheerleaders. To be clear, he was only suggesting this as a "bridge" energy source while other, actually clean technologies, get online. But nuclear fission has a bad rap for some fundamental reasons.
Our current nuclear power plants generate electricity by splitting heavy atoms. Unstable radioactive elements are brought together in a "controlled" reaction that splits atoms apart at a fundamental level. As a result, some mass is

---

almost everything he encountered after himself and there is an endless catalogue of Pikes creeks, and Pikevilles and Pikes-piss stops. The native name of this massive mountain is Tava- "the mountain of the sun" and had been for thousands of years prior. It should be again, if not on our maps, then in our hearts.
[125] "The Current War"(2019), "The Prestige"(2006), and of course: "Tesla"(2020)

destroyed, or rather, transformed into pure energy. Nuclear fission is the opposite of the reaction that occurs in the sun. Inside the furnace of Sol, atoms merge in the process of fusion, the building of matter. Fission is the shredding of atoms, the elemental destruction of matter. Einstein's $E=MC^2$ is on display perfectly here, just as it is in an atomic bomb blast. When matter converts into energy in this manner, the release of power is far greater than that stored in our fossil fuels' chemical bonds. Every gram of matter annihilated will release an amount of energy equal to its mass times the speed of light — squared! In a reactor, this energy is transferred as heat and is used to produce steam at extreme pressures. The live steam is then used to turn a turbine and make electricity. Lots of electricity!

The downside of nuclear power lies in its waste products. Spent nuclear fuel is by no estimation "clean." It is, in fact, about as dirty as dirty can get. Radioactive wastes, some with half-lives of thousands of years, remain after their heat is released, the turbine has spun, and the partygoers have all gone home[126]. A radioactive material's half-life is a measure of the persistence of its toxicity. It means that one-half of the radioactive material has decayed to a more stable form in the stated period. So, compounds like strontium-90 and cesium-137, which each have a half-life of 30 years, will still have half of their radioactivity, overall, after 30 years. Then they will have ¼ of the starting radioactivity after 60 years, 1/8 after 90 years, 1/16 after 120 years, etc. Plutonium-239 has a half-life of 24,000 years, and it is present, albeit in tiny amounts, in what is called high-level waste. One gram of it could kill ten million people, according to some sources. (Osterling, 1999) Currently, there is no post-processing of high-level waste in the United States whatsoever. (Agency, 2015)

So just how do they treat the toxic waste produced by nuclear power plants? They put them in swimming pools. Well, they are not precisely swimming pools, because you wouldn't want to swim in them. Instead, the spent rods and wastes end up stored in "spent fuel pools," ponds made of thick reinforced concrete lined with stainless steel sheeting. These pools are meant to cool the waste, which still has a lot of left-over reactivity. Just not enough to generate

---

[126] And are buried, and potentially petrified, to be strictly accurate.

high heat efficiently. The radioactive waste spends at least five years cooling in these pools before being encased in stainless steel casks, which are then wrapped in concrete shrouds. Note: nuclear decay is still happening in this waste, and the primary particles: alpha, beta, and gamma radiation, are still being produced. Alpha and beta radiation are relatively low energy particles and are barred by the water or the steel barrel alone. Gamma radiation, however, needs at least 1m (3.1 ft) of concrete to block it. These casks don't seem to have a 1m concrete outer shell because (wait for it) they don't actually exist. Theoretically, after the waste is "kegged," it is supposed to be moved to a stable underground storage facility and entombed for eternity. That is, in theory, because there aren't any existing permanent storage facilities at present, not a single one! No one wants nuclear waste in their backyard. So, this stuff sits in temporary storage facilities, on-site, waiting for someone to develop a good idea of what to do with it. (D'Arrigo, 2014)

In yet another killing joke that humanity has played on itself, this practice, bordering on insanity, has jumped the border and runs free to wreak its madness across the countryside. There is no "where" that is safe for nuclear waste storage. The pyramids are about the oldest and most stable structures ever erected by humanity, and at 4,000 years old, they would be woefully inadequate for this task, by a factor of 60. Putting this crap in the ground is another monumentally foolish idea. If this material gets into the water table by earthquake, flood, or act of terrorism, millions would die in a very un-peaceful manner. A one-hundredth of a microgram (one ten-millionth of a gram) dose of that plutonium-239 is enough to kill a human if ingested. That means 1 gram has the potential to kill 10 million people. (Osterling, 1999) Seriously. That's nothing to be messed around with and released, accidentally or otherwise, into the environment.

The only place that would be absolutely safe to dispose of this material would be to shoot it into outer space or at the sun. And that would pretty much negate both the energy production and economic gain that had been realized from it as an energy source in the first place. The cost of lifting a payload off of the planet's surface, both monetarily and in terms of energy, and into orbit is outrageous. And only one accident while trying this stunt could wipe out millions when the radioactive waste aerosols into the atmosphere after an explosion. Check out the term: "Videos of rockets exploding on takeoff" online and prepare to spend the entire afternoon or evening watching them go boom. Space cargo accidents are no longer a rare event, unfortunately. So, in our wisdom, we continue to do nothing about our atomic waste storage problems.

To be fair, a handful of people are working on devising methods of de-contaminating these materials and disposing of them permanently, Bill Gates for one[127]. He has been working on next-generation reactors that can take this spent fuel and re-use it to a nearly inert state. This process hasn't been exhibited yet, as this project is solely privately funded. Developing any such technology is stunningly expensive, and so, us techies are still awaiting a successful test. So, it's not that nothing is being done, just pretty close to nothing. Nuclear waste is inherently dangerous for a very, very long time. As humans, we don't have an outstanding track record for sustaining stable social structures for thousands of years. We're not that reliable of a species. Yet.

And then there are nuclear meltdowns.

A 9.0 earthquake and its subsequent tsunami hit the Fukushima Daiichi nuclear power complex in March 2011. Located within the coastal towns of Okuma and Futaba, in the Fukushima prefecture of Japan, this plant had six light water reactors. This complex, built on a bluff overlooking the ocean, was designed in the 1970s to survive both earthquakes and Tsunami. But, unfortunately, it was not designed to withstand the scenario which occurred four decades later: one of the most powerful earthquakes in recorded history.

When the tremors hit, the damage was minimal and automatic shutdowns began, as designed. But you can't just pull the plug on a nuclear reaction; it takes time, as it must be stepped down slowly. The reactor vessels, designed by General Electric, were of a boiling water type that had fallen from favor in the years since their construction, apparently for good reasons. They were basically stainless steel pressure cookers with a nuclear core.

Emergency generators were pumping water into the reactors to cool them

---

[127] TerraPower- a private nuclear design company in Bellevue, Washington. It is privately funded because there just doesn't seem to be a lot of status quo buy in on making nuclear waste safer. It seems many are just waiting for some sort of miracle to fix this problem.

rapidly when a 14 m (45 ft) wall of water hammered the installation, wiping out the batteries and drowning the generators. This strike stopped the pumping, which halted the cooling, which allowed the sizzling reactions to continue unabated. Units, 4, 5, and 6 were, thankfully, already shut down for planned maintenance. This left units 1, 2, and 3 vulnerable to meltdown with no way of looking inside to see what was happening. (Black, 2011) It took a couple of weeks of finger-crossing before it became apparent that reactors 1 and 3 were in meltdown. They had just not cooled down enough before the power cut out. But there was still hope. Reactors are designed to contain extreme heat and pressure for extended periods of time.

Then a hydrogen gas build-up caused reactor 2 to explode. Soon it became apparent that there were equally disastrous problems in the other two reactors. They cut vents in the containment vessels of the seemingly intact reactors to avert another explosion. And it did, by releasing a radioactive cloud into the atmosphere, which all share. (IAEA, 2011) The release included: Iodine-131; Cesium's- 134, and 137; xenon-133 and Nitrogen-16, which has a half-life of only 7 seconds but decays into a deadly gamma-ray burst. (Tokonami, 2012) These bursts were recorded on robot-mounted video cameras as bright flashes of light before the camera chips finally failed under the assault of the heavy radiation. An inspection robot and its camera had a half-life inside the reactor buildings of about 20 minutes.

The initial disaster led to the establishment of a 19 km (11 mi) evacuation radius, now reduced to a 12 km (7 mi) no-go zone. It is uncertain how many cancers might eventually result from this incident. Not only did this disaster affect the Japanese countryside, but the radioactive plume in the ocean made it almost to the United States' shores. What mutations it started under the Pacific Ocean waves is anyone's guess, but there's probably a Godzilla remake in the works. The sad takeaway from this disaster is that it could have been far worse. What if all six reactors had been running at full steam?

Humankind keeps having Three Mile Island Disasters, Chernobyls, and now Fukushima, and we keep accepting a few deaths and some zones of exclusion as a reasonable price to pay for our electricity. We continue to pretend: Well, one isn't bad, two is acceptable, three is actually pretty good considering how many plants we've built and operated for so long. [128] But why is this acceptable? We have alternatives and have had them for some time. Can't we just get over our fascination with tickling the dragon's tail? [129] We have yet to face the disaster of a spent fuel catastrophe or a terrorist bomb made of stolen waste material. But considering the risk, it's probably just a matter of time.

And so, now you know how earthlings do Energy at the beginning of the 21st century. They mine and drill and burn and spin to create the abundance of power on which all depend. They also damage and disrupt, pollute and kill and charge tons and tons of money for it. This endeavor is a huge part of why our atmosphere is filled with carbon dioxide, which is warming our planet and threatens finally to kill us all, along with a good portion of the species diversity with whom we share this finite world. We need access to energy that isn't polluting, extractive, or scarce. The universe is awash in energy, so why can't someone develop a source (or sources) that can supply our needs without the risks we now face?

They, we, can. Our future lies with renewable energies: solar power, wind power, and geothermal power (and eventually, nuclear fusion). If fossil fuels have taken tens of thousands to millions of years to form and are still, in essence, just forms of stored solar energy, why not cut to the chase and just power ourselves with the sun? Again, we can, and we will.

One of our primary Solar Energy collection technologies is called photovoltaics. Photovoltaics (PV) use intrinsic physical methods of extracting electricity directly from solar radiation. This technology is excellent, awesome, and in the geekiest sense, cool. A solar photovoltaic panel has no moving parts

---

[128] This is also nonsense: There have been over a dozen accidents, or criticalities, in the United States alone. How many have occurred under the Radar in Russia and China? The answer is certainly not: 0.

[129] "Tickling the Dragon's Tail" is an experiment that was conducted in the early days of atomic bomb design. The "Demon Core" was composed of two spheres of plutonium that could be brought together to create small, controllable criticalities (for fun and profit). Bringing the masses together was the tickling. At least 5 people died due to accidents while fiddling with the Demon Core- that was the dragon's tail smacking the poor humans down.

and does not require any conversion of energy sources like steam to motion to electricity. Instead, this reaction is pure physics in action.

Current PV solar panels work with two silicon layers, a chemically doped layer, and a pure layer. One layer will become the positive pole, and the other one the negative. Photons from the sun strike the silicon atoms and pop electrons to a higher energy state within those atoms; this is the excited state. As the density of the excited electrons rises, a potential is created between the two layers. This potential will have to find a balance. In doing so, a cascade of electrons will flow from the negative layer to the positive layer (Remember: the current flows in a reverse manner, from positive to negative). Tiny wires embedded between the two thin sheets "harvest" that flow of electrons and Eureka! You have electricity.

The atom is like a miniature solar system with electrons orbiting the core of protons and neutrons, called the nucleus. Just as with planets or satellites, some orbits are stable, and some are not. In the atomic world, atomic orbits are called shells. Some shells can hold only one electron, and some can contain many. Each shell, as it moves farther from the nucleus, represents a progressively higher energy state. Electrons can move from shell to shell, but there is an energy price. It takes energy to move upwards, and energy is released when an electron falls down a level. Within a solar cell, a photon supplies the energy for that upward mobility. The photon is a subatomic particle that is sunlight to our eyes.

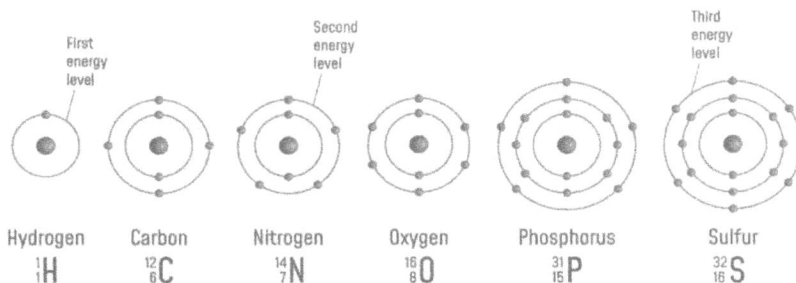

Diagram of various atoms with their electron energy states (shells)

Solar panels are a uniquely elegant method of harvesting the sun's power, the primary source of all energy within our planetary domain. The solar power available to our panel is roughly 1,000 watts per square meter, discounted by the panel's efficiency. When I first lived off-grid in a remote home powered solely by the sun, panel efficiencies were about 15%. Today's panels are in the 20+% range, and the future promises up to 40% efficiencies. That level of effectiveness would equate to an eventual harvest of ~400 watts per square meter of panel surface area in electrical energy, compliments of the sun. So just under 2 square meters (~86 square feet) could produce one horsepower for free.

So why doesn't solar energy power everything already? For precisely the same reasons, some people are still debating if global warming actually exists. There has been a concerted effort to slow, if not stop, the development of solar technologies. When James Earl Carter Jr. became the 39th President of the United States in 1977, he created the National Renewable Energy Labs (NREL) in Golden, Colorado. He also installed solar heating panels on the White House to usher in the coming era of clean energy. When Ronald Wilson Reagan became the 40th US President in 1981, he had the solar panels torn down and slashed the NREL budget. He and his ilk saw no need to waste money on these new-fangled devices. There was plenty of oil and coal. The atmosphere was enormous, and who needed pristine air, anyway? Besides, his friends were making a lot of money with things operating just the way they were. So, hobbled from the starting gate, solar power has crept along under its own virtues with scant federal subsidization.
Meanwhile, fossil fuel producers had and continue to have massive public subsidies. Even though they arguably don't need them. (Redmond, 2017) When you see the world as only money, the needs of that cash will outweigh all other considerations. And that's how Mr. Reagan and his "conservative" cronies saw the world. That money is an imaginary placeholder becomes irrelevant; to their way of thinking. In such minds, the devaluation of renewables was also an enjoyable task in and of itself, if only because of the discomfort it inflicted upon their "enemies." The further fact that people with differing opinions were not enemies was immaterial to them as well. These miscalculations have led to the continued devaluation of all proper actions and tangible objects as we keep discovering. Their mindset is why our estimations of value are so skewed and presently so damn wrong. Such a mentality illustrates why we must never allow ourselves to devolve into cruelty and

punishment, as they have. That path will never permit us to bridge our differences and come together to accomplish, well, anything. Such an attitude can never build. It can only consume. And it's currently doing a pretty good job of devouring everything right now.

Another type of solar power collection is solar thermal, and it's what President Carter was sporting on top of his temporary presidential residence. Solar thermal panels can be as simple as the ones my father made in the mid-nineteen seventies. He took a couple of old doors, nailed a hundred lidless tin cans to them, built sides, painted everything inside black, and finished it with a glass top. A simple fan would blow air in one side of the box, and the opposite end had a hole for an exhaust. He mounted the panels on the roof where sunlight could hit them. These panels could warm a room efficiently and almost for free. The only cost was for the fan and the Energy to power it. Modern solar thermal panels often use antifreeze tubes as the heat collection and transfer mechanism, and they can be industrial in scale. Several test sites use large arrays of mirrors to collect and focus sunlight to a central point. At that spot, materials, perhaps salt, are super-heated and used as thermal collectors for temperatures as hot as an arc welder. Recently, another Bill Gates (him again) energy project hit the news: Heliogen, a start-up that operated in secret for years, achieved success in generating >1,000 ° C (~2,000 ° F) using mirrors controlled by artificial intelligence. The attainment of this crowning achievement suggests that they can perform energy-intensive tasks, such as the production of glass, concrete, and steel, without fossil fuels. Common Salt, sodium chloride (Na Cl), melts at 801° C (1,474° F), just slightly under steel's melting temperature. This property makes it a great material to transfer heat, generate steam, and a promising substance for heat storage. Molten salt can be stored in the liquid state for days and then used in a heat exchanger to extract that heat when needed, making molten salt a potential thermal battery.

There is one other rather important reason solar power generation doesn't dominate our Energy production landscape yet. Solar cells only produce power when the sun shines on them, and the sun takes a nap every night as Gaia pirouettes on her axis. Our world has developed a need for constant power and has little tolerance for its disruption. If solar power only works for half a day, and then there are cloudy days, the general wisdom would say it is not a suitable mechanism upon which to base a global power system. And all things being equal, that might be a sensible viewpoint. But all things are not equal; there is an imperative need for clean, renewable power sources. So what is also needed is a reliable way to bank power for use when the sun isn't shining. While molten salt might be one form of power storage, we're going to need many more. Hold that thought.

You now know that inequalities in heat and humidity within the atmosphere generate the winds over vast areas. And you also know that the heat that bathes our world comes from the sun. So, wind power is yet another permutation of our primary renewable energy source, the sun. Wind power is based on a straightforward concept: build a windmill and let the spinning blades' rotational force turn a generator turbine. It has a lot of moving parts, but the power source is still free, and the wind always blows. Somewhere, most of the time, anyway. And that's its problem as well. Wind power is also intermittent; when the wind is blowing, you're golden. When the wind stops, you are adrift and without power, the bane of masted ships in bygone days.
The Greek poet Homer told us that when Ulysses and his crew were becalmed on their return to Ithaca, they unleashed an ox-skin bag containing all the winds of the world. Unfortunately, this gift from Aeolus, the god of the winds, backfired and only blew them farther off course when they loosened the knot. He had warned them to be cautious and balanced in their use of it, but they thought they knew better. Sound familiar? In the present day, magic windbags are out of stock. So, time once again to consider storage for times of cloudy skies and calm winds.

There is also a troublesome argument against wind power that has recently become quite popular. It goes like this: Windmills kill birds. And while that is true, the numbers of dead birds are relatively small, comparatively speaking. What these newfound bird-lovers seem to miss with their logic is that fossil fuels and nuclear power plants kill a *lot* more birds. A study in Europe went about trying to quantify bird deaths by the power source. And what they found was that wind turbines were killing around 7,000 birds annually in the EU. Meanwhile, nuclear power plants were killing 327,000 within the same time frame. And fossil fuel plants were responsible for 14,500,000 bird deaths each year within the study area. (Sovacool, 2009) Even when statistically adjusted for the greater prevalence of fossil-fueled and nuclear installations, windmill-caused deaths were far rarer.

This "windmills of death" argument is another excellent example of how debate works. The bird-killing argument is a negative team assertion proffered by people who conceivably don't care about bird mortality one whit. But they don't have to. This is just a disadvantage thrown at the concept of wind power. The argument has to have an internal logic, such as wind turbines = dead birds. It does not have to have any consistent reasoning, however. For example, "_Only_ wind turbines kill birds." While some Audubon Society readers are understandably upset by any bird dying, this argument is being used by people who could give a Rat's bare butt about wildlife. They just want to support the status quo (and perhaps fatten up their quarterly dividends). The blind support of the status quo is the very definition of a negative team argument. "I don't need to offer a solution; I just need to tear your solution down." Hahaha!

And speaking of the Audubon Society, they just released a damning report that predicts up to 2/3 of all North American bird species face extinction because of climate change. (Grand, 2019) This paper suggests a massive loss of species diversity that is practically assured if society continues down our present pathways. Here is the valid bird-killing argument. That bird-windmill contention was just a sham, and it should now be judged extinct.

So, there are some excellent renewable power sources, but they are intermittent. We need always-on power to keep reading actual books or swiping right after dark. Energy storage problems have always been with us, but never in the forefront as they are now. Our old methods of generating power are aging out of their usefulness, but they still have an ace up their sleeves. They run continuously. If enough energy can be stored while the sun shines or the wind blows to fill in the gaps, then we will have a proper replacement for our polluting power sources.

Our dear old friend Benjamin Franklin coined the term "battery" for an electrical energy storage device based on the construction of a "pile" made of multiple cells. A battery is a group, and this is how we got the name of an essential part of our clean-Energy revolution. Batteries store electrical power that can be discharged through wires connected to the poles whenever needed. Batteries can absorb energy slowly over time and release it for long periods of use. Now technology can engineer some batteries to absorb that power rapidly and then return it in a metered manner. Lengthy extension cords leading to everything that requires electrical power, all the time, are just a big drag. And that's essentially what our current electrical power grid is. This system is how they began electrical deliveries when Edison and Westinghouse were still buying their first mansions. Our present-day needs require convenient, portable electrical power storage and industrial-scale warehousing for massive amounts of electricity. And the solution to both problems lies in battery technology.

Alessandro Volta invented the first battery in 1800, which he called the voltaic pile. It was a stack of alternating copper and zinc disks, separated by saltwater brine-soaked cloths, that generated an electrical current. He believed the flow of electrons came from the differing metals, but the actual secret sauce was the brine, which is now called an electrolyte. How it worked was thus: First, the briny electrolyte would dissolve the zinc. But in order to dissolve, the zinc atoms will have to give up 2 electrons from their outer shell. This reaction turns the zinc atom into an ion that can readily travel through the wet cloth and meet with the copper plate's atoms. When the zinc ion meets the copper, it finds it easy to rob the copper atom of those two missing electrons—"Like stealin' candy from a baby." As the reaction repeats, the copper disc loses many electrons, and the zinc disk ends up with many extras from the dissolution process. As you know, once there is a surplus of electrons at one pole and a deficit on the other, current flows. This process is another lossy reaction as it consumes the assembly in a non-reversible manner. The voltaic pile creates an electro-motive flow, but it cannot be recharged. Once the reaction is done, it's done. And your workbench now has a stack of corroded junk sitting on it, requiring the wet clean-up kit.

In 1859, Gaston Plante' created the first rechargeable battery: the lead-acid battery. It worked so well folks still use it today. The reversible battery comprises a lead (Pb) anode and a lead dioxide (PbO_2) cathode bathed in sulfuric acid (H_2SO_4). The acid corrodes the lead anode and causes electrons to be released, creating lead sulfate ($PbSO_4$). The same corrosive reaction at the

PbO_2 cathode consumes electrons, creating our needed deficit; so, electricity flows. When a current is then pushed back into the battery poles, the chemical reaction reverses. The lead sulfate can again become lead and sulfuric acid, priming the battery for another discharge cycle. Now we have an efficient but not perfect reaction. Some Hydrogen from the H_2SO_4 is released in each cycle, consuming a bit of the electrolyte solution. So, while it's not flawless, it works well enough, and battery fluid is cheap and easy to "top off." Just adding distilled water will replace the hydrogen deficit, and we're good to go again for quite some time.

Next, let's jump to Lithium-ion batteries, the ones in our laptops and phones. The stages in between are many and sometimes subtle, but Li-ion batteries are a genuine breakthrough. Lithium (Li) is a metal with excellent energy storage to weight capacity, the best of all metals. What's more, lithium-ion batteries are made with a flexible polymer electrolyte, making them more useful for portable devices than a liquid electrolyte. Their excellent re-chargeability and low weight make them perfect for our modern society to use in everything from your Fitbit to your Tesla Roadster. In 2019, three scientists: John B. Goodenough, M. Stanley Whittingham, and Akira Yoshino, received the Nobel Prize in Chemistry for developing the Li-ion battery. A well-deserved honor that only took 30 years after their initial invention to its award.
The only teeny tiny problem with these batteries is that Lithium is a reactive alkali metal that can ignite in contact with oxygen or water. During the re-charge cycle, it can also grow dendrites, basically tiny stalactites of metal. These metal spikes can puncture the electrolyte barrier between the anode and cathode in a poorly designed cell. The result of this unfortunate reaction is a short out of the battery and the production of heat. This heat can be so intense that it melts the battery casing, and then the reactive metal can come in contact with air. Boom! This chain reaction explains how vapes can explode.

On the battery horizon is the solid-state battery. Solid-state batteries are, as the name suggests, wholly composed of solid materials that are stable through thousands of charge/discharge cycles. Several formulations exist, and time (as well as manufacturing efficiencies) will tell which one's make it into our lives and pockets first. A fascinating species of solid-state batteries can be made with carbon nanotubes. Carbon nanotubes are incredibly tiny structures composed of perfectly aligned carbon atoms that form hollow rods several billionths of a meter wide (nanometers). While not as pretty as a diamond, this carbon alignment may well become far more valuable to humanity. Their storage capacity is potentially ten times that of a lithium-ion battery by weight. Researchers at Rice University have developed a technique that uses nanotubes in a Lithium battery to increase storage capacity and workable life. (Williams, 2018) A pure nanotube battery seems to be something for the future at this point. Another excellent reason to stick around and not let civilization burn.

Our need for batteries is far more critical than just a smaller smartwatch or a longer movie watching time on our tablets. Since there are significant renewable energy sources that only work when the sun shines or the wind blows, next is needed industrial-scale electrical storage capacity. Clean energy will replace our current dirty methods only once the gap is closed and a system can provide constant, always-on power from intermittent sources. This goal is attainable and achievable. And it will be realized in multiple ways. Batteries are essential for storing electrical energy, but there are other methods of storing energy. One approach uses excess power when the sun is shining to pump water into elevated tanks that can later be released to turn a micro-hydro turbine. Windmill pumps could even achieve this same result without the need to generate electricity for the initial pumping. Drawing water like this is what those famous Dutch windmills were up to in all those old oil paintings. As mentioned earlier, molten salt can be stored for use at night to heat water to create steam to spin a turbine. Another experimental method stores compressed air to turn a generator whenever needed. Mechanical flywheels are excellent methods of storing kinetic energy and, while highly efficient, are rather costly. You could scale these methods large or small. But the most straightforward mechanism would be a cheap, reliable, large-capacity electrical battery. And all indicators suggest we will soon have them if we choose to stick around and fund their research.

Another beneficial, clean and abundant, if not technically renewable, power source that will be used more as the world moves to an emission-free Energy future is geothermal power.

Gaia's planetary core is a massive store of heat. This heat is a product not only of the gravitational pressures exerted on it, the compression of rock and magma, but also from the nuclear decay of radioactive elements in their proper place: at the planet's center and far away from us surface dwellers. This central core is composed of an iron-nickel alloy that is solid despite a temperature of 5,430° C (9,806° F), nearly as hot as the Sun's surface. The core is surrounded by a liquid outer core and then the plastic-like mantle. The crust floats over the mantel, and this life-protecting shell is thin, about 11 kilometers (7 miles) on average. It is thinner even than the atmosphere that protects us from outer space. The zones where the mantle and the crust meet are heated to a toasty 200° C (392° F), offering the potential for a massive energy harvest.

Volcanos, deep ocean vents, and hot springs are all examples of natural geothermal formations. Hot springs are my favorite geothermal manifestations, for not only are they fun and relaxing, but they also demonstrate natural bathing suit repellant properties. And they point us to a handy and clean energy source. The idea is simple: drill a deep well into the zone where the earth's crust is heated by the mantle and flood it with water. What comes out is live steam. We've just created an artificial geyser. Next, all you have to do is harvest its power.

When you think of countries that use Geothermal energy, you might often think of Iceland and its pioneering work with electrical generation and central heating. However: El Salvador, Kenya, the Philippines, New Zealand, and Costa Rica all produce over 15% of their energy needs using the planet's heat. This technique is a cost-effective and straightforward method of providing always-on power and could be implemented in many areas worldwide. There are hot springs and geothermal power applications even here in the Rocky Mountains, where the crust is thicker than average because of our mountains' massive uplifting.

In Mosca, Colorado, a deep geothermal well heats an area of interconnecting ponds and streams and is home to an alligator farm and Reptile Park. Why, you may ask, does anyone need an alligator farm in Colorado? I don't have a good answer for that one, except that it's kind of cool, and there is zero chance of the escape and the invasive colonization of alligators into the rest of the State.

We could construct different types of wells for various purposes. Everything from shallower wells for the gentle heating of homes and greenhouses to deep wells producing industrial-scale utility outputs. A single industrial well may lose its thermal differential after 30 years or so, as the local "harvest" of heat cools the rock and decreases production. But when you wait and leave it alone for another 30 years, that well will be ready to go back into production for another cycle. Such cycling could produce the human race's entire energy needs (Testor, 2015) until the sun increases in brightness enough to evaporate the oceans and steam clean the planet's surface, about a billion years from now. (Scudder, 2015) If we're not spacefarers by then, that's our own fault. The technology that currently exists for drilling oil wells could be directly translated into geothermal well construction. Here is an excellent source of future clean-energy jobs. Ponder, if you will, the reluctance of many to new industries that will replace fossil fuels while delivering essentially the same product and requiring the same, or an even larger, workforce. Why do we fight change when we know it is inevitable?

There are, of course, negative environmental impacts with geothermal energy production. Water scarcity is a big one; water needs to be available in large quantities to inject into these wells. The water might become contaminated by heavy metals and toxins from the bedrock and require decontamination. Hydrogen Sulfide (H_2S) release is also a concern. This gas smells like rotten eggs and can also be a precursor for acid rain. Remember, though, Gaia is incredibly ancient and has many tricks up her many sleeves. Her crust stabilized over 2 billion years ago, setting the stage for our potential thermal harvest of the core's heat. Before the uniform hardening of the surface, there were long periods of intense volcanism and crustal instability. But, again, life didn't care to wait and began issuing forth its first pseudopods over 4 billion years ago. And some early life had interesting metabolisms.

A few ancient life forms got their energy from chemosynthesis[130] or directly from geothermal heat. Organisms that live deep in the ocean next to crustal

---

[130] Chemosynthesis is the creation of organic compounds by organisms using the chemical reactions of inorganic compounds. Similar to photosynthesis but in the absence of sunlight; for example: Hydrogen Sulfide can be fixed with carbon dioxide to create amino acids with solid "poops" of sulfur left behind.

fissures still maintain these traits. These deep ocean vent bacteria can chemosynthetically break down hydrogen sulfide, creating solid sulfur and amino acids. Their metabolisms suggest that these or similar microorganisms might be used to detoxify the water used to harvest the planet's warmth. Again, investigating nature continues to provide us with the answers to our problems. If we refuse to operate in the ways we always have in the past and design our systems to include nature's agents; we can mitigate many of our problems before they grow out of hand. The neat thing about being the only creature known to recognize and utilize the concept of time is that we can foresee and ease our problems before they become catastrophes. If we decide we want to, that is. We need to conduct our lives to thrive within the ecosystem. And we need to act in more holistic ways, using the tools that nature has given us, to proceed sustainably. Human/microbe joint ventures will be a growing industry in the future, so invest now (if not your money, at least your attention and approval) and get in on the ground floor.

The Energy available from geothermal sources can be harvested 24/7 as it is always on. It could last a billion years or more. That's far longer than the human race will if we don't get our act together.

Our next renewable is also continuous: Hydroelectric power. Hydropower is the direct generation of electrical energy using water flows to spin turbines and usually involves constructing a dam. This type of electrical generation has been a source of clean energy for a few generations now, and while it is an established technology, you've now seen some cracks in its armor. But many of the woes considered earlier about dams can be mitigated if we use a different type of water-powered generator, a micro-hydro generator. Micro-hydro is the label for a class of small electricity-generating devices that, while using water flow as the energy source, doesn't require the construction of mega-dams. Micro-hydro systems are further defined as installations that produce between 5Kw and 100Kw[131] of power. This can often be accomplished with the creation of a small millpond and a drop (this is called the "head" in the water pumping world) in elevation of only a few meters. A small dam, often little more than an earthen berm, can impound a modest

---

[131] 5,000 Watts to 100,000 Watts of power.

amount of water and feed the overflow to a pipe. This pressurized water spins a small, high-efficiency turbine, and electrical power is the result. These small ponds can also act as lovely little skating rinks in the winter; and are another benefit of a multi-faceted, holistically balanced way of life. "Multi-faceted and holistically balanced" is also an appropriate description of the natural world Gaia has loaned us.

Micro-hydro is being used in "third world" countries and remote areas to produce modest amounts of power that can be used close to the electrical generating equipment. It often complements Solar power in remote outposts in that it will consistently generate power, just perhaps not as much as the PV array. People need less energy at night, so the templates offered by remote and temporary installations can show how to proceed, even at larger scales.

5Kw of electrical power is enough to run a home if your house is not an energy-hogging behemoth. I've certainly done it and had a backup to my entire off-the-grid home that was a 5Kw Honda generator. No sun? No problem, start the genny and let the chicks continue their peeping, and the kids their frolicking. Micro-hydro promises to help us in our diversified energy future, even as massive hydroelectric systems lose favor and potentially bury themselves.

Our final renewable power is a system that doesn't exist yet. It is nuclear fusion. Fusion is the power of the sun and the fundamental process of building matter. Where fusion creates more complex forms of matter, its sibling, fission, can only dismantle them. Fusion has been demonstrated since the 1950s but has never quite reached an energetic unity or even a self-sustaining reaction. The sun proves fusion power is possible. We're just still on the long journey to figure out how to do it here on the planet's surface. There is no radioactive waste from fusion. There is the potential to skip turbines and heat exchanges and produce electrical current directly from such a reaction. It could potentially create helium, which is in dwindling supply, for our MRI machines and birthday balloons. But the scientists aren't there yet. And if humans fail as a species and collapse, we never will be. We are so close to a sustainable and better future, held back primarily by our training and the "Old World Order" that is our status quo.

All of this brings us to our penultimate issues with Energy:

Who owns Energy?

How do we best harvest energy and distribute it?

The masters that Energy serves at present are the industrial utilities. Massive power generating, harvesting, and distribution systems require monolithic oversight and planning. But with recent developments in technology, this no longer needs to be the case. It is feasible to provide power to a home or business with solar, wind power, and even micro-hydro on site. Community-sized installations are also practical and even more efficient. This lack of efficiency inherent in our centralized power production is something that needs addressing.

A large electrical generating plant produces massive amounts of power distributed to a wide area via the grid. The electrical grid is essentially an interconnecting network of high-tension power lines and transformer stations that transport, step up or down voltage, and deliver it to your outlets (via a billing meter first, of course). Under this system, up to 50% of the power produced is lost in transmission to line resistance, faulty connections, and even local weather. Storms take power lines down, and people lose their vital power when they need it most. Electrical transmission lines are also a significant source of ignition for wildfires. They represent a massive amount of steel, copper, and other materials that might be used elsewhere or even, heaven forbid, left in the ground for future generations.

In Germany and some other European countries, they bury many of their power lines. This arrangement reduces damage from storms and increases efficiency a bit because of the constant temperatures and lowered magnetic field induction fostered by their burial. It is more expensive to construct initially, but maintenance costs are potentially far less due to decreased damage from falling trees and ice buildup. (P.S.C., 2011) Scandinavian folklore suggests it would be the best way to mitigate troll damage, as well. Utilities may choose buried lines to reduce aerial transmission system casualties, improve visual aesthetics, or minimize transmission losses. But it would still be part of a grid system under the status quo method of centralized power generation. If we could de-centralize power production, we would save considerable amounts of waste energy, material costs, and system complexity. Renewable energy allows us the chance to do just that.

De-centralized power generation will be the preferred method of Energy production in the 21st century and beyond. Nature does not work like the power grid. Nature is composed of millions of "islands" of habitats and micro climes. If any individual part fails, many others can take over the lost capacity. That way, the whole can continue to operate, often without skipping a beat. Such flexibility is not how our Energy infrastructure is currently built. There are pipelines that are thousands of miles long that pump resources from A to B, then C to D, etc. One tiny leak and the whole thing has to shut down. Ever-larger installations are built to create even more power to sell to even more customers. And while they build redundancy into the system, ask someone who just lived through a hurricane how well the utility system worked for them and how fast their power came back online. They will set you straight on how effective our current power grid is.

When your best source of power is a mega-plant that no one wants in their backyard, long lines are necessary. The loss of electricity because of these long power lines equates to a carbon dioxide release of nearly 1 billion tonnes/year (1.1 billion tons). Higher efficiencies and de-centralized power generation could reduce the waste by up to 511 million tonnes (562 million tons) of CO_2/year. These savings are equivalent to the greenhouse gas release of the entire chemical production industry worldwide. (Surana, 2019) And these savings would occur even if fossil fuel powered. But we have to outgrow fossil fuel as fast as possible. Still, it helps illustrate just how needlessly our current methods waste and pollute. De-centralized systems can go back up much faster than massive grid installations. And local people, with local knowledge, are always there to do the work when needed. Our current system is built on massive waste. People don't need to keep living that way. Smaller local systems would use fewer materials like steel and copper, lose less energy to transmission losses, and cost less to build and maintain.

Now it is easy to see that you could power a house with panels on the roof and a battery on the wall. Cities could power neighborhoods with solar arrays, some windmills, and any combination of micro-hydro, geothermal, and different power storage methods. Batteries can be centralized or in each building, giving them control over their own power back up. Here is a much more bio-mimetic system where multiple sources exist, each with strengths and weaknesses, combining to deliver the energy required when and where it is needed.

We need to consider this one last thing about Energy in our present state of societal development and cognitive evolution. And that is war. Fossil fuels exist in many parts of the planet. You could posit that oil and gas deposits exist where oceans once lay, coal exists where ancient forests stood. Raw materials are not everywhere, nor are they uniform in their deposition. Some countries have massive deposits; some have none. Because of their technological advancement, some countries consume more fuel, and some less developed countries use fewer resources. Do not confuse this with the idea that they require less; they just use less. While this could present an opportunity for sharing and collaboration between the haves and have nots, the needs and need nots, it is in practice a source of near-constant war.

Our War making cultures seem to love any excuse for a good incursion. Oil producers hoard and wrangle for increased prices. Petro-corporations act like nations and try to reduce the cost of extraction and competition by various practices, often unethical. The poorest nations generally get the least and bear the brunt of the ecological damage caused by these actions. (Eweje, 2006) When one country is at a higher level of economic development, and decides they need (want) the resources of a lesser state, they will use blockades, embargoes, and, finally, wars against the weaker. When the other society lacks basic economic or military power, they'll just get rolled over. In that case, one can hardly call it war, so let's call it what it is: theft and murder carried out by armies great and small.

Any pretext will work in today's world to start a war to steal another's resources. The United States did it in 2003 by openly lying about "Weapons of Mass Destruction" (WMDs) in Iraq. It was well known and reported at the time that there was no actual evidence of such terror weapons. In fact, no such WMDs were found because (hold your breath) they didn't exist. What did exist were oil fields, thousands of them. It was as if oilman turned Vice-President Dick Cheney had found his calling. All he needed was an excuse to get it, petty concerns like accuracy and fairness be damned.

Nevertheless, the war was real. We may never know the toll in terms of lives lost because that is the nature of war; who's counting? [132] America didn't win, and neither did they. But consider the destruction of regional infrastructure and the civilian dead, the shattered libraries, and priceless antiquities destroyed. The shared history of humanity squandered. But what the hell—gasoline stayed relatively cheap, and new fortunes were made repeatedly.

The rise of terrorism that resulted from these forays turned out to be good for business as well. There's always money to make selling arms to this faction or the next. And the United States of America is perhaps the biggest international arms dealer of them all. Societal instability will always allow for more profiteering. A person in a third-world conflict zone could have a net worth of less than $10, but some defense contractors might see the opportunity to make a couple of grand off of selling the tools to kill them. That's the world of today.

During the pandemic lockdowns in the United States, billionaires increased their collective worth by trillions of dollars, a billion dollars a day some say. So, this type of economic war is not something that just happens over there. The beneficiaries of the Middle East Wars were corporations, as has been the way for quite some time. (Jones, 2012) And again, many of our corporations just don't give a rat's bald bum, whether this is at all sustainable — the money's just too damn good.

The mixture of government power and corporate power, especially with some needless violence thrown in, is called fascism. Fascism is the status quo caught with its knickers down around the ankles. It is the opposite of egalitarian, fair, or just. It is a seductive ideology, especially for those who live in an imaginary world and want to stay there. No matter the cost to others, the planet, or life itself.

The sun's direct use as a power source will level the playing field, as the sun shines down everywhere most of the time. There is simply no need to invade another territory for their sunshine. Nor will it generate such excessive profits

---

[132] They say that history is written by the winners. But losers write it too. During the Vietnam war it is now documented that the U.S. Dept. of Defense intentionally over inflated the numbers of Vietnamese casualties, and under reported U.S. losses. Just to remind you- We lost that war.

that corporations will be able to continue buying governments.

It has recently been clarified that renewable energies have become more cost-efficient than coal and are staring down natural gas as well. It turns out that merely replacing coal-fired plants with renewable facilities will immediately save up to 75% of production costs. (Gimon, 2019) But, unfortunately, some producers will still build gas-fired plants or more nuclear stations if only to spite the world. They can do this because they have so danged much economic power. They can do this because huge decisions are made for you within our current status quo, not by you, regardless of what your constitutions say. And the inertia of all the lies told to discredit global warming and those who desire to make the changes required to fix it roars on like a freight train. As with many of our issues, our misuse of Energy systems stems from our ability to live in an imaginary world where humans are the first, and above all others. Above all, values, concerns, and consequences. Humankind may not actually have Dominion over the world, but we play an excellent masquerade of it.

Again, by emulating the ways of nature, we will establish a better system. It always was the sun, and always will be. It always was us, and it always will be, until it isn't. So much energy showers us every day, 256 million million watts per hour of solar radiation. We stand above a mighty, blazing hot dynamo. All that is needed is to harvest this power cleanly, and most of our carbon problems will be solved simultaneously. Renewable Energy fixes our carbon emission problem, so the next step is to remove the excess carbon left in the atmosphere. To accomplish that, finding the balance in all of our endeavors, just like ecosystems find balance, will close that loop. If we fail and continue as we're going, Nature will step in to restore the balance, and that equilibrium will be achieved by getting rid of the problem of us.

Benjamin Franklin was an energy revolutionary as much as he was a social one. Had he been born after Darwin, he would have no doubt been an "evolutionary" as well. And I believe he'd agree that:

The forward path of human evolution is the rejection of Energy Dominion. It is clean and fair energy for all.

CHAPTER NINE: FIRE

"Come not between the dragon and his wrath."
— **William Shakespeare**

"How can you rise, if you have not burned?"
— **Hiba Fatima Ahmad**

On August 20th, 1910, wildfires burning in Montana and Idaho combined when hit by hot, dry, hurricane-force winds. The joining together of multiple smaller fires into one large conflagration is termed a "complex fire." This complex became known simply as "The Big Burn." Over the next few days, 1.2 million hectares (3,000,000 ac.) of forest burned at moderate to high severity. It consumed 12,100 km² (4,700 square miles) of land in only two of those days. The fire spread to parts of Washington state and British Columbia as well. Eighty-seven people died, and the town of Taft, Montana, disappeared from the map completely. Wallace, Idaho, also burned to the ground but sprang back and remains a friendly town to this day. Devastation and carnage ruled the land that summer, and contemporary firefighters continue to study this fire. It is still listed as the largest wildfire in United States history. But it probably won't be for much longer.

President Teddy Roosevelt took this fire very seriously and rapidly elevated the fledgling US Forest Service into today's institution. He gave them a challenging and inflexible task: stop *all* forest fires within hours of ignition, whatever the monetary or human expense[133]. And that mandate would indeed

---

[133] By 10 am the next day, if possible.

have costs. Costs that continue and accrue to this day.

When the European immigrant population first set foot in the wild west, they had no clue that events such as The Big Burn were possible. The dwindling native population, of course, knew better. But if anyone even bothered to ask them, it's not recorded anywhere of note. Knowledge unshared is intelligence lost; knowledge unwanted is wisdom squandered.

In an epic act of hubris, a monumental exercise of man over nature, "The Old Lion," decreed that forest fire was a menace to be eradicated by any means necessary. Roosevelt was not the first to mistake his love of nature for a willingness to impose his misunderstandings about it upon its stewardship. Nor would he be the last. Many still follow in his footsteps. His actions charted the course for the grim realities the American West now faces.

Forests are burning worldwide. The loss of life, property and fragile environs is speeding up rapidly. Federal land managers oversee forty-six percent of US forests by area. (Tidwell, 2016) And as of this writing, over one-half of the US Forest Service annual budget is being consumed by the cost of fire suppression activities. Money, literally, up in smoke. Many project this trend to increase to nearly three-quarters of their funds by mid-century. Our current practice as land managers, in terms of fire, is reactionary. When a fire breaks out, firefighters run to suppress it, just like "The Rough Rider" instructed us. When a man has as many nicknames as old Teddy did, you simply follow his orders. All involved want to have a more proactive stance on forest fires, where mitigation and risk reduction projects are completed before the land ignites. But the budget for such activities is tiny, and it's getting smaller in the shadow of the growing cash drain of these increasingly large fires. As a result, fire is propagating relentlessly, increasing in size and destructive intensity. One day western US will experience another fire of the size of the Big Burn, and it will devastate the country. Population densities, the sheer number of structures, and the interface between the built world and the wildlands all but guarantees this. Until policymakers accept this as a reality and change how they manage our wildlands, this trend will continue.

Contrary to some suggestions, raking the forest floor has no appreciable effect on fire behavior. It is, instead, just another disruption of the ecological functions of the woodland community. One that is, at best, pointless. And, at worst, destructive in entirely new ways. Such a disturbance of the fragile duff has no logical purpose. So, the cynic might suggest, some will probably try it.[134]

From a Fire Manager's point of view, the forest floor's duff is fuel. From the Ecologist's point of view, it is life hanging on to its precarious foothold. Both perspectives are correct. We must be open to multiple points of view if we hope to understand what's happening in our environment. They say that fire was the first technology humanity mastered, but we need to question that assumption along with many others. For fire continues to refuse to recognize us as its master.

The forest has to burn. Fire is a foundational part of many forested ecosystems. It is not the judgment of a god for our sinful ways, as some may claim. Fire does not spare the faithful and punish the wicked. It is random and the manifestation of chaos bringing forth a new order. Fire is the natural system of selection and equilibrium within western US forests and many (but not all) forests around the world. These forests have always burned, and the original residents of this continent knew that. Indeed, many tribes practiced active forest management by setting fires to promote this cycle of renewal. The European settlers saw these actions as pointless vandalism, if they saw them at all. These practices had previously helped keep the forest in a dynamically stable condition and promoted healthy wildlife. What came as a harrowing shock to the transplants from across the ocean was common knowledge among the native populations. The forest must burn, and it has a schedule that needs to be met, else disease and starvation will come to call. Just as the apex predators like wolves are required to tune the herds, so too is fire the forest's beast of prey. Forest fires are transformational events, consuming waste and returning fertility to the ground. In a genuine sense, the frequent fire forest is the Phoenix of lore; its senescence celebrated in a fiery "death," followed by a triumphant rising from the ashes, renewed and rejuvenated.

But the European immigrants thought they knew better. When "god" gave us this land to lord over, it became our duty to protect it, at least in so much as that advanced our fortunes, of course. But the wildland is not a garden, and it is not there to satisfy anyone's ideas of beauty. Nor do artistic notions of what

---

[134] As Winston Churchill noted: "You can depend on Americans to do the right thing, once all other possibilities are exhausted."

the land should look like have any relevance in how it should be managed. That is the illusion of the newly minted private forest owner. And plenty of foresters share this view, and so do many of our thought leaders. It seems people just can't see the forest for the trees. Because the forest is not a thing, it is not a monolith. It is a system and can only thrive if its interconnectivity and dynamic nature are preserved.

When the western forest ecosystems operated under their natural, frequent fire regimes, their stands looked very different from today. There were fewer trees per hectare, having survived through multiple "selection" events to reach maturity. Those trees were more robust and fire-adapted by trial. Ultimately, our removal of one of the forest's prime drivers has led to an overpopulation of timber, which has led to malnourishment, insect infestations, diseases, and dehydration. This overcrowding stress leads to stand morbidity and mortality, just as it does in a crowded monocultural field. Our imaginations have inadvertently projected our ideas of fields and gardens onto our wildlands, an inaccurate and damaging miscalculation. Managers have selected tight forests for greater profits or aesthetic reasons all their own and subjected the entire ecosystem to stress. Our customs have adopted an all-or-nothing approach to forest husbandry and harvest. Humanity has exerted control over a natural process for no greater reason than that we could. Our imposition of human will has only led to more stressed forests and ecosystems. A stressed forest stand is little more than fuel for these chaotic fires that grow exponentially season after season.

There is chaos a-plenty in nature. And this explains how life has adapted to a planet that has experienced a multitude of instabilities and upheavals. Our geology is stable but often unpredictable. The universe is calm, but there are meteors. And then there's fire; fire from the skies, fire from the volcanos, and even from piles of decaying organic matter. Life can and has adapted to these occurrences and more. Humans just somehow feel we shouldn't have to. Mostly, we like to pretend we are above it all or that we are in control. Chaos reminds us that such a belief is just silly self-indulgence.

Chaos struggles against stability and its hidden curse — stagnation. Here is the wisdom of no wisdom, and it's what keeps life safe and ensures its future. Our impositions of order have merely wound a drive spring for greater, future turmoils. When Teddy stopped the fires, humans set the stage for the catastrophes we are only beginning to witness. Mark these words: we are only beginning to witness the devastation of catastrophic fire.

What the fire crews sees when they look at the forest today is a fuel bed. And they mark the coming fire in their minds, not in terms of if, but of when.

On a Saturday afternoon in June 2002, a Forest Service employee, with full forest fire training, and an equipped wildland fire engine, set some love letters alight in a campground fire ring, grieving over an ended relationship. It was a "red flag" day, meaning no sparks or ignitions of any kind were permitted on open land. Chainsaw work was discouraged, as was range shooting, and no smoking would be tolerated. Period. It was well understood that the forest was a powder keg waiting for a single ember to explode.

The wind lifted firebrands from this small pyre that found easy purchase on the arid duff, quickly igniting the Hayman Fire. Burning through heavy timber stressed by seven years of drought and drier than kiln-dried lumber, this blaze blew up like a volcano. On Sunday, June 9th, the winds kicked up to 82 kph (51 Mph), and the air humidity dropped to below 5% Rh. The fire raced, at times over 48 kph (30 mph). It ran 30 Km (19 mi) in a single afternoon, decimating over 24,000 ha (60,000 ac.) of heavy forest between lunch and the cocktail hour. Faced with such a conflagration, fire managers had no choice but to stand back and watch it burn. Locally, many of us call this the Monty Python tactic: "Run away!"

The Hayman became the largest wildfire in Colorado history (that record has now been broken). By the time of containment, on June 28th, it had consumed 56,000 ha (~138,000 ac.) of prime forest. That is approximately 530 km² (204 mi²). The initial estimates for the unassisted, natural return to its initial conditions for this forest were 1,000 years or longer. Without human intervention, many parts of the Hayman might never return to forest cover.

The US Forest Service had some budget for addressing post-fire restoration after the Hayman Fire; not a lot, but some. A small local non-profit stepped up to help coordinate volunteers, and the Coalition for the Upper South Platte

(CUSP)[135] was born. Under the leadership of their first full-time Executive Director, Carol Ekarius, CUSP forged ahead, planting tens of thousands of seedlings and building erosion control structures with thousands of volunteers. Here was a new type of community-based ecological organization. They were strictly non-partisan and non-advocacy; their sole role was to be the boots on the ground of restoration efforts and partner with whoever wanted to help. This common cause welcomed and was joined by bird watchers, ATV enthusiasts, through-hikers, fly fishermen and gun clubs alike. CUSP has now spent over two decades restoring the landscape, improving or removing roads where needed, planting trees, and experimenting with new techniques. Combined with Federal, State, and Foundation partners, their work meant this forest might now return in just over a century or two. Timescales are humbling when dealing with nature. What was done with one careless act, and transpired in just a few dozen days, will not entirely heal in our lifetimes. In aggregate, life has time on its side, individual lives, such as ours, not so much.

There is a period every year when fire occurrence is expected. It is called the fire season. And now, in some areas of the US West, that season threatening to consume the calendar from January right through December. When there is no pause to the rhythm of the fire, there is no safe harbor for life to reproduce or rest. Much like our damming of the pulse of seasonal floods, we have also inadvertently disrupted the timing of our fires, confounding their many functions, warping their positive effects into mechanisms of devastation.

The way a fire acts in a healthy forest is to stay close to the ground and wind its way through the duff and detritus (downed logs, dead or dying shrubs), killing weak trees and thinning out their density. Low flame lengths of 1-2 m (3'-8') selectively kill younger trees and fail to climb the larger ones who have either self-pruned lower dead branches or had those "ladder fuels" consumed by a previous burn. A low-to-moderate intensity event will release nutrients locked in the slow-to-decay wood and benefit overall forest health.

The ponderosa pine (*Pinus ponderosa*) is an excellent example of fire adaptation in nature. As the ponderosa grows, it develops a thick insulating

---

[135] I work for CUSP, a 501 (c)(3) non-profit that is now a template for place-based, collaborative ecological organizations. A much-needed cog in the wheel of environmental restoration and community cooperation.

bark. All living systems excrete waste products, and the ponderosa sends some of its into the bark, increasing its flame retardant properties. In addition, they auto-excise lower branches as they mature to eliminate any path for fire to creep from the ground to the living crown. The regen, baby trees flock around the older champions and try to protect one another when the fire comes. Only a few will survive to adulthood. The actions of fire create an open structure for the ponderosa stand. As a result, large trees will stand singularly while younger ones will thrive in clumps and groups, practicing safety in numbers. The forest's canopy then self-aligns in an open pattern where very few tree branches touch one another above the ground. This formation is a practical adaptation because the fire that moves from tree to tree, branch to branch, above the ground, is the most dangerous type of fire.

A crown fire is broadly defined as any fire that rises into the forest canopy and moves independently from the ground. This fire species occurred in the Hayman, the Waldo Canyon Fire[136], and the Camp Fire[137]. It is becoming an ever more common type of fire in the western United States. When a fire reaches the trees' crowns, and the trees are close enough to ignite their neighbors, the wind can drive it quickly. It also allows oxygenated air to enter the flames from both above and below. This catastrophic fire rages in an overgrown and moisture-stressed forest, running above and raining down an inferno onto the forest floor below. A crown fire essentially burns the landscape twice as it passes. This mechanism leads to the near-total mortality of the forest and the sterilization of the land. The crown fire is what has been recklessly encouraged by our acquiescence to Teddy Roosevelt's wishes.

In this scenario, the raging flame front pre-heats the forest before it, withering and desiccating the foliage in preparation for ignition. This preheating is crucial for extreme fire behavior and cannot as readily occur in a genuinely open forest structure, the wild forest, or forests pre-treated by thinning or prescribed fire.

---

[136] The 2nd most damaging fire in Colorado history in terms of monetary loss: 486 homes lost, insurance claims that reached over $453 million. The new most damaging fire in Colorado is the Marshall fire which was technically a grass fire that burned over a thousand buildings, did half a billion dollars in damage and happened in late December,2021.
[137] The greatest loss of life and homes in California history: 85 deaths, 18,800 structures lost, up to $30 billion in casualty loss.

The crown fire is chaos manifested. Catastrophic wildfires have so much power that they create their own weather patterns. The pyrocumulous cloud is characteristic of a cataclysmic wildfire. These plumes can rise to 12,000 m (40,000 ft) in altitude and create massive downdrafts of air to feed the inferno further. Fire whirls are tornado-like air vortexes composed of pure flame, and they can move like a jet force blast of flame, a searing gesture tracing carnage across the landscape.

This force of nature is akin to the Dragon, destroying with its withering breath and rising to the sky only to attack again and again until it consumes all of its fuel or the rains arrive to conquer it. This raging horror seems to be alive, acting with a singular purpose — the desire to consume everything in its path. There is no mechanism or tactic that that humans possess that can combat the dragon, only nature itself can stop it; by a change in the fuel base, the weather, or both.

The Pyrocumulus of the Waldo Canyon Fire, June 23,2012. Seen from 30 miles away in Divide, Colorado. Enter the Dragon -Photo by Jeff Ravage

Fire is a chemical reaction, several chemical reactions, to be precise. When heat is applied to a woody fuel, the energy causes the carbon/hydrogen bonds to separate in a process called pyrolysis, a destructive endothermic reaction.

Meaning it requires external heat. The jostling, loose atoms in the woody hydrocarbons can then combine with oxygen in the process of combustion. When hydrogen burns in the presence of oxygen, it creates water vapor (H_2O). When the carbon itself reacts with the O_2, it produces one of the carbon oxides, CO (carbon monoxide) or CO_2. These are exothermic reactions. They create heat, a lot of heat. This heat can then drive more pyrolysis, which births more combustion in a vicious and expanding chain reaction.

When combustion is clean, you will see little to no flame, and if you do, it's likely a blue color, like a natural gas (Methane, CH_4) stovetop. When the combustion is dirtier, you get flames in multiple colors. The often seen yellow/orange flame is created by the roasting of soot particles freed by an incomplete combustion. This glow is manifest in much the same way that photovoltaic panels create electricity. Only this reaction would be in reverse: the heat is the energy that makes an electron rise an orbital shell or two, and when they fall back down, they release a photon.

PV cells: photons in, energy out.

Flame: energy in, photons out.

Forest fires burn particularly dirty. There is moisture in the wood, itself already an imperfect fuel. The bark contains incombustible elements like silica and heavy metals, the trees' natural fireproofing. The needles contain nitrogen compounds, which, while not flammable, are still reactive in a conflagration. When the "shite" goes down, and a crown fire erupts, the amount of fuel available is enormous. As discussed earlier, the fire triangle comprises the three elements needed to sustain combustion: fuel, heat, and oxygen. The limiting leg of this chemical reaction, in a wildfire, will invariably become the availability of oxygen. The updraft of the rising heat creates a jet engine-like draw on the surrounding air. But there's so much heat and fuel that the local atmosphere's oxygen component can become exhausted before it even reaches mid-tree height. So, a zone of pyrolytic excitement is occurring, with no combustion possible yet. The atoms become frantic and collide to and fro, craving union but finding it denied. The column might collapse at this point to intercept more oxygen near the ground. Terrain that has not yet burned because the crown's flames have sprinted far in front of the ground fire. Or, the agitation and temperature gradients could continue to spin, searching for more components to burn and sucking the oxygen out of the soil itself. This process fuels fingers of fire that rake across the ground like fountain pens, drawing cryptic characters made of flame.

Or the whole thing could simply explode, a super-heated ball of fuel roaring upwards as it tears into the oxygenated atmosphere above. This reaction is the same phenomenon as the dreaded back-draft, where the super-heated gasses spontaneously explode when they finally find their oxygen source. Commonly, the updraft carries luminous fragments of bark and needles aloft where the explosion and wind drive them outward as ember storms. This rain of fire is one of the primary mechanisms for fire advancement, creating "spot" fires often far ahead of the advancing flame front. It is also the key process by which homes/structures ignite. Fancy log homes in the mountains are rarely designed to do much more than concentrate drifts of glowing embers into critical masses of ignition within their eves and cantilevers. Again, humanity lives our lives as if we were the only force of nature worth considering.

Ask a wildland firefighter about the things they've seen in a massive wildfire, and they'll tell you. Here is Hell incarnate. The pyrocumulous cloud is the laughter of the dragon as it disassembles our forest atom by atom. The fire's roar is the scream of the phoenix burned alive at the stake.

This type of fire is like a pottery kiln in intensity. It alters the mineral soil's chemical structure by fusing its surfaces. This melding eliminates the mineral particle's natural roughness that previously harbored bacteria and held moisture. As the trees burn, they emit organic gasses that aren't entirely consumed in the reduced oxygen atmosphere beneath the fire's main body. The blast of the furnace consumes the duff, in place,. Their freed organic compounds will then coalesce on that soil, condensing into a coat over the mineral grains, essentially becoming baked-on lacquer. These two phenomena combined result in extreme hydrophobicity. These hydrophobic soils shed water like a duck's back. They can no longer support plant or fungal communities because the ground now rejects the vital essence of life itself: water.

And then it gets worse. Since the soil can no longer hold water, when the rains come, and they will always come, the landscape has zero capacity to absorb it. The water rushes over the hillsides in torrents and creates devastating floods. The floods from the Waldo Canyon Fire in 2012 killed one person who was washed off of Highway 24 near Cascade, Colorado, and buried alive in the mudflow. He was just someone driving home from work when a cloudburst erased his future and condemned his last moments to unimaginable terror. It was not a scene from some horror film. It is the reality our pursuits have been busy building in the background, with the promise more terrors to come.

The denuded land left behind by a catastrophic fire has an altered albedo, as well. It is now highly reflective, compared to the forest that once stood there, and may still surround the open scar at some distance. This pattern causes storms to stall over the openings, increasing precipitation over the very places that now have the least capacity to absorb that moisture. As the drench falls upon the hydrophobic ground, it simply runs off, gaining velocity as it rushes downhill. It gathers ash and minerals and gains greater volume as it rolls. The hydraulic equation is simple: as water volume doubles, its force squares. And as the debris thickens this torrent, it adds heavier and heavier materials to itself. Until it births a freight train of mud, rocks, and even entire trees rushing downhill and crushing anything in its path. Flood damage to roads, homes, and infrastructure can be as bad or even worse than the fire itself in these scenarios. When there are no trees, plants, or mycelium to hold the soil together, combined with the soil's hydrophobicity, the sediment transport is catastrophic. The mixture that is created has the consistency and mass of liquid concrete. It will roll over anything in its path and flood the waterways below with choking sediment. These floods will invariably be flash floods and occur in minutes, giving very little time to prepare. This tragedy is not an occasional by-product of a catastrophic wildfire; it is a given.

Our best hope to reduce the dangers this reality presents is to implement mechanical forest mitigation and follow it with prescribed fire. Prescribed burning is the technique of applying fire to a landscape in reasonably good condition during times of low fire spread risk. This technique would mimic the practice of the original stewards of this land, except that they never let the situation get this far out of hand. Many tribes practiced this anthropogenic burning for forest health, ease of access, and even to improve grasslands and berry patches. Burning the land is a very cost-effective practice and is essential to the maintenance of a healthy forest. No artificial treatment can provide all the mechanical, chemical, and biological actions that fire can. Unfortunately, very few of our western forests are currently in a condition where this may be practiced safely. Without the pre-thinning of the forest beforehand, such practice is akin to drenching the land in kerosene and striking a match. The accumulated fuel being an explosive burden upon the land.

This can has been kicked so far down this road; there's no road left, just overgrowth and an abundance of dry fuel.

Prescribed fire is a controversial topic in the West. First, there are generations who have been told by Smokey the Bear that all fire is bad. Second, there is an accumulated risk factor that many in the public sector find uncomforting. And then there are escapes.

In 2013, a prescribed fire was performed in the mountains west of Denver, Colorado. This planned activity went as designed, and teams were in place to monitor any residual fire activity. It seemed completely safe. They set the fire in late March of that year, and there was up to 16 cm (6 in.) of snow on the ground. No post-burn activity had been seen for a couple of days. Then, extremely high winds kicked up on Saturday, March 24, disturbing embers hiding in the thick duff layers beneath the snow. These embers landed on dry fuel above the snow's surface and started the Lower North Fork Fire. By the time it finished feeding on this area, over 1,600 ha (4,000 ac.) of forestland had burned, 24 homes were lost, and three people had died. One of them was the wife of a colleague of mine. So I'm not too pleased by this outcome, either. The other two fatalities were an elderly couple. They died, the wife in the driveway and the husband just inside their front door. All because he had gone back inside for one more thing, one more second, when the dragon arrived. This fire resulted from a freak series of events: the wind, the presence of snow combined with a very low relative humidity, and the persistence of a buried pocket of embers. But it was also the result of a terrible human error. Crews with the Colorado State Forest Service had received a smoke report in the area on Friday, the 23rd, and, after a search of the locality, they stood down. They did not find the source of the smoke, nor did they call this report in to the Federal Fire Managers, as was protocol. The local Sheriff's dispatch was not alerted either. One forest manager made the call not to call, and so, this story has gone down in the logbook of infamy. Had anyone communicated this seemingly trivial bit of information, the result could very well have been the same. But it might have been very different. This mistake led to an entirely new set of protocols for the intentional application of fire to the land, as with every escaped fire. But even as the anger and hurt endure, we need to look to nature and hear what she is telling us. The forest has to burn. And the fact remains that prescribed fire is our best method of forest management, period. Because this is the system of nature herself, and the fire is an inviolate expression of the western US forest. The forest has to burn. If we think we still want forests.

In Montana, the Salish and Kootenai tribes have been practicing prescribed fire techniques, in many ways handed down by their ancestors, to great success. As a sovereign entity, they have taken the lead in the management of their own lands. And now, their fire crews spend a good portion of their time lighting fires in areas pre-thinned in preparation. Their Reservation Fire Management Plan (Steele, 2007) is an impressive and comprehensive document that could be applied widely. Since they have both a technical and cultural connection to their lands, their insights should be taken with extreme seriousness. No amount of training or experience can replace these practitioners' esteem for the land they manage. But when all of those characteristics are combined, as in this team of managers, the outcomes are just short of astounding.

I visited them recently while at a fire conference. An exciting thing that the reservation ecologist and FMO[138] showed us was how the application of fire spurred a regeneration of the Arrowleaf Balsamroot (*Balsamorhiza sagittata*), a historically significant plant for them. This plant had become scarce to non-existent for over half a century. But they knew from elder knowledge that it should be plentiful and was perhaps acting as an indicator of overall ecosystem health. After the action of fire was returned, the yellow blossoms of Arrowleaf Balsamroot filed the meadows once more. And it rebounded in quantities not seen in the memory of even their most senior of members. While this may seem trivial to some, it is information of the utmost importance if we want not just to restore but renew this world we have treated so poorly. Here is a demonstration of fire management's benefits in alignment with natural forces and performed with cultural gravity.

---

[138] Fire Management Officer

If humanity wants to keep our forests, they must be let to burn, for they are the Phoenix and their feathers are now quite ragged. People need to decide if we have the collective will to support the costs involved in the mechanical treatments necessary to make the forests safe for fire's return. While also understanding that collateral damage can occur. Many have moved into the wildlands out of a desire to live with nature. What many fail to understand is the true nature of the forest. Some don't even care, bringing their misunderstandings of safety and attitudes supremacy with them. The ecosystem's realities do not square with concepts like absolute ownership and Dominion, ideas that are hallmarks of our society's current identity. When people choose to live in a frequent fire forest, fire adaptation is their responsibility. The land can't adapt to us, it is us who must change. Only one thing is inevitable: if we do not choose to adapt and manage, the forest will continue to burn at severe levels. And more homes, lives, and our natural heritage will be lost, perhaps for good. It is estimated that in America, society is on a steady trajectory to lose over 70% of our US western forests (and up to 100% of our southwestern forests) by 2100. (Madowell, 2016)

While Land managers have known for some time that the increasingly devastating nature of fire and global warming are linked, firm evidence has been difficult to find. A recent study in California has directly connected the steady increase in summer temperatures since 1972 to the 8-fold increase in summer wildfire numbers (and their severity). (Williams, 2019) Scientists are reluctant to assert causal connections, even when their common sense screams at them, without a factual basis backed by data and statistical significance. This caution is the restrained attitude of professionals that propagandists, and some politicians have used to sow doubt amongst a public that doesn't easily see the difference between careful analysis and baseless opinions. Let there be no doubt that the human race has altered the climate, and it is steadily warming. And there is also no question that the cause is us: educated, technological societies. Societies that have the philosophical will and economic power to both disrupt the planet and deny our role in its decline.

Once a catastrophic fire has passed, the landscape is a black purgatory of matchstick tree skeletons. There is no soil left for plants to grow in. And even if there were, there would be few nutrients left in it to feed them. The intense heat having disassembled the nitrogenous compounds that formed the base of their nutrient pyramid. The fire expelled that nitrogen into the atmosphere, where it will remain out of reach, biologically speaking. Even if some tiny handfuls of soil with a bit of nutrients in them survived, the plants can't thrive without their mycorrhizal fungi to help their tiny roots absorb them. And even if there were some spores that made it out of the oven, they will probably all get washed away when the next rain arrives.

Fires naturally burn in a mosaic pattern, where groups of trees and patches of ground remain intact within the burn footprint. In a catastrophic fire, the mosaic is a pattern of either green or lifeless black. These green oases are the only seed sources remaining from which a barren landscape can restore itself. And in reality, that is often enough for natural regeneration. But it is not quite sufficient for the needs of humankind. It can take 80-100 years for sterilized soil to return to fertility. Then it can take 40-50 years for a tree to mature within that soil. The cones from these trees can only travel as far as the wind, gravity, or squirrels can carry them. (Bowd, 2019) These reasons explain why the forest can take centuries, even millennia, to return under the "humans do nothing" scenario. And that works for a billion-year-old living system, but for us? We need to fix things as fast as possible.

Post-fire restoration work, if funded, will often begin with the application of hay or woodchips mixed with grass seed dropped from helicopters. Hand crews, often with hardy volunteers, will fall dead trees and install them perpendicular to the slope to create log erosion barriers (LEBs) near critical infrastructures, such as roads. They might also install jute matting to keep switchbacks from washing out. This material will give seeded grasses a little purchase on the slope and a bit of organic matter from which to feed. Crews might apply similar hill slope treatments to sub-drainages with people living below them hoping to save their homes. All of this is expensive, and time-consuming, and not uniformly effective. The forest is fully evolved to self-heal after a fire, but not a catastrophic fire. Such occurrences in the wild can easily result in what is called landscape redirection. A redirected landscape can become grasslands or lead to desertification. The one thing it won't become again is a forest. We have teased the dragon, then unleashed it, and its wrath is terrible indeed.

Working with ecology to heal the damage done, many have been looking at the fungi to help speed regeneration. I began experimenting in 2014 using wild-collected native mushrooms to break down woodchips more rapidly than usual. Using training techniques, my crew and I adapted fungal strains to the specific medium we wished to decompose. What we found was amazing. Not only would the mushrooms consume raw wood, but the end product was a rich compost-like material that compared favorably to natural forest duff. This product was not some sterile gardening amendment; it contained an entire zoo of fungi, insects, and bacteria by the time a few years had passed. It is richer in nitrogen compounds than natural duff, but not overly so. (Ravage, 2023) It contained pretty much everything that was missing after the Dragon had had its way. And all that would be needed are those burned matchstick trees, a wood chipper, and some bags of mushroom spawn. It works, it would seem, because this is what nature would do, just sped up by our husbandry and focused by our technology. This process is still an experimental technique in a growing field of novel ecological restoration treatments: treatments that use ecology as technology.

As the climate warms and the effects of humanity's mistreatment of nature advances, there is another growing wildfire danger. Our grasslands are burning at high severity as well. In July 2018, the Martin Fire outside Elko, Nevada, consumed 176,000 ha (435,569 ac.) of rangeland. This 14 m (40 ft) wall of flames tore through grass and shrub that was only 1m (3.2 ft) tall. Driven by hot, high winds, this human ignited fire burned so intensely that it erased all the sage and native grasses, as well as many a grouse. The accelerant, in this case, was the bane of western ranchers: cheatgrass. Cheatgrass, Drooping Brome (*Bromus tectorum*), is an invasive species from Eurasia. This grass came to the continent in the 1800s in shipments of cattle feed. They made these shipments because the cattle didn't feed efficiently on our native grasses, since they were also an invasive species from Eurasia. The overgrazing of cows that is common on open western range lands gave the drooping Brome a significant advantage. It is tough and proliferates rapidly, filling in any void left by dead and dying natives. Cheatgrass also has a different life-cycle schedule than our native bromes and matures and dies back earlier, in June or early July. The timing of their cycle leaves hectares of dry brown tufts hanging in the wind just as fire season gets cranking. When it catches a spark, the land burns hot and fast. The severity of these cheatgrass fires threatens to redirect these already dry and fragile lands to deserts.

USDA soil scientists have been working on this problem for years and may have found a natural solution. Native soil bacteria can interrupt the brome's root hairs at a critical point in development. (Kennedy, 2002) This research into biological control took nearly 20 years to complete and is still being investigated by the EPA for safety and efficacy. Such work can seem to move at a crawl in a world that favors herbicides and quarterly profits over these careful studies. But this type of science presents the best chance of solving many of our problems instead of making them worse. Too many shortcuts got us into this predicament. We need steady, thoughtful action to get us out of it. This imbalance means we need to support funding for this type of work proactively.

Uncharacteristically devastating wildfire across all fuel types is becoming characteristic, a new normal, so to speak. The fires in Australia and New South Wales in 2019-2020 represent a template for what might soon become normal. Fires began appearing in June, which would've been expected if they had quickly burned out. Instead, however, these blazes continued for eight straight months, devastating fields, forests, cities, and rural outposts. The causes ranged from the natural, like lightning, to the pathetic.

As conditions worsened and they declared fire danger levels that had never before been seen, some "stable geniuses" thought it would be fun to experiment with some fire. No doubt, these individuals thought it was just a lark, another prank that was ultimately safe, because this world was made for them and their entertainment. "No harm, no foul, mate!" Except there was harm. These fires effectively shut down the country for months.

The Australian Prime Minister, Scott Morrison, a self-identified "religious conservative" and climate change denier, failed miserably in his initial response to this emergency. Many in the population viewed his actions with exasperation. He refused to acknowledge the role of fossil fuels and climate change in the increasingly frequent fires. Instead, he continued pushing a pro-coal agenda. He then went on vacation to Hawaii while his country burned and his populace died. Soon, anti-Morrison protests erupted along with the fires. Thousands of people took to the streets decrying his pro-carbon fuel policies and what many saw as cheating on his already weak pollution reduction commitments. He eventually apologized publicly for his shortcomings in addressing this catastrophe that had few precedents. COVID-19 tested him again, and it seems, wary of repeating the same mistakes, he finally got behind the science. A wise choice that many in the United States still refuse to make. These events became known collectively as "The Black Summer." The fires touched nearly every state in the country, the worst being New South Wales. 18.6 million hectares (46 million acres.; 186,000 km²; 72,000 mi ²) of that landscape charred, almost 6,000 homes and structures were destroyed, and over 30 people perished. Before the fires were done, they would pump 250 million tonnes (280 million tons) of carbon dioxide into the atmosphere. (Redfearn, 2019)

The fires were so devastating that some species, including the world-famous koalas, now face potential extinction. The smoke cloud crossed the Pacific Ocean and visited Chile and Argentina. These fires raged until rains arrived in February 2020. What would've happened had the rains not come is less a guess and more of a dire contemplation that all people who wish not to perish along with their countryside must make.

Part of the problem with these Australian fires stems from an invasive plant: Gorse. Gorse originally hails from Europe and the UK. This giant invader grows up to 3m (10 ft) in height and is covered by sharp spines. Its tissues have a high oil content that can rapidly accelerate fires and quickly spread ground flames to forest timber with their tall stature. Gorse will grow in dense thickets, essentially draping over the eucalyptus and other native trees. The intense crown fires encouraged by this plant also favor self-regeneration post-fire, when the barren soil gives them a head start on re-sprouting than the slower natives.

In southern Djaara county, Australia, a group has been working on combating the increased fire danger and distorted environments with a novel human-animal joint venture. The Goathand Co-operative seeks to "transition weedy, fire-prone forests into biodiverse grassy woodlands." Blackberries, as well as Gorse, plague their forests. These blackberries grow almost as high as the Gorse and are just as dense and spiny. When they overrun the forest floor, they displace not only the native plants but also the native animals such as wallabies and kangaroos. So they began what they characterized as "guerilla forestry," using goats in concert with hand removal to take on their weed infestation.

They initially remove the blackberry and Gorse by hand to remove the years of accumulated brown canes and shrubbery. Once the brown is knocked down to a manageable level, the goats go after the green shoots. Goats love prickly food. Spines, nettles, and thorns are all just flavor enhancers from their point of view. So they go after the stems, leaves, and flowers with equal gusto. Patrick Jones of the Co-operative narrates how such a human/goat collaboration proceeds in several promotional videos. He estimates that the initial clearing of highly infested lands can take up to a month to clear half a hectare (~1.2 ac). The following year's entry, however, will only require two weeks of goatly attention. They can clear the land next year in a week. Once the weeds are reduced and the ground well fertilized by the four-legged foresters, the native plants will regain their control of the land, and the forest will return to its natural trajectory. With proper husbandry, even our stock animals, such as sheep and goats, can help restore the land instead of overgrazing and destroying it, as has been our pattern for millennia.

Fires have moved across this land for millennia as well. It removed the debris and recycled minerals and nutrients back into the soil. It acted as a selective force, choosing which trees survived and which would perish. This force is needed to maintain the health and structure of the forest. A fire should return to any spot every 15 to 200 years in the American West depending on the elevation and aspect (north-facing, south-facing, etc.) and the predominant tree species composing any forest stand. If it occurs at the natural frequency, this fire will be of low-to-moderate intensity, minimally destructive, and promote the rapid regeneration of all life. Thus, I have portrayed the dragon as the predator of the forest. And like all predators, it is essential to maintain the balance and the health of the herd that is the western woodlands.

In ancient Chinese mythology, the dragon and the phoenix may be joined as lovers, partners which symbolize the Yin-Yang of nature. When balanced, they are the most formidable of unions, but when unbalanced, they become apocalyptic destruction and chaos. What an apt metaphor for the burning of our forests. The Yang energy of the Dragon is ambitious and outwardly expressing. Its chaotic form is the uncontrollable inferno brought upon the land by our actions. The Phoenix holds the Yin energy of wisdom and caring; she should nurture the forest back to life. Instead, the flames burn too hot, and the forest dies. Her husband has become abusive, and she is dying from the battering. We may lose 70% of our western US forests by the turn of the next century, killed by fire, disease, or insects. What will be left is grassland that cannot do the critical work of sequestering carbon from the atmosphere: that, or just more deserts.

Americans have loved our western forests to death. We have halted their natural cycle because it was inconvenient and somehow offensive to our human-centric aesthetic. The primary source of forest disturbance in the western US is fire. While earthquakes and landslides once were the secondary destructive forces, humanity's actions have replaced them. What began as a concerted effort to stop fires in the early 20th century has backfired on us. It may be helpful to think of these catastrophic fires as the forest's pent-up need to burn. The changing climate and the hotter, drier conditions that warming encourages simply speed up this destruction.

Westerners saw no wisdom in how the forest managed itself, and while they could've asked the remaining indigenous population how the forest worked, they chose not to. Again, racial misunderstandings and biases benefit no one. And they listened to no one and nothing but their overblown sense of self-importance. Commerce coveted the forests and wanted the resources they contained, so industry halted their natural trajectories of constant change. Change is the essence of the universe — just ask Heraclitus. In our present world, stopping the fires has only made for worse conflagrations. The dragon will not be denied, but he can be tamed. What we should've done and must now do is to adapt to fire. If people wish to live within the forest, it is they who must accommodate. Fire adaptation is the only way to co-exist within a forest, and it leads us to a humbling realization: it is us who must change, not nature.

It is humanity who now must evolve. Nature is already adapted.

CHAPTER TEN: THE FOREST

"The tree which moves some to tears of joy is in the eyes of others only a green thing that stands in the way."
— **William Blake**

In the late 1960s, lumber companies were interested in creating conifer plantations on the island of Puerto Rico. The conditions, both ecological and economic, seemed favorable. Land was cheap, labor was "a steal," and transportation by the sea offered quick deployment to many markets. The growing season was virtually year round, and other than the occasional hurricane, the situation seemed ideal. But when they planted conifers there, the trees topped out at around a meter (3.3 ft) in height, stalling their growth at an unmarketable size. This sub-Christmas tree height was useless from a marketing perspective and threatened their investments. So they hired scientists to sort out this disappointing result. Experiments were run to determine why their mighty trees remained stunted. It wasn't the soil's nutrients, which were, if anything, richer than their evergreen species required. There was plenty of sunshine and a pleasant climate that was neither too hot nor too cold. Everyone involved spent a great deal of time scratching their heads.

It wasn't until one scientist smuggled some soil with mycorrhizal fungi into the country, hidden inside a shoe, and spread that fungus to the ground that the saplings could thrive and reach their intended height. (Hacskaylo, 1967) (Vozzo, 1971) It turned out the island of Puerto Rico had had no suitable native mycorrhizas' prior to the shoe incident. The natural history of the trees had always involved a close association with mycorrhizal mushrooms. But no one had previously guessed how vital the two life-forms were to each other. This curious discovery allowed us a peek into just how elemental the fungi are to our forests and their health. It would take decades of continued research to understand just how inextricably intertwined the forest and its fungus are. So now when mycologists say that the fungi are foundational to our forests, that is precisely what they mean.

We all need to remember this tale and not forget what it's telling us about systems and the interconnection of life. But, unfortunately, too many still focus on singularities, such as a tree, measured in board feet. Or a crop calculated in bushels per acre and kilos of glyphosate per hectare. When the forest should be more accurately understood as a system, like a natural municipality, measured in biodiversity and interrelationships.

The forests were once very different affairs than our present experience of them. When European settlers first encountered the American continent, it was covered with about 674 million hectares (1.66 billion acres) of forested land. Land clearing for farming, mining, and settlement has reduced that coverage by nearly 30 percent over the centuries. Our mobile culture and desire to reorganize the landscape for our benefit have spread disease, invasive weeds, and non-native tree species, transfiguring the native woodlands into what one now encounters.

When foresters count trees for inventory and data collection, they routinely find 500- 750 trees per hectare (200-300 trees/acre). Pre-settlement estimates of tree stocking for these same areas were between 100- 300 trees per hectare (40-120 trees/acre). Sometimes, they now discover stocking rates of over 2400 trees/hectare (1000 trees/acre). With that many stems so cramped together, they can't grow very large, and their health, of course, suffers. They may become "lolly-pop" trees, with skinny trunks ten's of meters high and adorned only with tiny tufts of branches at the top. This configuration seals the canopy together in a continuous shield only a couple of meters thick and far above the forest floor. Often, little sunlight reaches the soil in this scenario, so there are no grasses, no flowers, and very little leaf litter for the squirrels to roll around in or forage through. This agitation of the forest floor by foragers is essential for a healthy forested system.

Today, one can get historical data about pre-settlement forests from old photographs, paintings, and stumps, which can persist at higher elevations for a century or longer. When our forebears began their exploitation of the forest, no one counted them the way foresters do today. They just cut down every tree they wanted. Then they moved on to the next one, caring little about the damage inflicted upon collaterals, trees they considered trash (the corporate term is unmerchantable). It was only after commerce had decimated the woodlands the first time that forest managers began exerting greater care in measurement and stock counts. Leaving the poorer specimens also had the unintended effect of selecting against the strongest and healthiest trees; it diminished the gene pool. It turns out that trashing the forests hobbled them into the future. Who would've guessed?

The ancient forest grew in intricate mosaics across the landscape. Many areas once exhibited an open canopy, which means the boughs of one tree only occasionally touched those of its neighbor. Temperate Savanna species, such as the ponderosa pine, lived on the dry southern slopes in parkland-like arrangements. They provided shade and shelter to the animals and grassy fields. The trees in the deep draws and northern slopes grew dense, and their canopy closure was tight. You didn't need to look for moss to know which direction was north; you only had to look at the trees and how they interacted with the understory for that revelation. This was a relief because calculating direction based simply on moss growth is a recipe for getting lost in many forest types.

The original people of this land now called America didn't get lost in their wild. They knew not only what was over this ridge, but the one after that, the next one, and so on. Perhaps they didn't know about plate tectonics and how it uplifted the land, but they knew how to navigate those folds and valleys and where they'd likely lead. They knew where the water flowed and how to follow its path, even from a distance. They knew this from experience and passed that information on through direct education and within their tales and songs. The natural variations of the land did not surprise them. What astonished them was the strange, unnatural conformity the Europeans brought with them.

The new immigrants looked at the giant forests as an intimidating and irredeemable evil. They carried this prejudice with them from centuries earlier when their ancestors waged war against Europe's own great forests. Upon arrival at Plymouth Rock, the Puritans would call the near-impenetrable curtain of woods the "Devil's last preserve" and seek its conquest as primary to the country's subjugation. Their weapons were the axe and saw the plow and horses; one of the first invasive species imported to the Americas.

But the Natives saw a different type of sin in the Immigrant's straight-line roads and their supposed divine spaces made of enclosed beams and shuttered windows. Beams ripped from the original sacred temples of the timber groves. The home preferred by the Goshawk (*Accipiter gentilis*) was such a space where towering spruces or massive Douglas fir would arch together, forming natural vaulted cathedrals. Chapels of nature, where light shafts pierced the dense undergrowth, alive with flower, fern, and fawn.

The settlers were correct in identifying the woodlands as chaotic, for it was chaos that drove the original order. These people, whose folklores overflowed with monsters and witches, all of whom could be traced back to the unholy forces of nature, could not conceive of the forest's true essence. The seemingly random interactions of light, life, nutrients, and fire acted almost like the Hindu goddess Shiva, destroying to make fertile; terrible, and relentless in her pursuit of renewal. And, then, nurturing and sacrificing for rebirth as well. It must have seemed very confusing for those who were on the path of losing their primal connection to Gaia's creation. Here, every life expressed with vigor and struggled to fill every niche, turning over every leaf, again and again. A forest formed by this rampant fecundity and subject to these selective pressures is stable. It finds its way and interacts with all of its residents in a healthy and sustainable cycle.

The paradox of life is that it creates its order through interactions with chaos. But this is not the perfected order we cherish in our imaginations. It is an order that allows all to prosper and perpetuate. The order of nature. The order that we have now almost lost, or perhaps surrendered.

When one surveys the forest now, tree species cohabit in uncharacteristic patterns. You can no longer be sure what elevation you are at just by identifying the foliage. Our activities have led to new stand configurations and randomly intermixed canopies. Commerce has created unique, hybrid, and engineered trees with no relationship to any other part of the forest. Thoughtless harvest has acted to select not for the fittest trees but for the weak and overlooked, those left behind by our enterprise's built-in gluttony. Dead and downed litter builds up faster than it can decay, making deep forests challenging to walk through and dangerous in a fire. This buildup does not contribute nutrients to the land because the decay rates are low, so scattered logs and limbs accumulate. These are nutrients locked away from life, simply awaiting the dragon's breath.

The pinnacle, climax form of the woodlands is the old growth forest (OGF). These stands are the direct connection of the forest to its past, perhaps its pre-historic past. It is quite difficult to define what is and isn't old growth these days. Part of the reason is that they were logged to near extinction for their enormous trunks and exotic flora. Seemingly, only areas of extreme inaccessibility survived the loggers' axe. Now, when we find them, they may be fragmented only a few hectares in size. Often, the only places they remain are on public lands, under the protection of the Forest Service or our National Park system. And the urbanization of these parks can lead to the slow decline of what was once an untouched forest structure. Their dense growth patterns can also make them easy prey for fire in our drying climate. And because some foresters don't believe they actually exist, we face another barrier when trying to define or improve them.

Old-Growth Forests are stands with a significant percentage of old, large trees, lacking clearly visible evidence of human activity. Often, they have never been logged. Their complex stand structure identifies them. There will be many snags and heavy ground loading. All stages of decay will be present and abundant. OGFs exhibit comparatively high biological diversity; with few invasive species and often rare native plant and fungal types and communities. Precursor stands diverse in tree size and age classes are required for their development. The OGF needs this pre-historic type of forest to emerge from. The type of stands that don't exist on lands managed by timber companies or their partners. Depending on dominant species, multi-layered canopies and natural gaps occur as multiple generations of trees die and create openings in the canopy. The openings of light then foster the undergrowth and younger trees beneath. These actions support many micro-biomes and multiple aged regeneration. The most phenomenal thing about them is they might remain at what is considered a terminal growth stage, stably, for hundreds of years.

So the Old Growth forest is the final stage of the forest. Or perhaps not since there seems to be a steady trajectory that they can find allowing them to survive like this for centuries. Like the established orbits where satellites can be continuously falling, while staying in place, almost in perpetuity. These old, large[139] trees can become an ecosystem of their own, supporting entire communities not present in the younger woodland stands. An example is the Agarikon mushroom (*Fomitopsis officinalis*). This massive conk grows on old conifers, Douglas fir preferentially. It needs to find purchase in a tree that is a century or older and then it can take another century for its fruiting body to mature. Such a life cycle is currently improbable in a human managed forest. And the discovery of antiviral chemicals in them has led to a race to find these rare mushrooms and take them for "naturopathic medicine" formulation. Once again, even those who may claim to hold dear the value of nature will seize its bounties for the slightest perception that it might benefit them. And as with rhino horn, there often doesn't need to be any solid basis upon which to make this decision to raid mother nature's cupboard. Perhaps the most important thing to remember about the old growth is that they contain the most complete genetic repository for the forest. So, if for no other reason humanity should endeavor to define and protect them as we would any ark of genetic diversity.

The forest's function, like that of any natural system, is multifaceted. Forests, like the oceans, act as the lungs of the planet. The massive expanses of vegetation absorb carbon dioxide from the atmosphere and exhale oxygen. They rip the carbon atom from the CO_2 molecule as a building block for their proteins, fibers, and sugars. They then release the O_2 molecule back into the atmosphere, where our metabolisms will do just the opposite. The energy for this process comes from sunlight captured by the photosynthetic process. Great tree trunks hold stored carbon that they have filtered over decades, perhaps centuries, from the atmosphere, as does every blade of grass and every columbine. The forest is a buffer; life keeps excess carbon tucked safely inside its tissues, stabilizing the climate for the good of all. Again, life holds its treasures within.

As trees mature and increase in size, the amount of carbon they can remove from the atmosphere increases as well. Many used to think that the younger

---

[139] In forestry we talk about old trees, large trees, and old,large trees. The difference is important.

trees, growing quickly, laid on carbon faster than the elder trees. That has been found to be an error. A 100 cm (39 in) diameter tree can accumulate the carbon present in an entire 10 cm (4in) tree every year. (Moomaw, 2019) Whereas it might take 1000 of those 10 cm trees to hold the carbon store of one grand daddy. As these champion trees age, they continue to increase in both size and vigor until they reach their end, which often comes quickly. The idea that they slow down in their golden years may come from a projection of how people age. And not how they naturally age, but how people age as residents of homes with central heating who own cars that carry us hither and yon, instead of using our legs and lungs. For aging need not require decline; it's often our lifestyles that manifest this waning of fitness. All the energy and vitality of life is within us. We are also a safe carbon store, as long as we stay healthy and alive.

Trees are amazing plants, monsters of the floral kingdom. They can tower up to 90 m (300 ft) and are home to thousands of species of insects, birds, and small mammals. A study in 2009 in the Bavarian National Forest Park gassed a 600-year-old tree with an insecticide to capture all the tiny creatures, the forest plankton, living within it and upon it. When the tarp was removed and the creatures counted, 2,041 individual animals were counted from 257 species. (Gossner, 2009) A brutal but illuminating study. Rest assured, new tenants began moving back in even before their truck had driven away. In the folds of this old, large tree, there were no doubt areas protected from the poisons. Perhaps beneath the bark, definitely entwined with the roots, because our tree is a community. And life is not playing games.

Our tree has three layers: the bark, the vascular layer, and the heartwood. The bark is the outer skin and may be thick or thin depending on the species, but it always acts as a protective layer for the living tissue beneath. It is also a shelter for those tiny forest plankton creatures. The vascular layer is the tree's living tissue and contains the xylem and phloem, which act similarly to our veins and arteries. One brings water up from the roots (xylem), and the other distributes the sugars and nutrients generated by the leaves (phloem). The heartwood is primarily composed of cellulose ($C_6H_{10}O_5$), hemicellulose, and lignin. The fibrous strands of cellulose are joined together by the crystalline lignin to form a robust and flexible composite that is responsible for the tree's strength, and therefore its height and majesty. The store of waste products called metabolites darken the central core of the heartwood. Metabolites are also sent outward to the bark, where they might help repel invaders who don't just want to live on the tree but wish, instead, to eat it. The excreta may also act as insulation or fireproofing, as mentioned before. Everything eats, and everything excretes, everything living anyway.

The interwoven matrix of these giants' roots, and their mycorrhizal companions, hold the fragile soils together and keep sediment from flowing downhill, anchoring the hillside in times of disturbance. This sediment, unfettered, could clog waterways and kill the aquatic life within. Held fast by this floral embrace, the land instead becomes a natural filter that regulates and fortifies the water that falls upon it. Herein lies the ultimate meaning of a watershed. It is the land upon which the water falls, moves over, and through. Healthy forested land is therefore essential for a clean, pristine water supply. Precisely the opposite of how, presently, agricultural land poisons and disrupts the water's flow.

The roots and their mycelial companions also store carbon, often nearly as much as you can find in the life above the ground. This system is as beautiful as it is vital. We react to its allure instinctively, for this is the anvil upon which our primal beings were forged. That connection is why we are drawn to it. But we need to act towards it now with care, having lost a direct linkage to it with our civilized ways. Humans can restore this connection, but only once we have corrected the balance within ourselves and are ready to restore the balance without as well.

The forest canopy shelters a vibrant, diverse world of plants and animals, each with a place and a function. Chipmunks and moles and worms and mushrooms all churn the soil, distributing nutrients and minerals to each other's benefit. My favorite forest pals, the Abert's squirrel (*Sciurus aberti*), form a triumvirate with the Ponderosa pine and the mushrooms that grow at their roots. When fall arrives, these tassel-eared rodents alter their diet to consume the fruiting mushrooms primarily, spreading the spores in their scat and ensuring the health of the next generation of ponderosa pines. The pines shelter and feed the squirrels with their cones. The mushrooms, in their turn, trade nutrients and carbon with the roots. Mycorrhizal mycelium protects the tree's root hairs from desiccation during periods of dry weather. That's why I like to say: "The tree is the squirrel, is the mushroom, is the tree." You don't need to understand every complex interaction of these interrelated lives completely. You just need to know that the function of their little system is as straightforward as this simple phrase.

Predators similarly interact with and regulate the populations of the herbivorous animals, maintaining balance. Humans are predators, which is part of our rightful place, as long as it remains balanced. Life is a balancing act, and it looks more like a pyramid of skilled acrobats than a scale with a product on one side and a coin on the other.

The forest is a city in its own way, and by that measure, the rainforest is a metropolis. The rainforest is a unique and essential component of our global ecosystem. These incredibly dense and diverse natural treasures exist in a very counter-intuitive manner. The soil of the rainforest is terribly shallow, much like the duff of our montane forests. It is also relatively poor in terms of nutrient content. This forest system is so complex and supercharged that nearly all available nutrients are involved with one life process or another; one animal, plant, or fungus. When something dies, it is scavenged or rots quickly, quite opposite the northern montane ecosystem. This process is recycling in overdrive. Plants grow on plants that grow on massive trees with shallow root systems that gain their fastness through sheer number. Superhighways of ants race back and forth, scouring the floor for the latest leaf scrap, the newest fatality. Kinkajous speed mid-canopy on their own elevated pathways, raining fertilizer on the plants below as they barely pause to pee. Monkeys and macaws scream and call, filling the air with an often deafening sound. The entire structure gains its physical strength through the sheer mass of the growth and the relentless speed of its nutrient overturn. This forest grows in warm and moist environments, so many of the limits experienced in our temperate woodlands just aren't there. Fire is not common, nor is it needed for the cycling of debris and nutrients. That all happens by virtue of a light speed decay cycle. The moisture content of the soils, leaves, and wood is high and therefore resistant to ignition. Their fire is cold fire—the metabolisms of all residents converting death back into life through digestion, both internal and external. The rainforest is always in a constant state of flux. Again, this is called disturbance. Where disturbance is high, but not devastating, life form diversity also tends to be rich.

The Amazon rainforest is a prime example, although rainforests exist worldwide around the Equator, between the tropics of Cancer and Capricorn. The size of the forested Amazon basin is 7.5 million Km^2 (2.9 million mi^2), over three-fourths the size of the entire United States of America. It is so dense you can't walk through much of it without a machete. Rivers and streams move through it like veins in a giant circulatory system. Many of them are undetectable from satellite images because of the heavily interwoven forest crown. This crown density promotes lifestyles that move through an elevated highway system, never needing or wanting to touch the ground. This system creates over 20% of the world's oxygen. Half of Gaia's known species live there. The rain forest is the perfect template for the system called permaculture. It is natural agriculture on a fierce scale and productive as it gets!

So how would you think humans would treat these amazing wonders of the world?

Take a guess.

Deforestation threatens the Amazon rainforest ecosystem as over 1% is clear-cut logged, burned, and converted into cropland or range for cattle every year. The problem here is that the soil is so shallow and nutrient-poor that a cleared area can only produce commercial crops for a few years before becoming depleted of all nutrients. Eighty percent of deforested land will never return to a forest ecosystem. Most will become scrubland, and some will convert to a desert in, again, ecosystem re-direction. At one percent a year, and with a 20% regeneration rate, it could all be gone in just over a century. If people let that happen. And that would probably mark the end of humanity as well.

Did you have a burger from a fast-food chain restaurant lately? If so, the chances are high that some of that cheap beef came to you via this destructive practice. Since every hamburger from many fast-food restaurants contains beef from hundreds of animals, it's a pretty sure bet you just ate some slash and burn goodness. Yum!

In 2019, the world was aghast as this slash and burn mentality went into warp overdrive. Jair Bolsonaro, the newly elected "conservative" president of Brazil, created a favorable climate for an acceleration in forest harvest. Seventy-six thousand individual fires were set, some legally, but many not.

This activity represented an 80% increase from the previous years' orgy of arson. Worse yet, they did not harvest many of the trees that were felled for use as lumber. They were just left to dry and then, if needed, burned again. (Borunda, 2019) This waste of resources is just a pointless release of greenhouse gasses, and a squander of wood that would be highly valuable if it were only located somewhere else. Such methods are cheaper when your ultimate goal is to use the cleared land for cattle production or to grow your GMO soybeans. The world's hunger for meat protein is growing, and this is what it looks like: the death of an ecosystem for a cheap hamburger with fries. So, when someone asks you: "Where's the Beef?" you now know the answer.

The practice of deforestation accounts for over 10% of annual global greenhouse gas emissions (UCS, 2015). Humankind can't keep throwing away the rainforests when they are one of the primary mechanisms of extracting carbon from the atmosphere and the producer of a significant amount of the oxygen we breathe. This ecosystem is the 2nd greatest carbon sponge on the planet[140] and one whose absorption of CO_2 is not potentially damaging. Its preservation is vital if we wish to increase carbon sequestration to clear our atmosphere of excess greenhouse gasses.

And then there's palm oil. Almost a third of all vegetable oil produced in our world currently is palm oil. The Oil Palm (*Elaeis guineensis*) is very productive, 5 to 8 times more productive per planted hectare than the next most productive oil crop (Singh R. , 2013). Most of its farming occurs in the Indonesian and Malaysian rainforests, but Brazil is also increasing its production of this desirable cash crop. Oil palm is a slash and burn crop, and it's in your food, cosmetics, and toiletries. It's in your Girl Scout cookies, for Pete's sake! The deforestation associated with palm oil is nothing less than a global catastrophe. It threatens the orangutan, the pygmy elephant, and a sizeable list of other species that lack their own marketing departments. And while we're speaking of Girl Scout cookies, in 2014 Rhiannon Tomtishen and Madison Vorva, two actual Girl Scouts, began a campaign to stop using unsustainable palm oil in the cookies they were asked to sell. Eventually, their

---

[140] Number 1 is the Oceans, and too much carbon dioxide absorbed into them leads to acidification and death, as we have seen.

letter-writing campaign led Kellogg's® to relent and begin phasing out slash and burn sources for their palm oils. This result is a positive outcome and demonstrates what the public can achieve if they put their minds to it. If two young women can do it, so can the rest of us. The dark side of this is that you will often be disappointed when you rely on corporate culture to do what they say they will, or to do it responsibly. Girl Scout cookies are still yet to reach 100% sustainability in their palm oil content, eight years and counting. Wilmar International, the globe's largest distributor of palm oil, also got on the ethically sourced bandwagon. They announced in 2013 that they would end the practice of deforestation as a method of farming. And then they did exactly the opposite and continued their slash and burn exercises, decimating an area twice the size of the city of Paris, France, between 2013 and 2018. Martua Sitorus, the then CEO, stood accused of creating a "ghost company" to insulate Wilmar from the continuing practice. When journalists discovered this ruse, he was forced out of his position. Whether his dismissal was for running the scam or getting caught is an unknown but legitimate concern for us all. The other unknown is whether this changed anything or was just more daytime drama meant to distract and soothe society until they inevitably forgot about it.
A common status quo view of loyalty is that someone will be blamed and sacrificed when the group is caught in a treacherous act. Rarely will it be the actual guilty party(ies). These organizations rapidly rotate sycophantic staff, much like a chess game played with living servants by the medieval French monarchy. Regimes that favor power over truth frequently surrender their pawns. There's always another patsy to be offered up to protect the masters. You can certainly think of other current examples of this practice.

Another type of forest that may not immediately come to mind occurs underwater along the western coast of the Americas, from Alaska south to Baja California. The giant kelp, or bull kelp, forest hugs many coastal areas and acts as a buffer against the waves and a safe harbor for oceanic diversity as it feeds off of nutrients washed up from the cold depths. Kelp is a macro-alga (*Phaeophyta sp.*) and grows much faster than the vascular trees of our land-bound forests. The kelp anchor in the shallows and extend a canopy that is a floating raft of fronds up to 25 meters (80 ft.) long. This mat is alive with fishes and crabs, sea otters, and nudibranchs. The kelp itself is a nutritious food source, and a vibrant food pyramid extends outward from it. Kelp forests occur around the world, and all have similar ecosystem functions.

As might be expected, these forests are rapidly disappearing, and the cause is climate change. Off the North American coast, masses of warm water started this loss, washing over the tidal zone and blocking the cold, nutrient-rich under swells. Once the imbalance began, around 2014, sea stars began dying off in ghastly manners, including the auto excision of their arms and the melting of their bodies. This turn of events then allowed sea urchins to reproduce without control, the sea stars having been their major predators.

The purple sea urchin (*Strongylocentrotus purpuratus*) eats green and brown algae exclusively. Once they figured out that the bouncer at the head of the buffet line had passed, they went hog-wild. Thus, from Mendicino to Marin counties in California, the kelp forests were almost completely consumed. It redirected them into what are now called urchin barrens. The purple sea urchin stripped two hundred miles of coastline of its bull kelp in less than a decade. This loss also decimated the abalone (*Haliotis sp.*) nursery, reinforcing the decline of sea otters who eat both the urchins and the gourmets' favorite mollusk. Fishing groups had already been at the vanguard, trying to eliminate sea otter competition before the sea stars suddenly declined. This bottom-up trophic cascade has led to a problematic situation where some locals reject the cure because, while the sea otters might help restore the kelp, they would undoubtedly affect abalone fishing at the same time. It is a similar battle to the one raging in the western US over the re-introduction of wolves. Humanity fights the very idea of predator/prey relationships, despite being one ourselves. Disconnected from the interrelationships that make up the ecosystems, many imagine that any predator other than us is competition. And this is because people have been taught that life is competition and don't recognize dynamic cooperation in many of its forms anymore. Unfortunately, such short-sighted viewpoints can present a hurdle too high in today's society.

All ecosystems developed intertwined and need to have all of their connections restored for their complete functions to resume.

Bull kelp grows fast and can incorporate carbon faster than most terrestrial plants. Scientists have estimated that the carbon-holding capacity of Australia's Great Southern reef forests ranges from 10.3 to 25 million tonnes

(11.3 - 27.5M tonnes) of carbon. Worldwide, kelp may capture 173 million metric tonnes of CO_2 per year. (Filbee-Dexter, 2020) In fact, kelp is so efficient at collecting carbon that it is the backbone of several carbon capture schemes. One idea is to create floating farms in the deep ocean, and then once the plants are mature, sink them into the briny deep. The kelp will have to be plunged over 1000 meters deep, where they will very slowly decay, locking their carbon away from the surface system. Hmmm.

Does that sound reminiscent of the dead zones? Even though the deep ocean floor seems like a desert to us, it is still alive. The decay of megatons of kelp will undoubtedly rob that benthic zone of preciously scarce oxygen, resulting in the death of any creatures that need O_2. That technique sounds a bit like sweeping a mess under the rug. So here's a better idea: let's restore our urchin barrens to kelp forests and let them get back to the job of sequestering carbon. In their proper place and function. Maybe some new forests can be established to expand their overall range a bit. Perhaps kelp can be farmed and folks can find new uses for it. People already eat it, why not use its fibers, or make kelp inks, perhaps there are pharmaceuticals produced by them. Waste should never be a primary part of our solutions if we want to restore the environment. And let's stop arguing with little sea otters and let them have the abalone for the next few decades or more. Do people always have to feel like they're winning? Even if it is causing us to lose?

An important and foundational part of every forest is the fungi. Many young foresters find themselves obsessed with trees and will readily tell you they have no interest in any of the other parts of the system. They aren't interested in the animals, nor the lesser plants, and certainly not the mushrooms. It's as if the mushrooms are just some sort of secondary parasite. They may imagine mushrooms are only attractive to silly 'shroom hunters out for a lark in the glen and vale. But they are certainly unworthy of the attention of serious land managers. That point of view is, again, the inability to see the forest for the trees. Fungi can make up almost 90% of the forest soil biomass in some ecosystems and accounts for 25% of *all* biomass on the planet. So such perspectives are, thankfully, passing by the wayside. The public needs to understand just how important each role is in the forest system; mushrooms doubly so. An intent that seeks to alter only one component of a forest, such as our current commercial timber management, will always fail in achieving its promise of restoring the ecosystem. Forest restoration must be a synergistic discipline that works with all the various life forms that compose any given environ.

The first colonizers of the land, hundreds of millions of years ago, are thought to have been the funga. (Heckman, 2001) They have had an extraordinary relationship with plants ever since. Plants have that astounding ability to harness the energy of the sun to manufacture proteins and sugars. And many a fungus can digest rock. Minerals are of vital importance to all life forms. People get ours through eating vegetables and meat, where they came from the plant matter previously consumed by that food animal[141]. And those plants got their minerals, primarily, from a fungus. All life is a system and a cycle, each part of which is just as crucial to the whole as every other part.

The prototypical fungus/plant relationship is the lichen. Lichen is a symbiotic organism composed of a fungus and an alga (or a cyanobacterium) that lives on rocks and bark and grows ever so slowly. They eat their mineral footing and absorb sunlight, rainwater, and the occasional draught of reindeer pee. They used to cover the land, the mountains, the plains, everywhere. In the untouched areas of the mountains, they still exist on the ground as indicators of pristine air quality. Lichen absorbs CO_2 from the atmosphere and uses that carbon to create their flesh. The initial absorption is by the algae, the phytobiont, which then exchanges it with the fungus, the mycobiont, in another beautiful example of mutualistic cooperation.

I have often seen people move to the mountains, rake up the ground lichen, and throw them away too often. They collect them in trash cans. Just like many may grab from around them any random thing, to either consume or discard, with nary a thought of what any of it means. Lichen rocks, many that have taken centuries to form, are harvested and sold in the cities and towns as moss rock for gardens and landscaping. They aren't moss, and the air quality that people find acceptable in our suburbs will inevitably prove fatal to these living fossils and their beautiful cloaks of color. This aesthetically driven slaughter will negate the lichen's centuries of work sequestering carbon and carelessly

---

[141] Some animals can get minerals from licking geological formations, or salted roadways. Fungi in their digestive tracts can help absorb these minerals. Digestion is a multi-species effort within our guts.

re-release it back into the environment, adding to the concentration of greenhouse gasses. A better idea might be to buy regular rocks and have an artist paint them. They would at least last longer, and few would be the wiser. If one just wants decorations, by all means, decorate, but please stop killing just for some artistic fancy.

This fungal cycle, which is ongoing in the mountain forest, is a foundational process. Here in the Rocky Mountains, the soil is often composed of decayed granite (DG) or other crudely deteriorated mineral-based materials on top of solid parent rock. Above this sand and gravel is a thin layer of duff. Duff is partially broken-down organic matter, primarily needles, stems, and leaves. There is little that resembles topsoil in the mountain forest. For the giant trees to survive in such a situation, they had to develop a relationship with the mycorrhizal fungi. Mycorrhizal fungi surround their root hairs, protect them from drying out, and exchange minerals and carbon for sugars. These fungi, often mushrooms, form extensive webs throughout the duff layer and share with not just one tree but with the entire stand. Suzzane Simard has documented that certain fungi will exchange carbon from older trees with the younger seedlings and saplings. (Simard, 1997) Her pioneering work in the forests of British Columbia showed an interaction between the trees and fungi that borders on nursing or even farming by this mycelial network.

The extent of the fungal mat in the soil is truly staggering. A single cubic centimeter of soil can contain 0.1 to 1 km (cubic inch: 1-3 mi) of fungal hyphae, the threads that make up the mycelial mat. This network represents a massive amount of carbon storage. Mycorrhizal fungi store 45%-70% of sequestered carbon in the boreal forest. (Clemmonsen, 2013) Carbon sequestration is the mechanism by which atmospheric carbon is removed from the air and reliably stored, as wood in trees, as stems in flowers, and by the fungi's mycelium. The fungal body is made up, to a large extent, of the organic compound: Chitin ($C_8H_{13}O_5N$). This material is the same substance crab and lobster shells are made of. And the astute reader will see the similarities in its chemical structure to cellulose, the body of the tree.
If we want to help restore ecological balance, we need healthy forests, and for that, we need healthy fungi. One hiccup in this realization is that we don't reliably know what a healthy forest fungal community should look like. Our

forests are quantifiably unhealthy at this stage, and data on what the mushroom community looked like one or two centuries ago are rare and fragmented at best.

My colleague, Dr. Andrew Wilson of the Denver Botanic Gardens, runs a survey of the extent and diversity Colorado's funga. The Colorado Fungal Diversity Survey collaborates with mycologists and citizens nationwide as part of the North American Fungal Diversity Survey. Using trained volunteers, they foray and collect, diligently documenting their finds. And then, they bring them to the lab and analyze their DNA. Using a mobile application: iNaturalist, they can pinpoint collections on a nationwide diversity map. They will toil for years to collect and identify a baseline, the first of its type in US history[142]. While this data is expected to be skewed because our forest function is highly stressed, humankind sorely needs this information. In 50-60 years, when over half of our local forests may be dead or dying, we will desperately want to know what fungi will be needed to bring them back to their proper glory. It is also expected that within that half-century, the importance of native fungi will become obvious to everyone.

The Morel is the quintessential gourmet mushroom and often presents itself after a fire, leading to swarms of collectors in the spring after a burn occurs. But what if no trees survive on which the Morel coexist? Morels are mycorrhizal mushrooms, and without trees, there will be no mushrooms. And as has been seen, the reverse is also true. Without mushrooms, there are no trees. This situation is not an intellectual chicken and egg exercise; this is the primary challenge of post-catastrophe forest restoration. Our current ethos of breaking things down to their parts to study function misses the point that single parts are often a misnomer. Many view the world this way out of habit and prejudice- for one part at a time is how humans construct things. But growing up and diversifying in harmony is how nature builds her mechanisms. Life is not a machine designed by some architect and then copied precisely from unit to unit. A bolt is not similar to the engine it is part of, like a cell is to an animal. Nor can that bolt turn into a screw when and if conditions require it. However, life can do and is both of those things. Life is pliant when needed

---

[142] If all of this sounds interesting, go to iNaturalist on the web and volunteer to help with this or any of a growing number of other projects.

and interconnected in every aspect. In nature, nothing stands alone. We continuously need to limber our thought processes when dealing with Gaia and her ways.

In western US forest management, many are currently trying to restore the forest to its pre-historic conditions. Why do this? Because the forest has adapted over millennia to exist in a stable state. They wish to return it to that state and, by extension, to its proper role and functions. They remove trees in a calculated pattern to reduce stress and competition for water and nutrients. When I do this, I think like a fire. Acting as the predator of the forest. I select those plants that a moderate intensity fire would likely consume and carefully choose the survivors. An overall reduction in tree density is required on most sites, as is an eye towards maintaining the diversity of both ages and sizes. With some practice and a bit of imagination, one can make choices that improve forest health and prepare the land for actual fire's return because fire must return to the frequent-fire-forest, as the name suggests.

This thinning improves the remaining trees' health and increases their resistance to disease, insects, and catastrophic wildfire. This process is labor-intensive and expensive. Not as expensive as cleaning up after a catastrophic fire, but costly still. In some areas, the trees that are removed have an economic value. They can make lumber or pulp for paper, for example. In other areas, such as where I work, there is little economic value to the wood. Trees grow slowly at elevation, so most of our removal is of small-diameter timber.

Trees are a renewable resource and can be managed sustainably, as long as there are trees.

Economics has eliminated the local timber industry in Colorado and several adjoining states. Well-meaning environmental groups have also aided in this elimination through the demonization of logging as an industry. And indeed, the logging techniques I saw in my youth were environmentally damaging, sometimes bordering on criminal. As are the tactics used in the rainforest and many parts of the world to this day. Included is the practice of farming "Franken trees," genetically modified organisms with no relationship or interconnection to the local ecosphere. This practice is common in the American Southeast. But this does not have to be the case. Today, many of our loggers are not only aware of the hows of timber management but also the ecological why's.

We can harvest timber and improve forest health at the same time if that is what we choose. Many now see the two as mutually beneficial and necessary. One way to save the forest is by logging. And many will soon be logging for forest health and carbon sequestration by retaining larger, older trees and removing smaller specimens. They will take their harvests the same way fire prunes the forest, selecting the weak and overcrowded for removal and keeping the strong. Loggers will use their chainsaws to increase diversity and not just mow down whatever stands before them and seems valuable. When the harvest is altered in this way, industry will have to alter post-processing to use these smaller logs. Advances in manufactured wood products and cross-laminated timber (CLT) will allow them to do just that.

Where there is no economical use for the removed material, a quandary is faced. Tons of wood have been removed with nowhere to take it. Currently, much of it gets sold as firewood for heating in rural areas, but those markets are near saturated. And our efforts are also taking a sequestered form of carbon (the logs) and burning them. This action just ends up releasing even more carbon into the atmosphere. If this waste wood is just left wood somewhere safe from a fire perspective and allowed to rot, the damage could be even worse. The bacterial decay of wood not only converts the cellulosic carbon into CO_2 but also creates methane (CH_4). That methane is vastly more damaging to the atmosphere than CO_2. While it is currently popular to designate recently harvested wood fuel as carbon neutral, it doesn't actually matter much. More carbon dioxide pumped into the atmosphere is still more carbon dioxide. Stop adding CO_2 into the wild and speed up its removal. That has to be our dual goals at this time in history. It can seem like a damned if you do, damned if you don't situation. Here, we should err on the side of don't. Don't add more CO_2 into the atmosphere, whenever possible.

It should be noted that perhaps the best thing that can be done with the wood we harvest is to make it into lumber and build a house (or a boat) out of it. Wood is made of cellulose, an extremely resilient natural fiber that can last for centuries if properly protected. I liken it to natural fiberglass because it is. Primarily composed of those long strands of cellulose, wood is a simple polysaccharide fiber; glued together with lignin. Lignin is a highly complex cross-linked and carbon-dense material that slowly crystallizes, an analog to fiberglass's acrylic resin. Combined, they make a strong yet slightly flexible composite perfect for construction, easily shaped, and stable for the long haul. Fifty percent of cellulose is carbon, so for every 2 tonnes of lumber used, ~1 tonne of carbon is locked away from the atmospheric carbon cycle. Were this carbon atom to join with two oxygen atoms to form CO_2, it would have a combined weight equal to 3.67 times the carbon. So, for every ton of wood sequestered by building our house, we have prevented 1.83 tons of CO_2 from re-entering our atmosphere. So here is an actual win/win situation, and it is just beginning to spur a renaissance in wood construction.

In Norway, wooden skyscrapers, such as the 280-foot-tall Mjøstårnet tower, are an intriguing new possibility emerging above the horizon. Experiments with large timber buildings are occurring in the United States as well. All-wood buildings that are more resilient to earthquakes than a concrete and steel monstrosity are being designed. Wooden car bodies now appear in design studios worldwide. Wooden boats are superior to fiberglass boats in both beauty and longevity. Much of our woody waste, in the form of sawdust, can be rejoined with resins to make high-tech composites and engineered woods. Or that sawdust can be used to grow tasty mushrooms.

In an exciting turn of events, transparent wood now seems feasible. What you may have seen as an interesting experiment in high school chemistry class now suggests windows can be made with an insulative value, and that can bend and never shatter. Not only is wood strong and renewable, but it is also inarguably gorgeous when worked and finished. And it never stinks or poisons the land, unlike fiberglass.

Only some of the material from any given tree will make it into a finished piece of lumber. Around half of the woody material will become "waste" that must somehow be used. This waste comes initially from slash, the limbs and tops of the tree, and then the sawmill loss. So, foresters are still up to their proverbial necks in branches and wood chips. I prefer to leave the slash and tops on-site and grind them back into the ground. The trees' green needles represent most of their nitrogen, the vital and limiting nutrients for plant life. This process retains the site's fertility and stops a destructive type of nutrient export we have inadvertently been enabling. This green material can represent over a century of nutrient accumulation by the forest stand, and many hate to see it hauled off and disposed of in any of the ways currently used. If burned, practically 100% of that nitrogen will convert back into atmospheric nitrogen, which, while greenhouse neutral, is no longer available for use by the photosynthetic cycle. And this is precisely what many do with this material. They make burn piles and torch them in the winter when the risk of spreading fire is low. However, global warming is making this more difficult as the safe window for burning becomes ever shorter, and the snowfall decreases in both frequency and depth. This difficulty is, strangely, a positive pressure on forest management. Burn piles just release more CO_2 into the air, and the heat pulse from large burn piles will sterilize the soil beneath them just as effectively as an extreme wildfire. Forest dead patches created by this process can last for half a century or longer, with no growth present or possible where large piles had burned. (Rhoades, 2015) The blaze of a large consolidated pile will not perform the job of recycling nutrients, either. Burning our wastes for no benefit or letting them rot are both horrible ideas arrived at by a lack of suitable alternatives.

There is another way. Our team initially conceived our fungal decay experiments to deal with this very problem. Fungal decay does not create methane. It does not cause piles to heat up like the composting process, so the piles will never spontaneously combust. Fungal decay is how nature recycles wood, albeit more slowly than the methods under development. Fungal decay appears to increase the carbon load (sequestration anyone?) in the finished soil[143]. And fungal decay has another benefit: it produces mushrooms, and many of the ones we use are quite delicious.

Our team collected wild mushrooms, called wood-rotters (*lignicolous saprophytes*), from the field and cultured them in the laboratory. These organisms are native and collected as close to the area of their intended use as possible. First, our downy mycelium is introduced to the materials we wish them to consume. Then, over multiple culturings, the amount and form of our desired materials are increased until they are feeding on exactly what they will encounter in our piles. At this point, they are considered trained. Next, mainly using standard mushroom farming techniques, little pieces of mushroom fuzz

---

[143] Hold that thought, because we can't prove it just yet. We are working on it, though.

are expanded into thousands of pounds of spawn. Those spawn blocks are taken to the field where they are inoculated into woodchip piles. Then nature takes its course. The mushrooms eat the waste, and our finished product is a compost-like material. This material is not the compost you buy in the store or make in your composter on the deck. This compost began as pure wood, with no addition of green organic matter. It does not need to be turned because the process of fungal digestion is self-regulating and aerobic. This process takes a few years, but this is an exponential acceleration over what would happen if they were left to nature's schedule. A fallen log or pile of logging slash in the Colorado montane[144] forest can linger for decades before it is wholly decomposed (Battaglia, 2018) (Wagener, 1972).

This technique is a perfect example of ecological restoration where ecology is used as a technology. This work presents a very low risk of unintended consequences with a potential for many applications, including post-fire landscape restoration and the revitalization of lost or depleted topsoil in agricultural settings.

If we sit on our hands and do nothing about this threat of warming-driven catastrophic wildfire, we may lose up to 70% of our western forests by the year 2100. This fact is made more distressing because our fires continue burning at uncharacteristically severe levels. Severe fires burn so intensely that there is no seed base left. There are no bacteria or fungi- no nutrients- just sterile gravel. Again, the worst-case scenario is redirection, where what was once a forest is now a completely new habitat (usually a grassland, maybe a desert). And if restoration is not used to address these situations, these lands will probably never be a forest again. The phrase: "not in a million years" is, sadly, quite accurate.

Eight million years ago, in Colorado, giant redwoods (*Sequoiadendron giganteum*) once grew. Since then, the changes have been too many and extreme for the Redwoods to survive. All that is left are massive stumps of petrified wood. In Central Europe, there were once Large leafed beaches (*Fagus grandifolia*), but when the ice age came, they could not survive the

---

[144] Montane- "of the mountains" the label we use for forests that occur between 2440 m (8.000') and 3050 m (10,000') above sea level.

extremely long winter, nor could they migrate over the Alps. So now they are only found in the southern-most parts of Europe. (Comes, 1998) It is well known that tree species have and will move worldwide as climates change and local weather patterns evolve. But this change has only previously occurred over millennial-scale time periods. The speed at which our global warming has altered our world is itself without precedent. As the climate warms, trees will be especially susceptible.

Trees can't just uproot themselves and move to a more hospitable spot. The "conservative" economist Alan Greenspan[145] used to be fond of saying: "People vote with their feet." Which is a clever way of saying that if people disagree with a policy, they can just move. A more cynical reading of his motto is: "Love it or Leave it." You might recognize that phrase. However, you might also note that "conservatives" rarely embrace this type of thinking when refugees attempt to vote with their feet and come to our advanced nations while fleeing turmoil in their homelands.

Regardless, trees can't vote with their feet; they don't have any. So as the climate warms and planting zones creep north, many of our trees will not survive. Cold hearty species have evolved to need a winter's deep freeze. This period of metabolic reset also acts as an effective insecticide against many potentially devastating pests. When the winters don't come, and the summers overheat, billions of trees will die. If we try to replace this loss after drought, fire, or disease with the same species that previously stocked these locations, we now often find that those seedlings will fail. (Stevens-Rumann, 2019) While people and animals will migrate worldwide to adapt our coming changes, our critically needed forests can't. So, we will have to help them.

The practices of assisted migration, and assisted colonization, supplement or even replace native species with species that might have a greater chance of survival in a new climatic regime. It may bring trees uphill from the plains or even from different parts of the world. This concept is still controversial in some circles. Most ecologists are reluctant to risk fostering new, technically

---

[145] The 13th chair of the Federal Reserve. Initially appointed by Ronald Reagan in 1987. His policies worsened the dot-com buble (which burst) and set the scene for the sub-prime mortgage collapse (which tanked the global economy). So, yeah, he was pretty much inocorrect in all his economic philosophies, "conservative" policies, so I'm told.

invasive species. But it's too late to pretend that drastic climatic changes are not occurring, as it is too late to avoid some of their disasters. This type of direct intervention into a species' native range is a practice that might have been considered decades ago. Unfortunately, it wasn't taken seriously enough when it might have been more effective. Now our hands will be forced by reality. Assisted migration is human-aided adaptation that takes the most aggressive form. It will be our forest scientists and ecologists who must decide what trees to plant, and where. And it's going to be a series of judgments that will be made without a lot of precedents. So: "Them's the breaks," as the saying goes. Humankind will need to choose carefully how to save our trees and grow our future.

Afforestation is the establishment of forests where there were none before. This term should not be confused with reforestation, replacing trees lost in a forested area by fires, disease, or over-harvest. This practice is the opposite of what has been done for centuries, where forests were cleared for fields or towns. Humans have pulled so much oil from the ground and released that carbon that the atmospheric CO_2 load on the system has increased by the equivalent of thousands of forests. Even if every forest were functioning at optimal capacity, there would still be an excess of carbon in our atmosphere to remove. The only practical way to deal with this is to increase the forested footprint planet-wide. If we've lost 30% of the America primeval forest, how much can we now return?

Akira Miyawaki has been pioneering the creation of "small forests" for several decades now and has trained hundreds, if not thousands, on how to do it correctly. Creating a forest is not establishing a tree farm, and many tree planting projects make this simple mistake. Planting hundreds of trees of the same species on one site may help, but it may have little effect, as well. It will never yield a self-sustaining woodland, because forests are not monocultures. To recreate a forest, you need to plant many species, not only trees but also shrubs and grasses and forbs. It may require the import of ground cover to start the process of soil establishment. It can even benefit from relocating dead and downed trees to act as windbreaks and soil micro-habitats. They then work as inoculants of bacteria and insects, all required to form a working habitat. You can think of it as an ecological "fecal transplant" where you inoculate from a forest to the field. This approach creates a multi-tiered complex that can attract and feed the many animals needed to develop an ecosystem. His method is the accelerated creation of a climax forest with all the differing canopy levels represented from the start. This is the assisted creation of the old-growth forest mentioned earlier. All the plants used should be native, and preferably, locally sourced. Creating a forest means starting all the environs and ecotypes required for all the animals that will become a part of this new ecosystem. Miyawaki establishes all the plants in a spatial arrangement that will become the multi-storied system that is a complete forest. He is a proponent of the rapid establishment of old growth type forests. His forests proliferate rapidly, establishing themselves as a community from the ground up, accelerating carbon sequestration. (Miyawaki, 1992) This procedure is creating a forest using techniques amazingly similar to those advocated by permaculture. Miyawaki's methods are not those most commonly used for reforestation projects currently. That would be to replant single tree species in a monoculture, or perhaps two species depending on the site's initial conditions. Since it is being done in an area that was recently forested, it is assumed that many of the cohorts needed to re-establish an ecosystem are either already there, or nearby. Nor is afforestation often considered as a method of carbon sequestration. Ecologists are currently wary of changing any habitat, even if it might improve what is now present. But these are the methods that will be used in the future for both the revitalization of our wild spaces and the recreation of our agricultural systems. We need to use all of our resources wisely and promote the fertility of nature to feed ourselves and every other creature on the face of Gaia. The Green Revolution was a bust. It must be retired immediately.

Forests are unique places in our hearts and minds. Forests are essential for many of our larger native animal species and the nesting grounds for birds and small mammals. They make up a significant component of Gaia's respiratory system. They are the source of our best renewable construction and writing materials. They are also so very fragile now. What was once looked upon as just more resources to exploit, materials used wastefully, forests are now seen as a major mechanism of our salvation. Mankind has inflicted damage on these woodlands with our harvests and even sometimes compounded that with ill-considered attempts at conservation.

Caring for the environment is a laudable sentiment. It is the primary concern for many of us. But there needs to be open conversations between all groups and a willingness to learn from each other. There is a difference between use and exploitation, and that's where the line should be drawn. Environmentalists have a crucial job to perform. They need to be at the table when decisions are made and the deals are done. They then need to make sure everyone carries out the projects in the agreed-upon manner. Harvest can improve the health of the forest just as easily as it can destroy it. It's the balance that most have lost sight of. When people refuse to cooperate on wild harvest issues, the result will often be: take everything or taking nothing. Often, neither choice is the right one. The problem is that our answers are not as simple as we've been trained to think they are. They never really were.

We can't live without using nature, and it can be improved by being used; disturbance is natural and essential. Nature is a give and take. We will need to learn that lesson now. We've been almost exclusively taking for far too long. Now is the time to take all that we've learned and apply it to healing the damage that the human campaign has inflicted. Now is the time to assist evolution for our planet and ourselves.

This is the our Green Evolution.

As we come to the end of this section of the book, it is important to take a moment to let it all soak in. Earth, air, fire and water were the original elements of the alchemists of yore. Civilization has come so far from their mindsets, and yet in some ways, we are still trapped there, in a stone walled timber roofed shack trapped by something that hasn't been able to keep up with our other advances. Held hostage by some medieval curse cooked up in unsanitary glassware, a part of our social order has remained frozen in a pattern of caste built by the ignorant superstitions of those very alchemical quacks. "All animals exist as objects for our possession, to do with as we see fit. All plants are ours to consume, contort and lethally control. Women must be owned by men lest they lose their way and again, get us evicted from our rightful domain. The planet itself is worth no more than sterile rock to be crushed to dust by the assault of our machines, pounded by our soles. Other men are to be vanquished because they want what you have and will take it, must take it. Because both of our gods demand it." Dominion haunts us and we will never be free until we can release its shackle on our still lingering primitive minds.

And of course we can do that. We've been doing it for centuries, however many fitful starts and reversals we have encountered. The human race has been practicing since before the Magna Carta with freeing ourselves from the masters and their philosophies, which were little more than codes of control. We've been getting better, as suffrage and civil rights have been won. The latter still being a work in progress. When some found the courage to bring into the open how diverse the human race really is, so far beyond the old black and white binaries humanity exercised that power again. To even be able to speak openly about the many genders and the near infinite ways people might conceivably bond is further proof that we are building both the strength and courage to change again. Because we have to change if we want to survive. And we can't fix the hundreds of real problems outlined in this text, without fixing the problem of Dominion. All we have to do is let it go and let life be. We're strong enough to do that. We've practiced long enough. It's time to end our childhood.

We will next look closely at what that thing is, this beast that screams in the darkness of human consciousness. What is it? Where did it come from? And hopefully, we'll see how to dispatch it with the light of knowledge. Our next step in evolution will be the greatest one so far, as all previous ones have been. And how many steps will there be beyond that? We'll first have to take it to find out. Humans are masters at "flipping the script" and that's what we must do again. This time we'll flip it back to living in harmony with nature and the living world that gave birth to us. If we do it right, we'll get to bring all our prizes with us. Our knowledge and wisdom, our compassion and our will power. And we can save technology as well, but only that part of it that can be used without destruction, untethered from greed.

CHAPTER ELEVEN: THE ONE

*"What we call man's power over Nature turns out to be a power
exercised by some men over other men with Nature as its instrument."*
— C. S. Lewis

On March 24th, 1989, the Exxon tanker Valdes struck the Bligh Reef of Prince William Sound, Alaska. It eventually spilled almost all of its contents, 260,000 bbl. (41,000 m³; 10.8 million gals.) of crude oil onto 2,100 km (1,300 mi) of pristine shoreline. Exxon responded slowly, facing eventual criticism for sitting on their hands for several days when the oil could've been cleaned up relatively easily had they addressed it promptly. Again, Dominion culture sees the planet as theirs, and therefore will resist any explanation (or apology) when they foul it. It's as if the words get stuck in their throats. Exxon's defense, when they finished formulating it, was: "We were waiting for clearance to use dispersant." (Andrews, 1989) But, once that permission was granted, the first air tanker drop of said dispersant missed the oil-soaked coastline entirely. And it only added to the environmental damage when its foam slimed an entire forest. Meanwhile, the spreading shroud of black death continued its relentless expansion across the once immaculate shores. Pounded by the waves and tides, it transformed into a crawling black froth of corruption and death.

Unschooled in the techniques now used by major corporations to address such accidents, Exxon made what may now seem like a rookie bungle. The corporation attempted to downplay the disaster in a blundering, unfocused manner. They forced reporters to assemble in Valdez, Alaska, a town lacking communications infrastructure and adequate housing. They may have hoped that the accident site's remoteness would lessen the impact of the sight of countless dead seabirds and otters. Or perhaps they thought that by inconveniencing the press, they might slow the spread of news. And maybe those pampered New York reporters would just give up before they saw the carnage, worried that the slog through clotted tar would damage their expensive shoes. Out of sight, being out of mind, from their perspective, after all. Meanwhile, the town of Anchorage was a scant 40 miles away by air. It possessed the hotels and communications infrastructure that the company itself used.

Whatever their hope, it was all dashed when the first lens caps came off the television cameras. Their carnage was laid bare to an American public who watched in horror from their kitchen tables. Because despite their best efforts, the word finally got out, and it was not flattering. Exxon's missteps would help create several new industries, the corporate image agency among them. They placed full-page ads in newspapers around the nation in an attempt to get ahead of public opinion. But they did not publish these first ads until some nine days after the accident, an indefensibly long pause considering the magnitude of this disaster. So rather than getting ahead of anything, they just added to the hurt and confusion. For what they meant to be a corporate display of contrition appeared more of a boast than an apology. Much like the language of Tagalog, there seemed to be no words for "we're sorry" in their corporate lexicon. Their initial reaction may have been clumsy, but it shaped up several years later when, as you shall see, it really mattered.

Mistakes and the intensification of this predicament by their corporate actions became the order of the day. This fiasco would become the worst oil spill disaster of its time, now eclipsed only by the Deepwater Horizon. And it would become the first time a major oil company was taken out behind the woodshed by the courts, which found them responsible for not only the cost of clean-up but over $500 million in actual damages. Once corporate officials admitted a drunken sea captain was in charge at the time of the accident, the courts found Exxon so flagrantly irresponsible that they inflicted an additional $5 billion in punitive damages. It was argued at the time that Exxon's executives just choked when faced with this unexpected tragedy. There was a cascade of failures and bad luck, to be sure, but it didn't seem that they had crossed that line of criminal intent. So there was no reason to suppose deeper insidious motives for their failures. Or so it appeared at the time.

Inaccessibility would finally work in Exxon's favor. Getting to the accident site was a significant trial. So, even though this was one of the biggest stories of the day, the cost of getting reporters on-site, and the endless complaints from many of them, led to a slow disconnect. And because the public turned their gaze away, much of that oil is still there, in cracks along the rocky shoreline and buried under the sediment a few inches below the seafloor's gravel surface. (Taylor, 2014) In the end, their corporate cynicism turned out to be a winning tactic. Because, damn it all, they got away with it.

The outcry did last long enough to lead to passage of the Oil Pollution Act of 1990 and local Alaska-specific regulations for tanker escorts. In addition, this one incident is credited with the Federal decision to ban any drilling in the Arctic National Refuge, a massive win for the environment but one that has been challenged continuously ever since.

Exxon chased that $5 billion through the courts and by 2009 reached a final settlement, having to pay only $500 million, a meager 11% of the original judgment. This outcome is not a startling, singular avoidance of justice. It was (and still is) business as usual when prosecuting large corporations for environmental crimes. Justice is slow, and the American attention span is short. You do the math.

And as it turns out, the captain was neither drunk nor even on duty at the time of the incident.

That's right, the third mate had the wheel on that fateful eve, and he may have been suffering from fatigue due to overwork. They had also turned a radar collision control system off, contrary to the written procedure, but perhaps in line with standard company practice. Having no alarm and navigating through pitch blackness, it would seem that he merely missed the impending collision, a simple human error. So the entire tale of employee negligence turned out to be just that: a story. An apocryphal account that many still believe to this day. Another steaming bowl of corporate BS served up for a captive audience. Strange, how the facts only came out after Exxon was in the clear. So, while the courts found them negligent, it was not as grossly indifferent as they actually appear to have been. It was their excessive production schedules and corner-cutting that put the crew into this situation. And it was their choice to blame a captain who was neither intoxicated nor on duty, effectively burning his career[146] in a blatantly indifferent manner for a bit of media camouflage and just to create plausible deniability. It's no more complicated than that.

---

[146] Joeseph Hazelwood, the Captain was acquitted of being intoxicated, but he was found guilty of "negligent discharge of oil" and fined $50,000 as well as given 1000 hours of community service. A harsher penalty than that given to Exxon Chairman Lawrence G. Rawl, or any board member. You can still find articles that incorrectly cite the captain's drunkenness to this day. So the scam worked.

Exxon did it because they could. Those in power will almost always get away with it because society holds institutions and their figureheads to different standards. Centuries of laws and customs, millennia of myths and folklores have built this code. Some people are allowed, no, expected to be above the people they supposedly serve. And above them is one whose power pretend is almost limitless. This chapter will turn our attention to how socially curated realities have helped shepherd us toward our appointment with extinction.
In our current societies, positions of authority often enjoy different rules than everyone and everything else. For example, in the USA, a police officer may kill a citizen in the line of duty and get off scot-free, even if the whole "line of duty" part is questionable. A politician can make a policy that harms or even kills thousands of people and yet bear no personal culpability for the damage they have inflicted. These are tough calls. It's a dirty job, and someone has to do them. So, they can assert that they deserve their singular privilege along with the weight of their decisions. It's lonely at the top, so those at the pinnacle need more: more power, more privilege, and, of course, more toys.

We, humans have an instilled obsession with "the One." One path, one leader, one country under one god. And it's a dangerous, yet seductive, heap of rubbish. Our myths and folklores tell of the superior man (it's almost always a man, go figure). Our fictions often reinforce this fantasy as well. And in a world where the line between fact and fiction is so intentionally blurred, that's a problem. The concept of an ultimate one is so interwoven into our cultures it is often hard to see it. One flag, one team, one proper display of subservience when in the presence of your betters. From these simple singularities arises a grand illusion of divine uniqueness. When there is some singular model of the perfect human, everyone else will be judged as less than. And whoever can pretend to fill the archetype of perfection can manage all sorts of mischief. But, as we all know, no one is perfect, and as we have seen, perfection is one of the most misunderstood myths to which humanity subscribes. A terminal myth that we must abandon if we wish to survive.

The true "one" was always composed of many, just as the multitudinous cells in your body produce the singular you. Precisely as trillions of atoms form a single object or a cluster of solar systems creates a galaxy. Our experienced worlds are ones of interconnected networks. Gaia is a vibrant ecosystem of varied creatures and physical interactions, all of which are necessary and function in chorus. Singular, monolithic thought is an indulgence in an old mindset from centuries before people began to understand this universe. And it is a pretentious fantasy that gave some of us more than everyone else. Repeatedly, perfect gods were created who control everything with a micro-management style and terrible wrath. This sort of fallacy will not serve us well in fixing the problems that face us.

We have inherited a masquerade where one sex is revered above all others. Many have exalted one skin color over the rest in the past, and many seem still to be clinging to that absurd point of view. Thankfully, cracks are now opening around these antiquated ideas. But even so, there is far to go in working out our relationship with reality itself. It seems almost coded in our very beings: this concept of a "One." Instead of actively combining different ideas to achieve balance, it often seems easier to choose one set of beliefs and adhere to them no matter what. But is this actually an aspect of our beings or just a clever trick we have taught ourselves? Synergy is the ability to combine ideas into new, harmonious forms. Our brains excel at this task, but only if they have been trained to do so, and exercised often.

Humans all too often make up our minds and resist changing them. Taking the path of least resistance many can ignore all data that opposes their prejudices. In those instances where we have trained ourselves to ignore contradictions, we can become blind to them. We'll just continue soldier onward, projecting a mental construct of reality upon our perceived universe. We will use our powerful imaginations to stop our built-in inventiveness, taking that power of mental synthesis and using it instead to hold contradictions side by side in our minds. Contradictions such as we (and our economies) can grow endlessly on a finite planet.

Communities have been organized based on the stories we've told ourselves about ourselves. But what if those stories were merely tales of how we'd like the world to be? What if our folklores are just a bunch of chamberpot water? What if our traditions are apocryphal, just empty myths after all? What if we are not looking at the world as it is but are instead looking at it through a distorted circus mirror warped by our powerful minds?

The legends of our dominant contemporary civilization are patriarchal, monotheist myths. They are about central powers and individual supremacy. Our myths are still about the *man* who becomes the king or kills the king and becomes the oppressed's emancipator. Such tales support the underlying prejudices of our current way of life. They are the myths of the status quo. They were designed to give license to the drive to rebel against injustice while almost ensuring that injustice will remain after each revolution. They promise us the power of the individual over the masses, the world, the universe. The provably false assertion is that the culture, planet, and universe can and must bend to our individual will. This is all just bad poetry that makes us feel that whatever happens, it will all work out to our advantage in the end. You can now see that is untrue.

Our studies of nature and ecology arm us with an irrefutable disproof of such a worldview. If humans continue to act as if we are privileged beings in a universe built for our convenience, all will soon perish. This idea of "the One" is primarily responsible for the actions that have led to this ultimate juncture in our species' history. If people stubbornly cling to their concepts of superiority, everyone will be laid low by reality itself. The only way to save our collective lives is to admit the human race is not in control and seek the balance that has been previously belittled—harmony with the natural world that created us. For Gaia has created us with her systems and chemistries, not via some supernatural process. That is the reality empires have run from so persistently with their tales and myths. We are real. We are mortal.

Our success as a species has led to many legends and stories about ourselves, usually entirely unrelated to fact. Therefore, this hollow foundation makes them practically useless as patterns upon which to base our actions within this real world. Moreover, our ability to adapt our immediate surroundings to suit ourselves has led to a backward projection of our psychology on the living systems around us. This is called Anthropomorphization: the projection of human traits and motives on animals and even physical structures. Everyone is guilty of this to some extent. It's something our minds do when trying to grasp unknown or complex situations and make some sense out of them.
Our sciences still don't know a lot about our consciousness, how it formed, or even how it works. Many of our prejudices and myths are probably artifacts of

how we process data to interact with the world. The brain makes shortcuts in perception, just as everyone makes shortcuts in our workloads. The amount of data that flows through an internet pipeline is staggering, yet it is a pittance compared to the sheer onslaught data that assaults our senses when walking out the front door. We are aware. There are thousands of interactions occurring around us, our minds can track them all. The squirrel catches the eye as it runs up the tree, simultaneously the dog barked, but it was barking at a cat sitting on the windowsill licking its paw. And still we automatically duck just before the newspaper hits us because the bicycle carrier was distracted by a backing car. It's always 186,000 miles an hour when the mind meets the world. The brain filters and sifts to stay on target and still gives the illusion of a seamless awareness. But it isn't complete, we are always taking shortcuts. The only thing we are always aware of is our self[147].

Our society currently lives under a delusion that the individual is greater than the whole and that it is greater than any other civilization. Many even believe that our empire is more powerful than the world that keeps us alive. Such an attitude is utter nonsense, of course. And it is dangerous nonsense. This belief is self-annihilating nonsense. Our myths have told us that all of this foolishness is truth, and so many believe without question. The masses expect a chosen one to arrive and confer his greatness upon us, relieving us of the labor of being great ourselves.

But there is greatness within each of us, and you need to see yours to survive. It just isn't singular greatness, an end-all, be-all pinnacle of human majesty. The individual's greatness is found when the self makes peace with itself. When you find you no longer need to horde or dominate to feel satisfied. This, along with recognizing our actual wants and needs, and the path towards achieving them. Even the greatest among us pales compared to the power of all of us combined. All but the psychotic have a social desire, a need for the group because humans are social beings. Make no mistake about it, the mentality of "the One", "the masters of the universe", "Dominion" is all a fever dream birthed of narcissistic psychosis. And they have fooled us into exalting them.

---

[147] Except when under the influence of the psychedelics, unique compounds that can actually lift awareness outside of identity. That's another things mushrooms have to offer us, but it is also at the very least another book.

Their trick has always been to fragment our singular significance so that "the One" can steal it all away. The trick has always been to fragment our combined significance so that "the One" can claim it for their own.

In the United States, has been a rise of twisted individualists who, in their anger against society and belief in their superiority, claim that they don't owe a damn thing to anyone else for anything at all. They call themselves sovereign citizens and assert that they deserve an independent State's power. Quite a few of them were involved in the January 6th, 2021 uprising at the US Capitol. This arrogance was on full display as US Capital insurrectionist Pauline Bauer claimed in federal court that it had no jurisdiction over her. Yet, she felt she had the right to judge capital crimes against others when she called for Speaker of the House Nancy Pelosi to hang. For the crime of, well, that was a little fuzzy. Let's assume she'll get back to us on that one. She'll have some time in prison to figure it out.

These "sovereign citizens" reject any outside control of their behavior and deny the rule of law and authority of the social collective. They are their own little compact, self-contained "Ones." They may drive Coal Burners.

Really? So they owe nothing to anyone? What about their mothers? What about the generations before them that worked to create a society that gives them the freedom to throw their temper tantrums? What about the teachers who taught them to form complete sentences to utter their selfish statements? What about the shirt on their backs? Did they grow the cotton or synthesize the nylon? No, they did not. The products of society require a society to produce, which may seem obvious, but it is apparently not, for some.

This sovereign citizen nonsense is just that: nonsense. Its ignorance only matches the pretension of such a position. And most times, the sociopathy of its adherents. But pull back the curtain, and they are just extremists of a very American point of view. Americans have been told that they have individual rights greater than their responsibility to the whole — the whole of society, the whole of all societies, the whole of the planet. This way many can believe that the greatest power is the superior man[148]. And so some might feel justified

---

[148] There is a source for this mental illness and it comes from a docturine of selfishness and is spread by pseudo-intellectual dipsticks. Commentators and politicians who worship money and may be follwers of sociopath Ayn Rand.

storming the seats of power because their candidate lost an election. Rational thought means little to a mob whose emotions have been stoked by anger and ancient tales of messiahs. Hundreds of them stormed the Capital while narrating into their iPhones as if they were the stars of some Hollywood production. The cohesion of society is weak with these ones. Throughout history, such enraged pawns have been used to disrupt nations, always for the benefit of a few and never, ultimately, to their own best interest or profit.

The products of a society require a functional society to create. The products of life require a planet and its entire ecosystem to produce, which hopefully now seems obvious.

We are all members of interrelated societies, and are greatest together. Alone, one would not nor could not survive. *One alone cannot reproduce* (unless you're a polyp, which you're not) *and therefore will soon perish.*

Nature is also a society. But of a far different sort, and interrelationships are its key feature. Multiple species fill multiple niches, each giving something that makes the whole possible. Fungi break down the minerals, and bacteria release nutrients from the dead. Plants photosynthesize, making complex carbohydrates from raw materials that others can then use. Animals eat plants, fungi, and each other, colonizing every imaginable habitat from the deep ocean trenches to the top of the highest mountains. Everything recycles, from the water by mechanical forces to the nutrients, which pass from one creature to another. Everything has risen together in both complexity and utility. For billions of years, life emerged and changed and experimented with itself and its environment. Perhaps we stand not on the shoulders of giants as much as on the caps and limbs of the fungal and floral kingdoms.

The phrase: "Ontogeny recapitulates Phylogeny" has a simple meaning: The individual's development reflects the evolution of those who came before them. When the human egg is fertilized within the mother, it begins as a single cell. It replicates and forms different structures as it develops. One cell, two cells, then the blastula: a hollow ball of cells. It has gills for a while and then resembles a chicken embryo. Limbs differentiate, and fingers form. All embryos are female until a certain point in their development. Then those with both an X and a Y chromosome generate masculine parts.[149] Always remember that as we grew in our mothers' womb, all of us were female until a genetic cue made roughly half of us grow testicles and penes. Some develop intermediate genitalia, and they used to be worshiped for their uniqueness; rare, but not unknown, or forsaken. The fetus will only begin to resemble a human baby at around four months of gestation. In essence, we all evolve within the womb. We start as a simple cell and will finish as the most complex creature on this Earth. The novice's path follows the path of all those who came before; it is not singular. Disbelief in evolution is, then, the denial of who we are for the sake of a myth about who we wish we were: divine and perfected.[150]

2.8 million years ago, what is now Africa's grasslands were much wetter and covered with forests. This area is thought to be the birthplace of the genus *Homo*, our genus. By 1 million BCE, that climate was drying, and our ancestors began migrating north. *Homo erectus* populated the Middle East and parts of Europe. Having come from the Equator, their skin had a rich level of melanin, the pigment that filters out high levels of UV radiation. This functions as the last layer of defense from solar radiation that began with the Ozone layer. The original humans had dark skin[151], which later became a trait looked down upon by the eventually light-skinned Europeans. Europeans who were but another transitory dominant culture in a long line of societies that all thought themselves the ultimate expression of humanity. (Hint: "the One") Even though the original humans were dark-skinned, racial prejudice remains today, despite any logic or rationale, to be honest. Science allows us, through careful study and a long steady course of learning, to understand this. People

---

[149] It is important to understand that X and Y chromosomes determine male/female development in humans, but there are many other ways to determine who makes eggs and who makes sperm in nature. In fact, there's more than X and Y in humans, but this is another topic that needs it's own book (or two)- if you're interested.

[150] Twenty minutes spent reading a feed on social media could be seen as absolute disproof of this conjecture.

[151] Certainly, there were occaisional albinos, but they are not the source of, or members of, the Caucasian color range. Some idiots are now claiming that Caucasians existed either contemporaneously, or even before, other skin color groups. It's just an apophrycal bundle of B.S. And it's a not-so-subtle form of racism.

waste a lot of time over old ideas and biases that are just not true, which may be disastrously wrong. Many still worship the "wisdom" of functionally ignorant peoples from thousands of years ago. That is not to say they were stupid; they were just missing a lot of information. Many do this readily because it allows them to justify and maintain a particular set of selfish worldviews, such as Dominion. Humans are all one species; we are brothers and sisters, cousins, aunts, uncles, mothers, and fathers to one another. Any other view of humanity is just ridiculously misinformed.

We do have cause to feel special. Humans are the most successful species in the history of Gaia. And because of study, both historical and scientific, people know quite a bit about this planet's story. And the story of us. The human race can adapt to, or alter for our purposes, almost every circumstance around us. People have changed the face of the planet and stepped off of it into space. This prestige is probably one of the main reasons we have this idea of "the One" so firmly ingrained within us. We stand above all other species in a few crucial ways. However, despite our strengths, we are still subject to the laws of physical reality. We are a great but still immature bunch. Many refuse to accept almost any limits on their actions. Many refuse to accept what is plainly in front of all of us. Life is a web of interconnected species and not just a pretty backdrop against which the human drama is to be played. We cannot continue to pretend that we can do anything we want and get away with it.

As humanity has grown up, there have been thousands of religions. Our complex minds reach out to understand everything around us, and where an immediate answer eludes us, one can find consolation in their gods. Gods can comfort us in troubling times. Religion can provide us with peace in the thought that things have a plan, and everything will work out as it's supposed to. However, only a very few of these theologies persist today. Those that do are primarily monotheistic and predominantly patriarchal. This situation is especially true in the western world and in the Middle-East. Their god, inevitably male, never changing and perfect, definably anthropomorphic, and concerned only with the human good: is all-powerful, ever-present, and in absolute control. Many believe that there is no need to do anything about climate change today because some god has everything under control. I do not intend to malign religion. Maybe these religions are all on to something, but perhaps they are not. Maybe there is something akin to spirituality within us. There certainly seems to be some type of spiritual drive built into our minds. But that drive seems to become distorted when spirituality becomes corporatized and turns into organized religions.

Science and religion need not be at odds with each other because they are very different human endeavors. Science is not a philosophy; religion is an ideology. When it comes to the direct, simple determination of fact, we have to side with the only framework we have successfully created to discover facts (and it's not the one that asserts absolute, infallible, and never-changing truths). Science can lead us to the future. But there is a place for spirituality there as well, for it seems to be an innate human drive and a powerful one indeed. However, there is no place for either if we can not survive.

One belief of many religions is that without a comprehension of "the God," humans can't be moral or fair. That we can never be ethical. This is the belief that we are children and need supernatural guidance. But ethics are not the result of some creed. The understanding of right and wrong doesn't come from an ethereal edict. And new anthropological research underlines that fact. Oliver Scott Curry's study into the anthropological roots of morality found that certain concepts are nearly universal. His "cooperation as morality" theory has given us an empirical measure of social morality worldwide by measuring 60 diverse cultures and their attitudes.

> "The present incarnation of the theory incorporates seven well-established types of cooperation—helping family, helping group, exchange, resolving conflicts through hawkish and dovish displays, dividing disputed resources, and respecting prior possession—and uses this framework to explain seven types of morality—obligations to family, group loyalty, reciprocity, bravery, respect, fairness, and property rights." (Curry, 2019)

The near-complete adherence to the social frameworks of cooperation, respect for others and their possessions, a desire for justice, and spontaneous displays of bravery make a pretty solid argument that what humans call morality is innate in our beings. This work stirred instant controversy and even included a comments section upon publication for dissenting viewpoints because statistical significance is excellent until it interferes with deeply held biases, even among some scientists. For our purposes, it should suffice to say that whatever our morality is, it certainly is not the possession of only one group of people, nor is it the gift of one unique god. Those concepts are plain to all or not (psychotics excepted). Here is the understanding that "Good" and "Evil" are concepts applicable to us, but not to nature. We are the force that can become corrupted and choose wanton destruction. The biggest lie has been that nature was the unconquered evil, and people are pawns, subject to its tyranny. Such a belief is the first step to turning people into pawns. Many creeds instructed their followers to bend nature to their will (and do so while denying the true salvation of knowledge) to make nature their tool. Such doctrines would be just and moral in what way?

Perhaps if you have all the power and answers and your eternal soul is saved, you don't need to take any responsibility for your actions. And many do act in this way. When one believes everything is preordained and you belong to the chosen people, any action you might take to advance your interests must be justified. This freedom from consequence is the ultimate gift of "the One". Accept it and you need never feel responsible for anything again. And this is the opposite of ethical behavior. It is the replacement of active discernment and empathy with a simple punch list of commandments. This covenant can then become a license to ignore the reality that is right there in your face. It explains so much cruel behavior exhibited by true believers in both love and war. And it is an extremely tempting viewpoint.

When such a worldview is replaced with the shared understanding that humans have damaged the ecosystem to the point of near collapse, and together we can fix it, we can establish a more stable foundation. We might synthesize these two (only seemingly) contradictory human pursuits. Science and religion should not be at odds; together, they could be the tools and the strength to use those tools that could lead to actual human salvation.

Older aboriginal faiths often were more flexible and nature-based. Their parables would often seem paradoxical. They challenged thought and inspired contemplation. They were tools for strengthening the mind. Such tales were coded with information on how to survive, who to trust, and when to act. Nowhere in these older creeds is found the command to conquer, dominate and control. But, again, it is when spirituality became corporatized that the troubles began.

A fascinating and dangerous modern expression of "the One" can be found in our current conception of the corporation. Corporations are business entities formed to operate in perpetuity. Thus, they have a certain amount of personhood, which allows for business continuity even after the founders have died, which makes some sense. This arrangement ensures shareholder protections that might not exist in an ownership or partnership scenario. Originally, corporations were granted these rights of an individual so that, for example, they could find protection from unreasonable searches and seizures, or simply have the ability to enforce contracts across jurisdictions. But the boundaries of their artificial lives have expanded considerably since the dawn of the 20th century.

The U. S. Supreme Court has been instrumental in granting more and more constitutional rights to these artificial entities. After the Civil War, corporations won protections under the 14th Amendment as if they were humans. And it was hard to argue that companies with a national scope would not benefit from equal protection, from state to state, as they conducted their trade. The dark side of this business as a person is the corporate entity's ability to insulate the actual, flesh and bone people who run it from legal liability or prosecution.

Let's imagine, for instance, that Exxon made a series of terrible decisions that led to an Exxon Valdes-type disaster. While the company can face criminal prosecution, fines, and potentially dissolution, the people who made those decisions would not. The sea captain, who was sober as a judge (and probably asleep at the time), received harsher penalties and stiffer fines than the CEO, who merely got slightly embarrassed. In these situations, the shareholders might lose some value of their investments, but even if the company was found to be an ongoing criminal enterprise, they could potentially escape prosecution altogether. A pretty good deal, all the rights, none of the responsibility. The power of "the One."

In 2010, corporations received the Pinocchio treatment from the Blue Fairy that is our Supreme Court and became actual boys and girls. Tap, tap, and with the swish of a wand, here's some real magic for you. Citizens United v. Federal Election Commission granted corporations full First Amendment protections. Citizens United was a Political Action Committee (PAC) whose sole function was to advance "conservative" issues. It still exists and is funded by large corporations and, notably, the Koch Brothers[152]. And while they were united, there were no actual private citizens involved in their organization. As mentioned earlier in this text, this decision allows corporations to donate unlimited money to political campaigns. Before this, corporations had to stay out of elections. Except for initiatives that directly affected them. Now corporations have the right to free speech, and now their money is speech. Let no one or no state infringe on their unlimited political spending. However, just because a court says the moon is made of cheese, does not make it so.

There's another chilling offshoot to these First Amendment rights — The right to the free practice of religion. Corporations can now be true believers. Wow! It's about time! Now they can discriminate against people based on whatever criteria they choose, as long as it offends their sincerely held religious beliefs. They can deny their workers reproductive health care if it makes them doctrinally uncomfortable. They can fire LGBT folks at will and deny them any rights of redress. For some reason, there has yet to be an instance where a corporation lowered their prices to decrease profits because of their deeply held beliefs. Until they demonstrate some universal love for humanity in their custom, we'll have to conclude that these religious corporations are merely cons, designed to skirt the law.

The Framers founded the United States of America as a secular nation. They formed the Constitution to protect everyone equally. The Supreme Court has opened this new loophole to let religious interests in, potentially violating the First Amendment of the US Constitution. And they did it with the First Amendment of the Constitution. They must feel like such clever boys and girls. Shamelessness defines such sects, which need to discriminate against those weaker than themselves to find validation.

By design, no one should discriminate against anyone for anything that is a legal activity or status. And that pissed some people off. Hate, prejudice, and vindictiveness are integral parts of the status quo. It's the flash that distracts the populace while they empty the tills. It is a perk of power, letting the mighty crush the weak without having to break a sweat. And now it's legal- if you're a

---

[152] Look them up if you don't know who they are. Hint: oil money- lots of it, insane amounts of it.

corporation, that is. Don't try this at home unless you have a ton of white-shoe lawyers on staff. Non-profits, churches, as well as pharmaceutical companies are all (or can be) corporations. And they can all discriminate in ever-growing ways against whomsoever they choose, as long as those people offend their theology.

Corporations are not people. They are groups of people. That realization should be a given. If they want an individual's rights, they need to accept the responsibilities of one as well. When they break the law, they should all go to jail, just like the rest of us. When they wreak environmental havoc with disastrous spills and melt-downs, they should have to pay, at least some, out of their own pockets. That would stop a lot of corporate skullduggery on the spot. But that is not what the status quo desires. Instead, they want to continue the creep of rights and power invested in these imaginary constructs. Corporate status protects the insider and allows for all sorts of unethical and corrupt exercises. The citizens need to put the brakes on this and put corporations back in their place. There's nothing inherently wrong with the existence of corporations, but there seem to be nearly endless problems with their cultures. These ethical issues point to the current conflicts of interest when corporations attempt to do proper science. Corporations are piles of bias, legally protected, and patent approved. They have proved repeatedly that the truth is secondary to their continued immortal existence and profit-taking. They are conglomerates of individuals hiding behind the protection of a legally invented homunculus. And that's perhaps the nicest thing you could say about them. Capitalism turned sour when the profit motive was seen as justification for any action that increased that profit, any action whatsoever. Lying in court, pinning the blame on scapegoats, media manipulation, and group-think cultures became the norm. Corruption becomes so ingrained that it is no longer recognized as such. It's just jealous protection of the "brand."

In 1992, Stella Liebeck spilled hot coffee on her lap in a McDonald's drive-thru and burned not only her lap. She burned the phrase "frivolous lawsuit" into our collective awareness. You've probably heard of this case and have an opinion about it. That is by design. The truth here is far murkier than a styrofoam cup of Joe with cream, however. Being the absolute opposite of an ambulance chaser, Mrs. Liebeck initially approached McDonald's after her accident. Humbly, she asked that they merely cover her medical costs, but they scoffed at the very idea. Admitting wrong, any wrong, was against their corporate ethics.

So, she sued them. That's the American way, and everyone's right in such a situation. And so, the corporation counter-attacked her. And they did so so brutally that most of us remember it to this day.

Mrs. Liebeck was a passenger in her grandson's car when she asked for a coffee with the cream added at a McDonald's drive-thru. They didn't add cream; it came in single-use plastic deposit- fill-seal containers. Deposit-fill-seal containers are manufactured objects, a type of blister pack that is easier to manufacture than to use (by humans). She'd have to add it herself; that was the store policy. Her grandson pulled over and parked so she could open the lid and pour in the cream. The cup exploded when she pulled off the top and covered her lap in coffee just below boiling point: 88° C (190° F). The coffee came in a single-use Styrofoam cup, with a single-use plastic lid, both of which were engineered to snap together easily once and only once. The force required to separate them to add the cream was greater than the force needed to crush the cup. Indeed, the cup had minimal structural integrity without the lid attached. She suffered third-degree burns on her legs and crotch. The damage was so horrible she required skin grafts on her thighs and genitals. They immediately scoffed at her injuries, but the corporate officers did want to see the photographs.

These painful procedures were the source of the medical bills that McDonald's refused to pay. Instead, they preferred to make her sue them, choosing to spend millions to shame her, and slander her name, then to just do the right thing. And all of this had happened before. But the McD corporation had always bluffed their way out of responsibility previously. (Gerlin, 1994) Remember that: It had happened before, over 700 times, actually. But they'd always litigated their way out of any responsibility for their coffee's blistering temperature, or just ignored it until it went away. Uncertain that they could win in court, McDonald's began a campaign to discredit Stella in the public eye. If that worked, they might stack any future jury in their favor. With a pool of public opinion firmly against this "money-grubbing old woman," they could potentially win even though the facts were solidly against them. And gosh dang! It almost worked!

Many people remember it to this day as some sort of gross injustice and a crime against the poor, innocent hamburger clown. And a whole segment of society was so upset about this issue that they actively voted to constrain future claims against large corporations, effectively reducing their own rights, should they find themselves in a similar situation.

In the end, McDonald's lost in court because, as one juror said: "(the company displayed) callous disregard for the safety of the people." The jury knew the facts, having sat through the entire case, having had a front-row seat to McDonald's trickery and dishonesty. The publicity campaign that was raging in the outside world had not settled upon their ears, as it did with the public at large. She received $3 million from the jury. McDonald's quietly lowered their coffee holding temperatures worldwide. But they did not stop in their persecution. They wanted a warning to anyone else who might have the audacity to sue them in the future.

This case was cited in congressional hearings, on numerous television shows, and almost always in an intentionally inaccurate manner. Even after the corporation lost, they pursued the case in the public arena. Only this time, they could do it without the need for any facts or fairness. Legislatures passed bills in many jurisdictions to protect our poor corporations. George W. Bush (#43) ran for president on a platform that included a firm stance against frivolous lawsuits. What that really meant was protecting the powerful from culpability for crimes against the growing powerless. And America cheered, sharing a victory with those who sought to diminish them. This tale illustrates not only how corporations operate. It reveals how they believe they must operate. Vigorously defend everything, regardless of fault. When threatened, attack back. This habit alone should clearly demonstrate why corporations have no business in politics and must be highly suspect when wishing to conduct science. Corporations have shown that the less regulation enforced over them, the more narcissistic and sociopathic they become. Humility is never an option for such a mindset. And so, they lie and cheat and deceive, and humanity faces extinction.

There is this (incorrect) belief that capitalism is some sort of magically perfect economic system. It is said that markets are self-regulating and will always, in the long run, bend towards universal benefit. This is, of course, hog-dip. There is, presently, a perfect model of un-regulated capitalism and that is the drug cartels. Here, you have people so jealously guarding their brands that they have their own armies. They will happily kill any who oppose them and corrupt entire governments for profit. They will mix ever more addictive substances into their products to trap more people into an endless cycle of consumption. A cartel can alter any compound with any adulterant at any time for greater expedience or profit, without repercussion. That's an absolutely FREE market. It might be a good time to remind ourselves that the government is an extension of us. And its efficacy pretty well reflects how dysfunctional (or united) we as a population are.

Corporations are not people.

Science is not a philosophy. It is merely a method of discovery. Not all scientists understand this; however, many have their own problems with "the One." They think there is only one way to think, act, or be. Science is best when unfettered by bias. To ask a question scientifically is to accept the possibility that you may be wrong. Science is about being wrong, being wrong until you eliminate every other possibility than right. Science is humble.[153] Science, when properly conducted has no will. It has, like nature, the will of no will. And that's why it works.

Like some religions, corporations state absolute truth and seek to enforce it. Science looks for truth by trial and error, with little to no preconceived notions of what that truth may be. The power of science is that it starts from the simplest of premises: "We do not know what we're doing or what is going on" and proceeds from there. The discovered truth is only as stable as the next experiment that might re-write it. While perhaps not everything is possible, it should be worth considering until shown to be in error. In reality, such an attitude is tough to achieve, so everyone owes an incredible debt to those who walk this line.

All of these reasons are why science needs to be under the protection of the people. Science can and should be overseen by universities and specialized non-profits, organizations that are more than qualified to function transparently

---

[153] Although sometimes it can be sarcastic, as when debating flat earthers. "So, are all planets flat, or just this one?"

and ethically. When profit is removed as the goal, then the pursuit of truth can come to the fore. Corporations can still sponsor the research, which could restore much funding to charitable and academic institutions. This might lead to, heaven forfend, a reduction in tuition costs for future students.

Many non-profits, especially those who venture into the fields of science and the environment, are currently suffering. Politicians who are out to just score points can defund them easily. Grant funding is so regulated that it can be impossible even to afford to apply for many funds. And then those who support the common good may be slandered and belittled by those who stand to benefit from the loss of this sector of the economy. Many in the for-profit sector see non-profits as competition. A united citizenry could compel them to cooperate.

Corporations can still achieve their benefits. They can co-patent products produced and tested in this manner. But gone will be scientific trials run by people with MBAs in marketing and funded by accountants that demand immediate results. Slower testing overseen by ethical scientists, whose most important goal is to get the results right, can improve outcomes. And lab work performed by trained people who are not under the pressure of threats or job insecurity will be of a higher caliber than that which has given us countless questionable pharmaceuticals. Drugs that eventually breed litigation faster than relief.

In 2021, California congressional representative Katie Porter released a damning report on the impact large pharmaceutical companies have had on the nation's health. Entitled: "Big Profits, how big pharma takeovers destroy innovation and harm patients." It outlines how these mega-pharmaceutical corporations have taken to a course of profiteering over all other concerns. Between 2010 and 2020, the number of large drug companies in the USA decreased from 60 to 10, a six-fold decrease. As these monopolies consolidated, they spent more money acquiring smaller, genuinely innovative companies than doing any research on their own. In addition, they would frequently take the new products they'd bought and halt their release to eliminate competition with drugs they already had on the market. It didn't matter whether the new drug was more effective or cheaper. All that mattered was that they would give each product a chance to squeeze its maximum profit from the marketplace.

And those drugs that were on the market saw price increases year after year, sometimes by as much as 1000 percent. She documents how some appropriated companies had their vibrant cultures of science and exploration crushed by their new corporate overlords. The number of new, needed drugs released every year has been on a steep decline. And there is no end in sight.

> "from 2000 to 2008, nearly 2,000 drugs in discovery were discontinued, and the vast majority of these decisions were made not because of any issues with the science but for "strategic" or "financial" reasons" (Porter, 2021)

The massive corporate tax cuts that occurred in 2017 were supposed to free corporations to become more innovative. These windfalls could have been used to give raises to scientists, general employees, or for new facilities. However, the money was not spent creating new, life-saving drugs as they had previously promised. Instead, it was used by many of these corporations to buy back stock and increase the wealth of their executives and stockholders. But that probably doesn't come as much of a surprise. When corporate culture becomes one of rampant greed, no good will come from it. Corporate misconduct is on the rise at the beginning of the 21st century, and until we the people step in, there is no reason to assume it will get any better. When profit is the only thing left of value, then everything else is made valueless.
This poisoned corporate culture needs to change. Many seem to be moving in that direction already, as executives burned out and disillusioned by their former masters' schemes move to make new ways of doing business. But the power of some multinationals is tremendous, and their desire to do better seems lacking. If they don't want to come along with us to the future, society can compel them with regulations. We could impose self-control, self-awareness, and social responsibility on them. Or finally, the people could bring them to task for their misanthropic behavior and break them apart. Corporations are not people. It might become prudent to remind them they have no inherent right to liberty or the pursuit of, well, anything. Those privileges belong to actual human beings. If they want to keep their artificial personhoods, then they better act like people and not over-privileged, vicious brats. You can put them on notice by voting out the politicians who protect them, regardless of supposed party affiliation.

Our final troubling "One" is currency, money, the economy. It is crucial to begin by stating the obvious: money is not real. Currency has grown in importance as societies have advanced. A great tool to assist in the growth of markets, cash was intended to be a floating variable of exchange for tangible goods, services, and property. These goods, services, and properties were always the real valuables. Money was once an object of value itself, such as silver or gold. But the limited availability of these actual items became a burden on economies where production, labor and innovation outstripped the availability of scarce metals. Therefore, a more flexible form of exchange was desirable, as it could stimulate growth and not constrain it. New currency acted as a stand-in for value as various commodities moved from hand to hand. Hence, our current money is not to be viewed as a possession itself, but as a social instrument, owned by all and lent only to support easier trade. In The United States, all currency has the title: "Federal Reserve Note" on them; that's because the money belongs to the government, and therefore it belongs to the people. It exists solely to aid in the trade of those goods and services. But what has happened instead is that currency has become an end unto itself and not just a means. The accumulation of wealth has become an industry wholly divorced from any actual product or labor. Money is traded and lent, multiplied at each stage, and consolidated in the hands of people who may do little more than math on balance sheets. Some just write programs to slice milli-pennies from light speed computerized transactions and dump them into offshore accounts. While this is illegal mostly, it is allowable for certain corporations to do just that. These pursuits have become an encumbrance upon our economies. When the people who do the work and create things are bound by a scarcity of currency, now held by those whose contributions may or may not benefit society, that excess of money may be used to flood elections with cash and feed a less than honest messaging machine.

Today, the symbol has become more valuable than the actual thing, and this must stop. That it is somehow more economically efficient to ship raw materials halfway around the world to be assembled by extremely low-waged workers and then back to the domestic market is insane. The actual cost in terms of energy and time (and carbon pollution) is obviously greater, but our obsession with money makes such waste seem like a wise idea.

Money has a place, but that place is not primary. It was created as a tool and it should be used as such.
The bottom line isn't.

Economists know the foundational power of the economy is belief. Society believes this system works, so it does (more or less). The problem is, our current economies are not working so well at present. This is despite the fact that stock exchanges are soaring and GDPs are on the rise. We now often find ourselves not doing what we know must be done because of the belief that we lack the means ($). This excuse is more hogwash. The expertise is there. The human power to get the job done is as well. If the economy collapsed tomorrow, the sun would still rise, everyone would still be here. The crops would continue growing. The water would still be flowing. The only thing missing would be the piles of rectangular pieces of paper. Paper you can't eat, as Alanis Obomsawin[154] so astutely noted.

The Covid-19 pandemic brought this fallacy into sharp focus. There was no capacity, it seemed, in our current Monopoly game to address the simple need to pause for a time. Our economists were having panic attacks when faced with a reality that required us to sit still for a few months or face the disaster of perhaps millions of deaths, deaths that could create actual economic meltdowns worldwide. Everyone could be okay if they would just stay put for a while and act calm. But the accountant's balance books were about to explode. Television pundits were almost wetting their slacks, one moment predicting destruction and the next pleading for patience. Our sun was still rising, and people were still available to work in some capacity. This crisis reduced the need for our society's constant driving about and its subsequent fuel expenditures and carbon discharges. Most people were willing to accept lockdowns and social distancing. Some, for purely political reasons, were not. So the pandemics results were less disastrous than they could've been but more deadly than they needed to be. From the top-down, poor leadership made

---

[154] Alanis Obomsawin of a Canadian first nation, the Abenaki from the Odanak reserve, became a documentary filmmaker of note. Her complete quote was: "When the last tree is cut, the last fish is caught, and the last river is polluted; when to breathe the air is sickening, you will realize, too late, that wealth is not in bank accounts and that you can't eat money." And while it is often attributes as a generic "Native American Proverb" Alanis actually said it in a meeting in 1972.

matters worse and never quite answered the question of how to best deal with worldwide disasters. As a result, the long-term effects of this disaster on the world's economies remain to be seen.

New challenges face us in our future, and any money spent to keep our nations afloat during the pandemic was only so much paper in the wind. We got to see that acting with contemplation and restraint didn't doom us after all. But, unfortunately, we also saw the disasters created by pretending that "business as usual" was more important than public health. By the time the dust settles, millions will have died, and many needlessly in our technologically advanced world.

So, what now? Will our societies figure out how to keep the artificial flow of numbers on balance sheets happy, or will humanity curl up into a ball and give up? Essentially, this is a psychological conundrum: returning value to goods and services and relearning that currency is merely a placeholder. Can our societies let go of our attachment to cash? Can they stigmatize or even halt its hoarding? Now that it is understand that cash is a shared social asset, there should be no trepidation about re-allocating it if it's being used against the greater good. Especially those gains which were gotten unethically, or even illegally. Public awareness will be at the core of our new economies, and if they have the will to let currencies flow instead of being held by the few, then the value of the placeholder can begin working for us instead of against us. People need to realize that money is being held hostage by the few to keep the rest of us bound. The hoarding of capital is analogous to the acceleration of entropy in our systems of trade. The more money held out of the system, the more that system falls into chaos and disorder. Our economic systems are not true thermodynamic systems, therefore they do not need chaos to balance order. They can be kept in balance by fairness and wisdom.

All money belongs to everyone. It's just a tool and not a thing at all. When the economy is freed and capital flows, we'll have the financial means necessary to save ourselves and the world. It is the metaphorical equivalent of removing the dams from our wild rivers. Only this block is mental and ideological.

Save the world; we must, so let's do it. If our economists tell us it can't be done we can fire them and hire some who say we can. It is that simple.

America used to love Alan Greenspan, even though his theories were almost always wrong. Trickle-down economics was pure BS, yet it worked for a few. Our economies are trickle-up currently. Money flows from the working classes into the pockets of the ultra-rich, where it stays. Corporate capitalists, who tend to school and feed like piranha, overdraw everything and cause repetitive crashes with their frenzies. This forces the government to step in and bail everything out with a fresh injection of cash that can once again begin it's upward concentration. We've been repeating this pattern for quite sometime now. Our society could fix that, if everyone wished, they could create an economy that mirrors the thermodynamic system, a system that actually reflects reality. Then such a system could be enacted; it almost seems we have just preferred not to.

The economy must work for all of us; otherwise, it has no use to any of us. There's criminal justice, but very little talk of economic justice, at least not by the status quo. What has been used as a tool of conquest and subjugation can become one of fairness and equality. It is, after all, something people just made up. It is not a force of nature, a law of physics, or a meteor racing towards us. Therefore, the economics of survival is perhaps a more straightforward fix than it at first may seem. But it will require a step beyond the mindset of "the One" and into the realization of the many. It also requires societies stop treating life as a parlor game and face reality.

The thermodynamic economy is one where everything balances. Every action has an equal reaction. Energy is neither created nor destroyed. It just changes form. Profit does not exist in nature unless you realize that your profit is the privilege to live. The paradox is that profit seems to motivate people well. But just like birthday presents and ice cream can influence children, it does so for only a short time. Once a person gets more money, they may find even more wealth is desired, and even yet, that cannot fill the void, creating a cycle of the empty pursuit of riches. Wealth can not fulfill us, because it isn't real. It is suitable as a tool and a customary exchange, like a handshake, and nothing more. Adults can find more meaningful motivations, perhaps the greatest of which is the freedom to live in peace, the opportunity to find companionship. The actual economy, the interrelationship of life, the exchange of gasses and nutrients are how all life has survived to this point. We would be wise to learn from nature on this subject.

If our unwillingness to cooperate destroys the human race, it will be the saddest display of failure in the history of Gaia.

We must then look to nature as the model of the effective use of energies and resources. Everything recycles, usually through multi-stepped activities by different organisms and mechanical processes. Recycling must be the foundation of our economies, not just an attribute of them. Every truck that moves to a landfill is an opportunity lost, tangible wealth squandered. When everyone finally realizes that goods and services are the valuables in our economy, that they are the valuables, it will be understood that waste destroys value. Our cultural obsession with: "Everything, all the time" probably isn't even a core desire of our beings. It is just an entrained habit many have fallen into that only benefits a few while stressing out the majority. A more holistic, slower lifestyle that is more firmly rooted in the ebb and flow of the seasons might increase personal feelings of well-being and health more than a thousand therapists or all the pharmaceutical drugs ever concocted. Humans evolved as active beings that interacted intimately with the world around us. By becoming that again, joy and wonder will replace many of our fears and anxieties. We evolved to live in amazement, and every creature seeks joy.

This job of planetary steward is the one job we are qualified to perform. The position of taking care of each other is the greatest employment to be had. The real conservatives are, like the original peoples of this land, humble and grateful for life. It is a conservative human value to live within one's means and keep the planet safe for future generations to come. It is a conservative value never to waste nor incite. This responsibility can be summed up in the seven generations philosophy[155] of the Iroquois and other peoples of this land:

> "If you ask me what is the most important thing that I have learned about being a Haudenosaunee, it's the idea that we are connected to a community, but a community that transcends time.
> We're connected to the first Indians who walked on this earth, the very first ones, however long ago that was. But we're also connected to

---

[155] The seven generations philosophy is simple: all decisions should be made not only for the benefit of the present generation, but for seven generations into the future. Very different than a corporate "this quarter, next quarter" type of motive.

> *those Indians who aren't even born yet, who are going to walk this*
> *earth. And our job in the middle is to bridge that gap. You take the*
> *inheritance from the past, you add to it, your ideas and your thinking,*
> *and you bundle it up and shoot it to the future. And there is a different*
> *kind of responsibility. That is not just about me, my pride, and my ego,*
> *it's about all that other stuff. We inherit a duty. We inherit a*
> *responsibility. And that's pretty well drummed into our heads. Don't*
> *just come here expecting to benefit. You come here to work hard so*
> *that the future can enjoy that benefit."*
> -Rick Hill Sr. (Tuscarora) Chair, Haudenosaunee Standing Committee
> on NAGPRA[156]

The Haudenosaunee is a confederation of six Iroquois nations who practice the past peoples' ways (and those of the present and the future). They demonstrate a proper conservative way of thinking and so we should surrender this mantle to them considering the failure of those who in-authentically claim that title presently. Those who are concerned with fairness in our society, and those who work for ecological restoration and justice, will be seen as the true conservatives in the future. For without them, there will be no future.

You are the end of a chain; from the very first life on this planet is the unbroken lineage of Gaia, leading up to us. You are billions of years old, by this reckoning. The matter that makes up your physical body is older still, having come from the stars. There are many branches on the tree of life. They all hail back to the root. And they all bathe in the sun's light. We all live on one world, and it's the only one we've got. This order is our immortality, and it is our birthright. There is only one human race. We are all royalty, for we are the descendants of the original "One," the first cell to organize in the broth of Gaia's primordial seas. All life is based on the molecule DNA. In unity, we survive; singularly, we will fall.

Now we can finally see the ultimate meaning of "the One." The one is the many seen from a new perspective, the perspective of a thermodynamic system. A system is a function. The group is the singularity. From a systems point of view, many discreet beings, trillions of lives, formed the past. A

---

[156] Native American Graves Protection and Repatriation Act

similar number has joined in expressing the present. Within the teeming masses of life lies our future, as well. If we still choose to have a future. The One is the unbroken chain from the first cell to you. It is all the life present in the now. An interconnected web of life gave us this now. And the continuation of this interwoven system is what we must provide for the future. If we can live out this simple understanding, we will have evolved.

The Green Evolution offers us the conscious choice to evolve.

CHAPTER TWELVE: TRAINING

"Education is the cheap defense of the nations."
— **Edmund Burke**

L emmings are small, vole-like animals of the Genus *Lemmus*. They are a cute little group of fuzz balls that look like a cross between a hamster and a guinea pig. Lemmings have a curious reputation for running en masse off of cliffs, following a leader to their mutual deaths. This peculiar habit has made them infamous, even though most people couldn't pick one out of a police line-up. The curious thing is: they don't actually do that. No animal takes its own life, with one unfortunate exception. Some may sacrifice themselves to protect their colony or may simply expire from grief after their most loved one dies. But, in general, if an animal chooses its own death, there's an altruistic reason behind it all. Natural interactions, even those that appear to be competitive, are fundamentally cooperative. Even death is a step forward for life.

Walt Disney produced a documentary called "White Wilderness" in 1958 and created this myth about lemmings. An apocryphal tale that has clung to our collective imaginations like sticky fly-paper. The film crew, working in Alberta (where no lemming species is native, by the way), staged a mass suicide of imported animals merely for "cinematic effect." They didn't need any research to back up this claim of animal mass hysteria. All they needed was a turntable, some tight lenses, and a complete lack of respect for accuracy. (Mikkelson, 1997) An absence of concern for animal life was helpful as well. The ASPCA[157] was not required to be around when they made their little G-rated "snuff-film" in the 1950s, as they are today.

So the Disney crew, who demonstrably knew little to nothing about wildlife, felt qualified to make "educational" films about them. "Who cares?" they may have said; "if nature doesn't act like this, perhaps it would be better if it did". It would certainly be more entertaining. Here is an example of ignorance transforming into myth before our eyes. Little did they know they were setting

---

[157] American Society for the Prevention of Cruelty to Animals. Pretty much every movie that uses animals now, has a representative on set to insure this sort of thing doesn't happen.

up the human race to actualize their folktale. They probably wouldn't have cared; Walt was making money hand over fist.

Humans strive to make the world in our image. And we use stories, fables, and folklore to do so. Mass suicide is not a problem in the natural world, as only one animal practices this behavior — the one that reads and writes books. Additionally, the only creature capable of deceit is the human being, and they do it well because they have trained themselves to do it well. Training ourselves is an essential task. But in the absence of actual knowledge and wisdom, training can become mis-education that can directly lead to failure. For mis-training can give one the impression that everything is fine and rosy, right until the cliff's edge.

When a winegrower attaches a vine to a trellis, they are training the plant to grow in a specific manner. If done correctly, it will improve the plants' access to light, maximize growth potential, and support the weight of a heavy harvest with minimal damage to the vine itself. That is, if the grower knows what they're doing. A novice could easily misuse the same techniques and hasten the failure of the crop. Humans have interacted with nature in such a manner for thousands of years. The march of civilization has capitalized on our knowledge of how things grow and interactively nudged them along for our benefit. Generations built upon these successes until humanity created a complex web of strategies and tools that, while increasing our control, led to an ever more unstable enterprise until we reached the pinnacle of the Green Revolution.

For centuries people have exploited areas, exhausted their resources, and then moved on to the next. Since we did not recognize that life was a group endeavor, we also failed to notice that our actions were breaking nature's interrelationships. Unaware, humanity planted the seeds of its ultimate undoing along the way, like Johnny Appleseeds of calamity. We failed to realize that all life forms had diverse interactions with each other and their environment. If there were groups or individuals that understood this, they were generally either ostracized or expelled. Gaia was cleared of the earth-based ideologies, cleansed by the swords of those who worshiped more jealous gods. Early secular investigators of nature experimented and collected, seemingly with no concern for consequence. So, while humanity could've used our growing husbandry powers to benefit all of life, the march of progress focused primarily on our wants. Not upon our needs, but primarily upon our wants.

Our perceived mastery of the universe made us oblivious to the reality that nature gave birth to us, and not the other way around. Some felt free to make up any explanation for the world they saw before them, and those who might disagree could not speak otherwise. Because they could not speak. They could only chirp and trill and howl. Unable to understand their messages, human cultures were free to birth to gods and give them absolute power. They then assumed that power would trickle-down to serve them as well. That was a severe error in judgment.

We have trained ourselves to ignore the plain truth. We have been too successful without an actual plan of what we wanted or where we're going. Humanity has grown empires with little or no goal beyond proving to our myths that we could. We are the dog that caught the car. So what are we going to do with it? Shall we rip it apart, pee on it and watch it rust? Shall we remain heedlessly unaware that this planet and its ecosystem can take us somewhere?

No, we will not. Not any longer.

It is hard to conceive of nature's wondrous interactions if you have rarely set foot in it or only seen it in some staged documentary. Even reading a book such as this is no substitute for lying in the grass on the side of a mountain, watching the clouds drift by, or the marmots play. Instead, most of us live indoors, have food delivered to them, and think that life is about concepts like safety and pleasure. With the freedom to judge and little fear of being judged in return, too many of us sit safely in front of our screens and pretend we are in control.

We are not in control. We never were in control and never will be. But we can live within the world's balance as it has evolved, and we can nudge its systems to work to our advantage. We can use our husbandry skills to better the world. But only if we do it as a good partner would, paying close attention to the other's needs. We can even heal the damage we have inflicted. We can glide over the waves of existence to significant benefit, but try to overpower the ocean, and we will drown. Just ask any surfer. They will set you straight.

Humanity cannot re-create the world in our own image, which is really what we've been up to these many centuries. We've been the children who were given the keys to the candy factory, and the results are just what one would expect. It is time to grow up and become what the human race should be—the stewards of this planet, not tinpot dictators. Eventually, all dictators are deposed, and rarely in a tidy or peaceful manner.

In the most fundamental of ways, the climate crisis is a crisis in education, under education to be precise. Our primary method of training humans in our modern world is our educational systems. All developed nations have them, and most developing ones do as well. Yet, in the United States, many consider ours a damaged system. And for different, sometimes opposing reasons.
This injury began shortly after breaking the solar panels on the White House roof when President #40 chose next to break our school systems. Ronald Reagan took a study on America's education system (Panel, 1983) and came away with the conclusion he'd probably reached before he even read it (if he read it, that is). Our problem was our damn teachers; Liberal[158], bleeding-heart, weak teachers. Teachers who were indoctrinating our children into radical, permissive ideas, such as justice and fairness. Instead of instructing them to swear allegiance to our perfect capitalist patriarchal system and fight any who might question it.
Reagan had already decimated public education in California when he was governor before his elevation to national office. He slashed State funding of schools which caused the closure of many poorer schools and the depreciation of all the others. As President, he tasked his education secretary, Bill Bennett[159], to fix the problem of an education system that was failing to produce enough young "conservatives" nationwide. Their answer- was: "Time to get strict with these teachers and their students as well." First, they famously declared ketchup a vegetable to reduce the waste (cost) of school lunches.

---

[158] Dictionary definition: "a supporter of a political and social philosophy that promotes individual rights, civil liberties, democracy, and free enterprise." But when I capitalize it, it is a label used as a derogatory by those who believe in none of the things listed in that definition.
[159] "Bully Bill Bennett" -according to Jonathan Kozol, a veteran education activist and researcher. Bill wanted to eliminate financial aid for college stundents, suggesting if they owned a stereo they could afford college.

They next began categorizing students, America's children, the same way people graded potatoes on a conveyor belt. "Got to get tough with these little bastards!"

Predictably, none of their reforms helped improve our educational system's state. And that is because teaching students facts, with all their messy ramifications, leads to more open-minded citizens. Open-minded citizens are hard to lie to and challenging to coerce. The dreaded liberalism was just a by-product of teaching critical thought, unfettered curiosity, and empathy. The only way to instill the status quo's desired ideology would be to distort the instruction of reality and suppress dissent. And that was, and still is, the "conservative" goal for public education—that and funneling public dollars into private pockets.

When George W. Bush Jr (#43) enacted "No Child Left Behind" to answer the lagging achievement initiated by his predecessors' ineptitude, he created a system of testing and measures that mimicked those used in tractor factories to increase productivity. Year after year, his requirements of rising test scores, like a banker's projection sheets, set the stage for our present horror film scenario. Teachers' wages were linked to production (test scores). Schools could be penalized by reducing their budgets when their numbers fell, precisely the opposite of how one should address a systemic weakness. This general demonization of the profession had what many would see as the desired effect; the depreciation and erosion of the American education system. (Smith, 2018).

Rudimentary skills and weak critical thinking abilities favor employers who want easy to control (and cheap to pay) employees. And it also favored a political movement that hoped to enrich a very few and prey on the ignorance they had instilled in society. People who struggle and lack a complete understanding of their governance will easily fall for jingoistic slogans and partisan manipulations. They might even elect grossly incompetent leaders who promote fantastically preposterous notions. For example: "The country needs more oil, not less! Those egghead scientists are lying to you so that they can line their pockets with 'easy' grant dollars." They might interpret increasing gun violence in our schools as normal or tolerable; and not the result of the system's stresses upon the teacher, student, and administrator. Combine this with a horribly open system of gun sales and possession, and everyone sees the results on the nightly news. Today: "This is America," as Childish Gambino so astutely observed.

Our education system is in trouble. When teachers need to work multiple jobs just to show up to the classroom and buy the supplies their students need to learn, something is wrong. The inability of some in society to see these people's heroism results from this system's corruption by those with self-serving goals. Betsy DeVos, the US Secretary of Education under President #45, continually demonized public education. She saw it as just another pot of money to raid that perennial "conservative" goal: the privatization of everything and ownership of the world by an elite class. Her vision was one of cynicism, disrespect for democracy and a disregard for the plight of her charges. The true goals of today's "conservatives".

> *"If you don't live in an area with good public schools, you can move to a different place,"*- Betsy DeVos.

 So, if you're poor, and so is your school system, just move. Can I hear a: "Love it or leave it"? How about a "Vote with your feet?" How about strive to make all schools better? It is important to note that this woman hailed from an incredibly affluent family, and many saw her as out of touch with the public's everyday experience. She attended expensive private schools and married into the outrageously wealthy "Amway ™" [160] family.

> *"My family is the largest single contributor of soft money to the national Republican Party."*—Betsy DeVos.

And it seems those contributions bought her a Cabinet position in what will go down as the most corrupt and unethical administration in US history (so far). Betsy's ultimate goal was seemingly to de-fund public schools and replace them with for-profit private schools. (Spewsic, 2017) She would call it: "School Choice." But make no mistake, there is a desire to funnel money from public education into religious schools that can discriminate and teach

---

[160] Amway, the company that perfected multi-level marketing. Basically, a pyramid scheme that preyed on poor folks and enriched a very few with their colorful catalogs of crap.

fallacious doctrines (thanks to courts that allow this sort of thing), such as "creation science" or racial purity. Such schools could mass produce the long sought, culturally manipulated "conservatives" some interests have desired. Or, perhaps they'll just grab that cash for their private schools that may not have a public schools' accountability. An accountability to the public that once made our nation's educational systems second to none.

Ultimately, it is all about de-funding our local schools and making them 2nd (or 3rd) rate institutions for those with no other choice (read: those who are not wealthy). The goal has been to create a new serf class. And recreate a ruling class that can continue these private societies' love affair with centralized power and unbridled consumption by virtue of their wealth and privilege. Instead, it seems, they have only created a new type of pitchfork brandishing villager.

In 2020, the courts found Betsy in contempt for failing to cancel student loan repayments to people who had been defrauded by for-profit schools that were little more than scam factories. When ordered by the court, she refused to comply, sending people into debt collection and lower courts over obligations that had effectively already been waived. Why would she do this? It certainly couldn't have been to punish people for complaining about being ripped off, could it? Could it just have been to feel them squirm as she ground down her sworn charges with her spikey heels? Well, she sure smiled a lot while ignoring the court's order, and she certainly chose to next double-down on the whole caper.

Because Betsy then brought before the Supreme Court a case that allowed public funds to be transferred to religious schools, seemingly in violation of the first amendment. Schools that *may* discriminate based on sex (or possibly race) and could teach, as facts, things that are known not to be so (example: African slaves were just being tested by the Lord so that they would become better people[161]). She attempted to fulfill the "conservative" goal of elevating opinions to facts, thereby confusing the two. This intentionally incited confusion is how the status quo operates before the lights come on and they scatter: degrade, manipulate, subjugate, repeat. And she certainly made herself the face of that ideology.

---

[161] That's in one of the textbooks from a famous Christian Universtiy named after "Who's your Uncle?" Jones.

There is a way to fix education, and it's not just to pay teachers more or respect them for what they do. However, both of those would be a good start. We must stop treating the entire system and all its parts as cogs in a machine. Learning is a natural drive within humans, and they will learn whether we teach them or not. They just might not be learning things that will serve themselves or society as a whole. They might never learn how to determine fact from fiction.

Learning is a human instinct; training is the standard method of addressing that innate inclination. When people engage in the desire to learn with factual and diverse training, we all end up with a well-informed and capable populace. One that can be trusted to do the right thing and not destroy the planet out of folly, or to fulfill some ancient myth.

In nature, in reality, creatures mature in definable stages, but chronological age can not strictly predict those stages. Some children develop faster, both physically and mentally. The same applies to lambs, bats, and geckos. Some individuals grow slower. This attribute is not a bug; it is a feature of life. It is, again, nature hedging its bets with diversity. Numeric categories are not great tools to apply to stages of psychological or intellectual development. That is just an attempted imposition of our wishes upon life itself. Here is the accountant's desire to simplify the world by defining it solely with numbers. Our obsession with quarterly profits has blinded us to what is needed for true sustainability.

When you look at light after it passes through a prism, you'll see the spectrum: red, orange, yellow, green, blue and finally violet. But that's because you've been educated in color. The spectrum is not an array; it is a continuum. Where one places their pure colors could depend greatly on the physical structure of their optical receptors, or their societal conditioning. It could also have a lot to do with the color of the light coming from their sun. The take away is, wherever you plant your flag, or mark your first, second or third rank, is always going to be somewhat arbitrary. An educated person can take such an idea and find engagement in contemplating it. While those who lack a quality education will just scratch their heads.

American schools currently score and rank by numerical values in our schools and have very little discretion left to value our children in other ways. If a third-grader gets a 40% on a test, that child is going to feel like an abject failure. It's hard to pull yourself up by your bootstraps when the concrete, empirical measure of yourself is shown to be lacking. It is even more difficult when you have no boots. Such a child could become depressed and withdrawn. They could act out and reject the entire system, setting themselves up for stagnation and failure. Which, should be noted, sets up society for stagnation and failure as well.

As it's written, the law defines where you need to be (academically speaking) and when. If you are not there, where are you? And what of the few who are faster at this learning and find the subject matter boring or pedantic? They might withdraw or revolt as well. Our systems aren't just trying to force round pegs into square holes. There are star-shaped pegs, oblong pegs, and rectangular pegs, all told to go through that square hole. Ouch! When the system's primary method of dealing with nonconformity is punishment, might not those individuals who have been rejected turn to punishing others as well? Is this what we're really teaching the children?

There is another way, and not so strangely, it models nature more than our imposed assumptions and our ranked ordering of the world. The Northern Cass School District in North Dakota has begun an excellent deconstruction of our old system by eliminating school grades. (Berdick, 2018) Not grades on tests, but by removing grade levels K-12. No longer are you failing 5th grade because you are not in the 5th grade. This system allows students to move through their studies at their own pace. Some kids get maths faster than others, some are slow to grasp English. No problem, you work at the pace that is you. The stigma of being held back is eliminated because schools are no longer ranking by grade level. The child is simply somewhere on a continuum, always moving forward, never backward. No child would ever be held back a grade.

Sure, they will still conduct physical education and field trips in age-appropriate groups. And perhaps a person of the same age as your student will be their tutor on a subject they are having difficulty with. Because the tutor has already mastered it. How refreshing and de-stressing that would be. When educators stop assigning machine numbers to children, they free them to be what they are: diverse individuals. If you still don't get algebra by 8th grade, no problem, get it when you get it. Lowell Wood was a poor student in his early years. He often held the lowest score for many tests in elementary school. He persevered and entered UCLA at 16. He now holds the worlds record for the number of patents awarded to a single individual, 1,761. He took the crown from Thomas Edison in 2015, surpassing the formers lifetime achievement in patents received. And unlike Edison, he actually invented all of his patents[162].

The climate crisis is a crisis in education. The scorn of intellectualism and profiteering off of colleges to make advanced degrees a form of either privilege or servitude; was the Reagan plan. He was undereducated and a bit dim himself, but look at what he achieved (or don't, in the case of those who worship him[163]). An educated public is dangerous to those who want to mollify the masses with lies. It's better to have a voter base who has never read Orwell's "1984" because they might understand the doublespeak that is currently working so well against them. And if they've never read the entire book, they may misunderstand it differently because they only heard of it from a talk-radio anger merchant.

When a powerful group can convince the general population to vote and act without thought or even curiosity, they set themselves up for a special status to win debates with no facts and cheat the public right out in the open. They can act like brutes and intimidate and bully opponents who can't act reciprocally. This assault on education was intentional and coordinated with the calculated result of getting people to believe that the reality of climate change was still in question. Sure, they intended to confuse more issues than just that one, but global warming was always a primary topic of this disinformation campaign.

---

[162] Edison was a pretty "hands on guy" at Edison labs. But many of his "inventions" were actually the work of scientists and engineers in his employ. Warren De La Rue, for example, invented the incandescent light bulb.TE's team just imporoved the filament and got the patent. Edison excelled at business and corporations still use his "take your employees patents and put your name on it" trick. The first motion picture camera was invented by the Lumiére brothers in France prior to 1895. Edison is purported to have gotten his hands on one and patented it in the USA in 1897. He is, to this day, attributed with the invention of the movie camera. He was not.

[163] Reagan tried to destroy renewable energy and despised the environment. He traded with an "enemy" to fund right wing militias, all in order to overthrow democratically elected governments. He supported Dictators. He made light of the rape and murder of nuns in Central America. His administration held the record for most felony indictments for corruption until recently (30+ is the number to beat!). The list goes on- look it up.

There is no question about human-made climate change, only bluffs, tantrums, distortions, and lies. Yet, it remains a question in some minds and a tool for extended debate by others. The desired outcome of all of this educational manipulation was many people's inability to tell the difference between reality and pure bullshit.

Education is critical for the survival of humanity. There is no future where we can succeed by ignoring facts. There is no going back to hunter-gatherer societies. We've already messed up the ecosystem too severely for that. We won't be able to survive without technology, and that means we need educated people. *There is No Future* if we continue to believe instead of learning how to know. Our minds are our greatest tools, but like any tool, if we don't keep them sharp, they're no better than that dull chisel. The staggering, incomprehensible problem with the status quo's tactic of dumb-ing down the nation is that it had no actual plan, no final programmed outcome (other than the death of everyone involved?) So, it is interesting to wonder if the proponents of this anti-intellectual strategy also pulled a fast one on themselves. One look at an Ivy League graduate who, while elected President (#45), can't speak or read above an 8th-grade level, will suggest the answer is: Yes. Watch Betsy DeVos's congressional confirmation hearing to discover for yourself what it looks like when power is awarded like a participation trophy. So accustomed to privilege and the receipt of bounties, Ms. DeVos seemed to have just assumed she was entitled to whatever boon her "conservative" allies were about to bestow upon her. Despite her obvious lack of qualifications or even a basic understanding of the job offered her. A job whose particulars she could not articulate.

If a tyrant burned all books on science and religion, religious texts would return in one form or another in a thousand years. But the books of science would return, within that time, almost word for word. That is the difference between belief and fact. Facts do not require belief. Sadly, beliefs do not require facts, either. So, while they may seem similar, a solidly educated person can see the difference in a heartbeat. As with water, many want education to be recognized as a human right, not just a privilege of businesses that want a semi-literate and pliable workforce.

CHAPTER TWELVE: TRAINING

Organized religions were pioneers in the education of youth, in monasteries and academies, throughout our early civilized periods. Once, education was confined only to the aristocracy and the priesthood. By holding the keys of knowledge, they could exert unwavering control over the populace. This hierarchy explains why their tales warn so emphatically about the dangers of knowledge. They desired that it be only for the few. They would not teach women to read, nor slaves or those of lower caste. Power was for the knowledgeable. Knowledge was for the powerful. Wisdom was rarely of much use to either of them. Some wish to return to those days, but they can't. The planet won't survive the insult.

Now the Green Revolution has transformed many of our agricultural practices beyond training and into subjugation. Teams can now rip an organism apart at the genetic level and re-combine it with another, creating something novel. This little trick can be performed by almost anyone and requires no particular level of ethical development from whatever person or corporation undertakes this exercise. It is even easier than making methamphetamine in the home lab, because that requires a rudimentary understanding of chemistry. Bio-hackers test their edited gene sets on themselves and willing others with no controls and often no clue. A gene drive is an engineered organism that has been given heritable traits and then released into the wild. Not only can they alter the entire genome of their species, but that is the intention. These hackers can and have built these germ lines in sheds and garages. Corporations have made them in labs and released them into the ecosystem. The killer mosquitos mentioned earlier and the Monsanto plants designed to thrive under massive doses of weed-killer are just two examples. Once a novel organism has been released, there is no way to take it back. Nor is there any regulation to stop them, except for self-regulation, and that is one character trait people are still talking about more than exercising. So instead, we dilute the natural ecosystem and sue others when our creations go astray. Well, corporations do anyway.

Since the natural world has faced and overcome extinction-level events before, it is wise to look towards her for the "wisdom" that has made this possible. You don't need to look far because it is within all of us, from the King to a cockroach. Inside us is DNA, the singular life form on this planet. Strings of data encoded as nucleic base pairs make up our chromosomes. The chromosomes are bundles of these sugar/phosphate strands that contain the information required to build our bodies and define our beings. Thousands of genes make up each chromosome, and some are turned on or expressed, and some are turned off. Now is the genetic age, where advances in our understanding of these chemicals are occurring at a dizzying pace. It gives us powerful tools to alter all lives, our own included. But the means to save our ecosystems is within us; we don't need to hack genomes with CRISPR[164] to save anything.

Biohacking and the gene drives created by it are potentially dangerous and deserve intense scrutiny. Computer models now show that the potential harm to the ecosystem from these practices could be devastating. (Nobel, 2019) We have to take this Molotov cocktail out of the children's hands. Another sane regulation we could place on ourselves is to make the materials required to conduct gene-splicing controlled substances. Not illegal, but available only to those with a license and the integrity to use them wisely.

When science first began classifying creatures by their expressed types, how they look, act, and who they mated with, they created a tree that categorized life by these recognizable traits. DNA analysis has rocked this tree to its roots. As a result, scientists have discovered that some animals that were thought to be closely related are not, and life forms that seemed very different are, sometimes, next of kin. The discipline of taxonomy reveals this, and an explosion of genotypic reorganization is currently underway.
Taxonomists specialize in the naming and categorization of species. They have followed the rules laid out by Linnaeus[165] to classify life and its interrelationships. Taxonomy is a precise and intricate field of study. The recognition and recording of tiny differences are of the utmost importance to these scientists, and only the most studious and careful need to apply. They

---

[164] Clustered Regularly Interspaced Short Palindromic Repeats (in a DNA sequence). A method for locating specific coding sequences and editing them.
[165] Carl Linnaeus 1707-1778 - A Swedish botanist who standardized the categorization of life forms with the binomial naming of species. The structure of a binomial name is **Genus** followed by **species** (ex: *Homo sapiens*). The genus is a related group and the species is a singular identity.

express their work in museums of specimens and as libraries of texts. When they began sequencing DNA, a lot of their previous work got defenestrated[166]. As a result, they revealed a surprising new world to us. DNA is a great way to classify relationships in a precise and empirical manner. By reading the code of life, you can see connections in greater focus than just measuring wings and counting digits. Now is an exciting time to be a taxonomist as many long-hidden secrets are being discovered daily. And new, exciting names are being created by these folks, such as *Heteropoda davidbowie*, a tiny spider with Ziggy Stardust's shock of orange hair.

A recent, significant discovery concerns the expression and suppression of genetic traits. As evolution occurs, the old code is not just thrown away; it might be needed later when conditions change. A good example is the pigment melanin in human skin, so let's look at it a bit more closely. Melanin is the only pigment that humans possess. And some of us have more of it than others. Alleles are small bits of genetic code that exhibit measurable differences in physical traits when they vary. Dominant and recessive alleles in our skin's genetic sequence control our many skin shades.

In lands where the sunshine is more intense and relentless, the body can produce more melanin to protect the skin tissues from ultraviolet radiation damage. This trait expresses as darker skin. In climes with more cloudy coverage, shorter days, or both, the skin's genes might suppress melanin's production and become lighter. Some ultraviolet absorption is desirable, as it leads to the synthesis of Vitamin D. This is not a rapid change; however, it generally takes many generations to complete such a shift. Place a light-skinned person in Africa, and they will remain light-skinned. Place a dark-skinned person in Norway, and they are still dark-skinned. The only common individual variation of melanin production is minimal, and it is called tanning. So, in general, the allotment of melanin is fixed within an individual, the result of which alleles are expressed. And indeed, skin color is an example of genetic expression and is a measure of both which alleles are attendant and how they manifest.

All human skin is the same color. You can easily prove this by looking at the

---

[166] Literally: thrown out the window. Even if you have to wait a lifetime for the opportunity to use this word, it's worth it.

negative of a photograph with multiple people of mixed "races" in it. They will all have the same strange greenish/blue skin tint. This is because human skin color is a by-product of the skin's melanin content and not the result of different pigments, such as you would find in fish or tropical frogs. We see different races in humans by skin color and minor differences in eyelid and nose shape because *we have been trained to do so*. But, to our fearless companions, dogs, we all look the same. This controversy over skin color would be laughable if not used so tragically by so many and for so long. Just ask your local taxonomist; they'll set you straight about the human race (there is only one).

A standard function of speciation taught in school is: "Acquired traits are not inherited." This mouthful means that if your dad had his nose broken, you would not be born with a crooked nose. That is correct, but with one new twist. Genetic expressions are inheritable. Suppose your great-great-great grandpa moved from Egypt to Iceland, and his family's skin hue began lightening as it passed down the line, generation after generation. In that case, you can still potentially inherit an expression of his darker skin, perhaps centuries later. It's not necessarily a mutation when a darker-skinned child is born to light-skinned parents (or vice versa); it might just be an expression. Somewhere in your thousands of genes, there could still be one that might express as more melanin. This expression is just usually turned off if you are Caucasian. From time to time, it can turn back on. While this could be the plot of a television sitcom, it could very well play out as a sad tragedy in today's world.

The study, not just of genetic codes but their expressions, is the discipline of epigenetics.
These new understandings would never have been possible without the genetic revolution and its tools. We have yet to discover everything this new science has to show us, but even the glimpses into life we have gained so far are eye-opening. In a disturbing example of an epigenetic expression, a new study has shown that a mother's exposure to air pollution could lead to an activation of asthma disease in her offspring. This asthma can then carry to their offspring for up to three generations hence. The trigger in this investigation was the mother's exposure to diesel exhaust while pregnant. This exposure then caused a change in her immune system-linked dendritic lung cells. After birth, her

baby then exhibited the same modification in their dendritic lung cells. A change associated with early onset asthma. (Gregory, 2017) This mechanism is concerning because our systems pollute so wantonly that we may be inadvertently weakening our children in ways not yet understood. Perhaps there are built-in mechanisms that will suppress a species that actively damages the world for everyone else. The science of epigenetics is still in its infancy and may soon reveal more pleasant and painful possibilities for the future of humanity.

This cause-and-effect cycle became crystal clear when during the Black Summer wildfires in Australia, there was an uptick in premature babies. Scientists estimated that the smoke in many parts of the country made simple breathing the equivalent of smoking 30+ cigarettes a day. One woman, whose child was born weeks early, delivered an atrophied, gray, and grainy placenta moments after the baby. (Burke H. , 2021) The midwife had never seen such a diseased organ before in a successful birth. By the time this child reaches adulthood, we'll undoubtedly have many more examples of this syndrome. But will we have a greater understanding of its long-term repercussions?

Epigenetics also suggests a new tool in the toolbox for adaptation and evolution. Nature seems to have an ace of spades up her sleeve just in case things go haywire, and a radically novel series of events threatens life at its core. When the excrement impacts the rotating blades of future chaos, there is a chance that at least some individuals might have their great great great grandpappy's old gene set that existed in a past circumstance. Those adaptations, even ones long discarded for their irrelevance, could now be just the thing needed to survive. What was old can be new again. And these things happen. Meteors can hit. A new pathogen might mutate out of the swamps. Some smart-assed hominids might alter the environment so much that the entire ecosystem collapses. Nature has a Plan B, and C, and so on. Those survival plans don't necessarily include us. But life, as a whole, has strategies to adapt.

The Francevillian biota illustrates what we might be creating for our future. This clan of multicellular creatures existed up to 2.3 billion years ago and they can no longer be found on our tree of life. They were discovered, in 2014, in fossil form. Had that not happened, we'd be blissfully unaware of their existence. Their entire trunk of life died off because of rising oxygen levels, because of cyanobacteria getting way too successful, previously on planet Earth.

At the end of the Archean epoch[167], anaerobic life ruled the planet. The oceans were a brown-orange, filled with iron oxide (rust), which was lucky for the residents since oxygen was toxic to them. The rust bound all the oxygen and kept it safely away from the fragile life forms. Chemosynthetic metabolisms were the norm, so oxygen had no place in their biochemistry. Then the cyanobacteria learned a new trick: photosynthesis. Using the energy of sunlight, they found they could break water apart into hydrogen and oxygen. The hydrogen was useful for energy production, and the oxygen was a waste product that they essentially farted out into the ancient sky. Later, photosynthesis would settle on the more efficient splitting of carbon dioxide instead of water. Life works with what it's got, as we now know. So once the cyanobacteria began releasing oxygen, a deadly pollutant to their contemporary life, the world turned a completely different corner. Oxygen levels skyrocketed, and the great oxygenation event led to the extinction of the Francevillian biota, which set the stage for us. It is estimated that up to 90% of the then extant life forms perished. Had they not, we probably wouldn't be having this conversation. (El Albani, 2014) Their downfall was oxygen, how paradoxical that ours might be carbon dioxide.

As a species, our chief strength is our ability to manipulate matter and create based on pure imagination and thought. One of our strategies used to be our strength in numbers, but fragmented by our foolish preconceptions and social training, that power is working against us presently. Our imagination is also working against us because many imagine our future will be just fine. Many can imagine that their deity will come and save them[168]. Such beliefs are absolutely not helpful. Fortunately, we can also train ourselves to look at the

---

[167] The Archean epoch: 4 BYA- 2.5 BYA, the period after the planetary crust had stabilized somewhat, and the first life arose.
[168] Some believe they will be lifted from the earth by aliens and taken to a new world. From whence they can watch our planet burn. Good Luck with that.

conditions before us pragmatically, use our brains, and cooperate our way to a solution.

And perhaps we can train other life forms to help heal the damage we've inflicted as well.

When we train mushrooms in the lab, we're trying to stimulate their genetic memory of novel foodstuffs they might have once encountered. So, perhaps it could've been millions of years ago that they ate rich organic muck, or fragmented cellulose, instead of the clean, fresh wood inside of a log. Perhaps it was only ten thousand years ago. You can liken the *Pleurotus'* (Oyster Mushrooms) genetic memory to a card-catalog of chemical processes. I like to joke that *Pleurotus* can grow on mayonnaise because they probably can. They are usually the first mushrooms hobbyists cultivate[169] because they are so hardy and forgiving. Somewhere in their genetic makeup, they potentially have the recipe for the digestive juices and growth patterns needed for whatever task might be at hand (or filament, as the case may be).

We start our mushrooms much like everyone else, cloning onto a media made of carbohydrates and sugars. Once established they are moved to jars where the woodchips in question have been mixed with just a bit of carbohydrates. We time their run through the media and sequentially select those strains that develop fastest. Through rapid lab generations, the proportion of the material of interest is increased, in our case, the proportion of wood chips. this technique is termed sequential conditioning. And this type of conditioning has acted upon all species, usually over great lengths of time.

But it doesn't stop at wood chips; these mushrooms are quite capable of digesting petroleum by-products and other toxic crap like PCBs and PAHs. It's just a bunch of carbon/hydrogen compounds to them, and they can release the energy in those molecular bonds for growth. The side effect is the destruction of those toxic chemicals. And they can do this just as quickly as they decay stable cellulose. They use their cold fire to dismantle hydrocarbon compounds in a similar but less random and devastating manner than the hot form.

The mushrooms we work with are all wood rotters. They live inside of deadwood and are the primary recyclers in the forest. A pile of wood chips is

---

[169] That and another incredibly easy to cultivate species named: Psylocibe cubensis.

very similar, yet also very different from their standard fare of logs. Wood chips have a lot more surface area than a tree trunk, and this is a potential benefit because it allows the mycelium to envelope more food bits faster. The pitfall is that this represents more surface area for bacteria and other, possibly parasitic fungi, to invade and compete for the same material. Or perhaps these contaminant organisms merely want to eat the mushrooms themselves. While some of these secondary organisms can't process raw wood as effectively or exclusively as our mushrooms, they can consume the mushroom's filamentous body. Infection is a much more significant threat in a pile of chips than in the pure and unsullied heartwood of a dead tree. That's why training is essential. Our mushrooms have to remember the strategies and chemistries they once used on a foodstuff that was similar to the ones we now want them to consume. They need to remember their challengers and how to defeat them as well. Fortunately, these ancient organisms seem capable of switching alleles "on the fly". Training these organisms and then adding them to a substance in the wild is called Bio-augmentation.

Using mycoremediation is one way many of us envision accelerating our ecological restorations: working with the existing ecology of life to save us all. If a mushroom could conceive of such a thought, I have little doubt they would be fine with working with us if it meant avoiding yet another extinction event (they must get old). So you don't always need complex technologies or deep genetic manipulations. The genes needed are already there; they just need a little coaxing to stimulate their expression. The potential of human/microorganism collaborations could be one of the most significant technological breakthroughs yet on this planet. It would close the circle and harken back to our first microbial partnership: wine. The creation of that delicious ambrosia required not only the training of the vine but the taming of yeast, another form of fungus. It would make Gaia proud to see her children finally playing together instead of quarreling.

Another example of Bio-augmentation is seen in the work with coral. Scientists around the seas are working not only with genetic editing but with good old-fashioned heat tempering to develop coral bodies adapted to the rising ocean temperatures. The corals have seen warmer seas in their ancestral past as well. They just can't reliably adapt fast enough to keep up with our climate meddling ways. Nor can they do it with the added stress of their neighbors dying and rotting next to them. These techniques can help them. People can train them. While this type of accelerated adaptation works well with simpler organisms (fungi, polyps), it is possible to do the same thing with more complex life forms as well. Selective breeding was serviceable to create the crop lines that could survive our inaugural onslaught of toxic fertilizers and insecticides[170]. Remember: The Green Revolution occurred before the era of our genetic breakthroughs. They created their original pesticide-resistant lines by simply cross-breeding and training food crops. It can happen again, and our brother and sister organic farmers are doing just that.

Sometimes, it's just as easy as bringing back the ancient strains, called heritage crops, from before our hubris told us we could make agriculture perfect. Combine those hardy old crop types with permaculture, and we might be on to something. This type of work needs to be massive and widespread. Science needs to serve the people and the planet. Not the profit motive and the greed of a ruling class. And it needs to do so now!

Our lab is the world. So, our experiments must be safe and straightforward. The scientists must be of the people, trained in the rigors and disciplines of science. They must be free from the indoctrination and restraints of the current corporate culture. We have to try, and we will sometimes fail, but in other cases, we will find out how to best work with nature to save ourselves.

Science is not a philosophy. You don't need some magical blessing to perform it. You don't need to adopt any esoteric belief system to use it, either. Science is a method of inquiry, and if followed, it allows us to increase our knowledge and share it. Humans share while corporations horde. (That's how you know they are not people.)

Let's talk about our salmon friends again, now that there is a bit more context. Overfishing and the loss of access to their native spawning waters has devastated our ocean-going buddies. Salmon are an important economic species, so we've put much thought into helping them cope with the damming

---

[170] It was the increase of toxins initiated by the Green Revolution that, in turn, created greater tolerance in the insects or weeds, that led to the need for even greater toxin levels in the pesticides, and greater resistance in the food crops. This is an endless cycle if we choose to continue on such a path. Well, actually, it will end if we continue as we're headed.

of their rivers and streams. The human cure to this problem was breeding them wholesale to make up for the loss of native spawning grounds. People have also made fish ladders and salmon cannons to help them get past the roadblocks put in their way. These solutions are a very human-centric way of looking at things. People use stairways all the time. They shoot other people out of cannons for entertainment. What fun! Not so much if you're a fish. Hopefully, you will never see a female Salmon stripped of her eggs or her beau bled of his milt[171]. But suffice it to say: we've found a way to aid their reproduction that is both unnecessarily cruel and fatal. This artificial fertilization occurs in five-gallon pickle buckets, assisted with the stir of a paint mixer. How romantic! Next, the now-viable eggs are soaked in an iodine solution to kill fungus (which was introduced with our paint mixer). Then they get incubated until they hatch. Billions of baby salmon are produced annually, worldwide, for release back into the wild as a fortification of the continually dwindling wild population. This effort represents an economy worth over 3 billion dollars each year in the United States alone (Charbonneau, 2010). This market also equates to thousands of jobs, millions of fingerlings released, and the continued decline of wild salmon.

The release of these captive-raised fish has not only failed to preserve the wild fishes; it is aiding in their decline and potential extinction. And it's all because folks treat these animals like machines. They are "farmed" in intensive installations, raised in holding ponds, and dumped into streams and rivers. These (s)pawns have been raised without the selective pressures of nature. They have been fed artificial foods, rarely do they see a fly or midge in their first year of life. The natural world's scents, sounds, and textures are never encountered; the epigenetic triggers these interactions would've flicked do not activate. So ultimately, this is just the mass-producing a domesticated fish. Fish that fare as well as a city-raised child, dumped into the middle of the woods, would. If they happen to survive in the wild, we have only succeeded in diluting the dwindling native gene pool. (Murphy, 2019) Once again, our earnest attempts to aid nature have blown up in our faces.

Nature is a system. We have to learn this lesson. You can't rip a part of it away and expect the same functions to continue. The salmon are keystone species in

---

[171] Semen

their web of life. They are the base of many animal and plant communities. The problem is not that there aren't enough salmon. The problem is that humans dammed the rivers, polluted the streams, and shattered their life cycles. We don't need to farm them. We just need to get out of their damn way and let them heal themselves. And this means removing the damn dams we put in their paths. This removal is already occurring on more minor scales, and the results are what you'd expect if you've read this far. Those populations are returning on their own. Nature knows how to fix herself.

As we learn this elementary lesson, we need to be careful not to continue making matters worse. Sometimes human intervention will be needed, as with forests that can't recover without us. And sometimes, just letting nature be is the correct answer. Put back the Beavers and step away. Tear down impediments to spawning routes, and bow out. Perhaps the fastest advance we can make is just to *stop breaking nature* and let Gaia heal. It's all about balance. Complex interactions don't take well to simple solutions.

Take less and give more—practice self-control.

We need to examine our status quo critically, and judge what parts work for everyone and what features do not. When it is understood that everyone means everyone: man, woman, child, fish, bird, bug, and worm that creepeth and crawls upon the visage of Gaia, we'll be off to a good start. There will be no understanding until this simple truth is accepted, the ecosystem is a system, a system we a part of. We are not above it, or beside it, but up to our necks in it. And no part is greater than another, even though some roles may have greater agency than others. The living world is pretty much at our mercy, although, as has been seen, when it breaks, we'll end up at its mercy as well.

Our institutions and belief systems are artificial and do not possess any de facto power over life. Try as you like; you can't rhetorically outmaneuver the coming climate catastrophe. You can't buy off a hurricane. You can't bluff a pandemic[172].

If our institutions and beliefs do not serve all of us, what good are they? They have been used to confer almost unfathomable privilege on some, at the expense of nearly everyone and everything else. When economic systems are created that are little more than board games with winners and losers, we

---

[172] Not for lack of trying, it seems. See: 2020 pandemic in any future encyclopedia.

inflict the same damage on ourselves and our societies as have been inflicted on Gaia. Many live longer lives, nearly bereft of meaning, chasing a brass ring that will never fulfill us as much as the freedom to watch a single sunset or to hold another's hand. What will it take for our winners to realize they are losing?

Our systems can be re-imagined to mimic the ways that the ecosystem functions, but only if we accept we can't have everything all the time. If we'd stop running endlessly towards some ephemeral goal (which we often can't quite verbalize), we will find stillness, and be happier and healthier.

This crumbling world has not been made for us. Instead, our cultures inhabit an imaginary landscape designed by an idea of what could, should, or might be. Our obsessions have created a world for deities and mythological heroes, neither of which actually exists. Our imaginary construct has little resemblance to what the people want or need. They want community. We need each other. Almost everyone wants to be free to dream and achieve and feel settled in ourselves, secure in our future. However, our current status quo doesn't quite offer us these things, so we accept something less.

A world defined by ideas that may or may not even be real is nowhere to live your life. The game humanity keeps playing has no endgame. It doesn't even have a goal unless material wealth is a goal. But massive wealth is not a human goal; it is the goal of those mythical beings and gods our cultures created and then convinced the masses to emulate. Ask a billionaire if they are indeed happier than the average Joe, and if they are honest, they'll admit they are not. They will probably concede that staggering wealth has made them less happy. Great wealth brings a multitude of headaches and heartaches. We have trained ourselves to be deluded by the promise of wealth, and it's supposed freedom. Instead, this pursuit has made us miserable in every sense of that word. True richness belongs to everyone. It is the completeness you feel when you hug your children and laugh with your friends. It is the abundance of Gaia, which is to be shared by everyone, all inclusively.

We need to start doing the right things for the right reasons. The economy of nature, of the universe, is the thermodynamic economy. Every action has an equal and opposite reaction. Energy is neither created nor destroyed. All books balance at the end of the day, week, month, eternity. Profit is what you feel, not what you get. Profit is the swell of satisfaction in your chest that you have done well, and all signs point to a future where you will continue to do so. The profit of nature is that you get to live. Everything else you feel and achieve is then a value-added bonus. There is no artificial way to attain this feeling. You can't buy it, manufacture it or steal it.

The real deal is self-independence, social interdependence, and non-coerced cooperation. Currently, profit is greed, and greed is the desire to get more than you give. Nature shows us that the secret of life is that you only attain true profit when you give more than you get. This understanding is the true reward of the hero. It is wisdom.

Here is the real lesson of life hidden from sight by our words and conventions because it is a disruptive force against the status quo. Practice it, and you'll find it's true. So go ahead, you can do it. Give away your RV to a homeless family displaced by a fire or hurricane. It's probably just sitting there gathering dust. Convert your water-hogging lawn into a garden and share with you neighbors. Give away an evening a week to help feed the homeless or read to the elderly. Attend public meetings and let your voices join together. Find the common ground on which we all stand, together.

The utterly massive scale of our problems is beyond our current economies' power to address. There is not enough money to mitigate every forest stand and restore every coral reef. We can't afford to pay subsistence farmers to stop burning the rainforest down for a scant couple of years of production. We are caught in a positive feedback loop, where people need money and money requires consumption. Waste is good under such a paradigm, and even pointless effort is encouraged. Like lemmings in a wheel, entire civilizations spin and sweat, going nowhere.

So how do we fix our world considering these economic issues?

We volunteer.

We labor, do massive amounts of work, for free. Give back what nature has given us—everything you can. But it's not really working for free; the wages of volunteerism are completion and satisfaction, two more things you can't buy. The profit of volunteerism is the new people you'll meet who will also survive because you all gave a damn. When a mountain community joins together to cut down trees, haul away the slash and restore the forest, it works.

From the perspective of someone who works in this field, it doesn't work well any other way. At the end of the day, bathed in dirt and sweat, neighbors sit together and laugh. They heal and re-create the connections that were once a given for human communities. People rediscover what they've lost, realizing that people have so much more in common than those things that separate us. Our separation from each other results from our training as instructed by an economic system that prefers competition over cooperation, an economic system with no greater ultimate goal than that of a petri dish of bacteria.

Volunteering to help the forest or clean up a river; is the act of retraining ourselves.

Volunteerism doesn't stop there. It happens in the city when communities address polluted soil, or construct a new playground on a vacant lot. It happens when citizen scientists go on "bio-blitzes" to scour an area and document all the plants, insects, birds, and reptiles that currently exist there. This data is valuable information that, again, is just not economically practical to collect any other way. The fungal diversity projects are formed of volunteers in Colorado and nationwide. Hundreds of volunteers support the Billion Oyster Project. Coral studies on assisted evolution and heat training require volunteers to sustain their viability. CUSP uses volunteers to restore ruts from logging or plant trees on burn scars. These are environmental champions, citizen scientists.

Volunteering is the social pact, the bond between us that needs to be exercised to remain healthy. Corporations need to do all of this as well. They need to give back what they have taken for so long. Theirs is an almost incalculable debt owed by those who took extravagant profits, often by proffering elaborate lies. And they will give back freely when they too realize the gain from doing so is their continued existence. The same profit we'll all get to share.

Humans have created a system that promotes standing by until some can pay you, making so much human power unavailable. Could that be by design? We are a social species, and social species rely on cooperation to thrive. Somehow, that framework was shifted in favor of a system of "tit for a tat." Why? Because it rewards some more than others, and they fooled us into thinking that someone will someday be us. It's a con job.

When you volunteer, you are training yourself to understand that there are greater rewards than just money. People don't raise children to make money. They do it because this is how humans will continue. It is our drive to survive. All humans are family; we have just repressed that fact. Families are diverse and not always harmonious. There's always someone contrary and often insufferable, but they are still family. That is a pretty good thumbnail description of the human race.

The pursuit of the fulfillment of our dreams of how life could be has effectively blocked our experience of how it is. And we have chosen this path. Now we must choose a new one.

The Green Evolution is the evolution of life into a state where it can once again proceed both autonomously and with the stewardship of the only species capable of comprehending its interconnectedness and wonder. That species is us. Now is the first time we know of that a species has expanded its own consciousness in order to survive. Perhaps this is a vital evolutionary step and the genuine acceptance of the fruit of knowledge that just might let us save the garden we are in danger of losing.

The Green Evolution is the evolution of humanity.

CHAPTER THIRTEEN: THE GREEN EVOLUTION UNBOUND

"We are all connected; To each other, biologically. To the earth, chemically. To the rest of the universe atomically." — **Neil deGrasse Tyson**

" "The Secret Life of Plants," by Peter Tompkins and Christopher Bird, was published in 1973. The book contained a collection of experiments performed by Cleve Backster, a former FBI polygraph expert. Mr. Backster hooked up plants to a polygraph and registered galvanic skin response (GSR) reactions to stimuli, both pleasant and "painful." The responses were remarkable, repeatable, and seemingly dependable. His experiments seemed to show that plants could not only react to stimuli but were aware. This potential phyto-consciousness was a mind-bending idea that occurred during a time of heightened ecological interest in the USA. Many botanists and scientific philosophers immediately condemned it as rubbish. The very idea of plants having an awareness was an insult. An insult to what? Not to the scientific method, for sure. But to the biases of the scientists themselves. Decades of follow-up experiments have shown not only the plausibility of Backster's original work but of even greater flora capacities than he ever dreamed, as you shall soon see.

I enrolled in an experimental psychology class at the University of Wyoming in the early 1980s and had the chance to reproduce one of these experiments. It would be one of the simplest since my freshman resources were slim and institutional acceptance of the subject was even slimmer. I hooked up a philodendron[173] to a polygraph and tried to see if a plant could recognize a person, or more accurately, an assailant. I hooked the plant up to the GSR electrodes and left it alone in a room. Ten students were selected and given a random order to enter the plant chamber. Each one signed in on a sheet, the only record of their entry order. The Lab Assistant gave them a folded piece of paper as they entered the room. I asked each student to touch or pet the plant

---

[173] I believe that philodendron retired to a nice office window, but we fell out of touch long ago.

and then unfold their paper. If their paper was blank, they threw it away and left. Conversations with the plant were optional. One scrap had a large black dot on it. That person was to tear up a leaf with prejudice, to grind it into mush. After leaving the room, the students could not interact. Instead, they quietly had to await a new order from another bag that randomly assigned them a different entry sequence. And then they paraded past the plant once again.

No touching was involved in the second round. The random entry orders were used to establish a blind testing structure. This design helps eliminate biases, intentional or otherwise, within the experiment.

The plant reacted. I had marked each visit's entry and exit points on the rolling graph paper with a black pen. The moving needle traced arcs strongly when initially assaulted and then again when the mugger came back into its room. I went back through the sign-in sheets to verify the plant assailant's identity by correlating the entry orders. But I could quickly tell that person they were the attacker, much to their amazement. To everyone's surprise, myself included. It immediately psyched everyone. Something seemed to be going on here.

We reran the test promptly, and the results were the same with one minor difference. The plant was still pretty shaky when its original molester touched it, but to a lesser extent. It was still quite obvious who did the second deed, judging by the graph swing's magnitude. The plant knew what was up. Amazing.

These results impressed my instructor. We had pulled off a novel experiment, shown initiative by cobbling together the equipment needed, and followed through on a good experimental design. It was the class favorite, if only because of its novelty. The instructor took me aside and said I'd be getting a good grade, but she hoped I understood the results were meaningless. Because, of course, plants can't have awareness.

This is what scientific bias looks like.

In 2006, during the 4th season of the edu-tainment show "Mythbusters," they performed this same experiment. They based their design on Cleve Backster's notes, just as I had. And, to their shock and amazement, they got the same results. And this result perplexed them. So rather than trying to wrap their heads around it, they punted. Unable to reconcile their experimental results with their preconceived certainty that plants were mere objects, they did the very thing good scientists don't do. They altered the experimental parameters to get the results they wanted.

The group ditched the GSR test and found an Electro-Encephalograph machine from a local hospital. An EEG is a tool used to measure the electrical currents in the brain and record them. It measured electricity induced through nerve cells. But plants don't have nerve cells. So while changes in the leaf's chemistry or the pressure of the plant's cellular cytoplasm, or perhaps increases in surface moisture related to its respiration, could explain a change in galvanic resistance, there's no way a leaf is going to generate an electrical current. And that suited the Mythbusters just fine. They got a null response from the EEG and got to mark the experiment with their cheesy "Busted" stamp. Instead, what they actually busted was their credibility. Seeking the result you desire is not science. It is, by definition, pseudo-science.

So this is how bias spreads- through the assertion of impartiality where the opposite is actually true. Science with an agenda is a dangerous thing. Our society's agendas, railroading both the scientists and their science, are the real problem and the source of our current troubles.

Corporations force their will upon science, farmers, and the citizenry without the constitutional protections each class has from government bullying. This arrangement is why these artificial entities have been granted such power. Because now they can do what elected officials can't in a "free society." They can enforce the rule of the minority with the imaginary, yet nearly endless, power of "the One." This is how psychopaths design and implement their rule. They've had their run, it's time to retire them and end it.

By now, our pattern should be obvious: we're held hostage by our myths about ourselves, and many keep acting as if they are just about to prove true and miraculously save us. Trapped in an addictive cycle with petroleum, rampant consumption, and the illusion that we (and our wants) can grow limitlessly. Too many believe in human Dominion, and are obsessed with an unrealistic view of our powers and a seemingly boundless sense of self-importance. Until now, the chief business of humankind has been to consume this planet and its inhabitants. That, and to stare into our many reflections with awe. Our leaders have told us it was our right, our engineers have shown us how to do it ever more efficiently, and our religions have commanded that it is our imperative duty to do so (many of them, anyway). We have lied to each other about what we've been about, cutting corners at every junction. And we've lied to ourselves that everything will all be okay. Nothing will be okay unless we now act with restraint and purpose.

Gaia has us dead to rights. Our actions initiated the mechanisms of our extinction. And they are being amplified by the feedback apparatus of this world's systems. While news to many of us, the carbon cycle is one such system that we are breaking yet intimately rely upon for survival. Humans are around 60% water; we are also about 20% carbon (by mass). All life on Gaia is carbon-based. All life reflects the planet itself. How life from another system might present itself is a mystery to us. Even if life exists elsewhere is a mystery, a question that haunts us deeply. A question we will never answer if we don't choose to stick around.

The energy that fosters all of our lives is the sun's energy, fueled by primordial forces, not the energy stored by life forms in their carbon bonds. It is not a destructive act when suns burn, but a creative one where more complex elements are created instead of ash. The forge of the Sun is the factory of matter. Its reactions can seem to run in opposition to the law of entropy, the second law of thermodynamics. Entropy demands that the result of every energetic reaction is an increase in disorganization. But with the production of the higher elements, suns are setting the table for a new order to arise. Entropy directs the arrow of time (the order of processes). It clearly delineates before and after in terms of energy movement and the increase of disorder. Yet, here on Gaia, life forms are becoming more complex before our eyes, increasing order with each energetic process of life. Life rides on the tide of the universes' downward energy flow, rising on its fall. The universe's drive towards disorganization amplifies life's physical order. Stars build matter instead of just consuming it. Perhaps what is occurring in them is another type of life[174], but, again, we'll have to survive in order to answer such a question.

---

[174] Oh boy! More controversy! Fire is often used an an example of pseudo-life, an example of the difficulties of defining life. But fire deconstructs, it moves in harmony with entropy. Fusion acts in opposition to entropy, as does life. But hey, it's just a

Until then, you just need to remember that you are the child of ancient suns. We are the children of Gaia as well. Our current behavior would fail to impress either parent with how we've taken their billions of years of toil for granted.

So, here we stand on the edge of a Green Evolution that will reverse humanity's dependence on the mining of the dead for their fossilized energy and allow us to use the energy of Sol directly. All can now clearly see that the systems of life are intrinsically interconnected. As I've said: "The tree is the squirrel is the mushroom is the tree." And now also, "The sun is the planet is the life upon it." And this is a practical way of grasping the essence of our mortal existence. These simple ideas might be the most crucial realizations for you to now make. Within the planetary construct, if you destroy one system, you disrupt others, eventually all the others. Then, finally, the life you ruin will be your own. And that's precisely what's happening.

Societies have used up the best, easiest resources. They've eaten all the low-hanging fruit. Consuming without replenishing, our actions accelerate life's drift towards entropy on a local scale. For while life is self-healing and the ecosystem's balance is self-correcting, we've been at war against both. The wounds our actions inflict are too deep and come too rapidly for self-healing to keep apace.

And unfortunately, many seem content with this self-fulfilling prophecy of extinction. If our pursuit of immortality and deification ends with human extinction, then no one will be left to point out the folly of that pursuit. Then the deceived and deluded will be at peace in their myths, at the cost of the best traits humanity has developed: intellect, imagination, and sacrifice. Finally, the Dominion mindset will have reached its desired perfection: the perfection of ruin.

There is another way, the path that led to us in the first place. There is a beaten trail through the primeval forests that remains, although fainter beneath our trampling. This way is the course of nature. It is the will of no will. This is the intention that drives discovery without pre-conceptions, pure research. The flora, fauna, and funga that share this planet with us can't conceptualize the way humans do, so they can contribute to the future in whatever way we ask.

---

metaphor and a thought experiment folks. Or is it? We won't be able to answer this question with flint knives and bear skins.

Our refusal to acquiesce, our stubborn need to control, has taken us from the path.

While our improper intention has led to our rapid expansion, it has also cost us our proper place in the intricate network of life. Now is the time to change course again and deliver back unto life the gifts we have won during this detour. We can bring our sciences and skills to bear on the injuries we have inflicted and heal this world.

In 2019, The International Renewable Energy foundation released a report that found the near elimination of carbon release by the transportation and energy sectors could return between $3 and $7 for every $1 spent to achieve it by 2050 (IRENA, 2019). Such a massive return, in such a short time, makes one wonder why this is not headline news. The report spells out how adaptation will be a windfall for the global economy and all people's living standards. Replacing our infant systems, breaking our juvenile cycles, and adopting mature attitudes towards life are pretty hopeful potential outcomes of our current difficulties. Ending this insane moratorium on sharing ideas and technologies that is presently called our economy would accelerate this process even more. Shattering the lies of profit motives and replacing them with the acceptance that life's ultimate profit is life itself would heal our hearts. We will count wealth, not as bars of gold or deeds of trust, but as the bounty of satisfaction, the quantity of friends, and the number of laughs per day, things that bring real peace and foster true security. When partners surround you, you don't need vaults of riches to ensure your safety and well-being. Old age holds no uncertainties when the lives you have nurtured in your arms are there to foster you in your times of need. When your value is seen not as another competitor, but as a necessary link in the chain mail of human society.

New energy economies will replace the old oil-based ones and provide better and safer jobs. This is clear and backed by multitudinous studies and a growing wave of human activism. Those who have failed to use the head start given by their foreknowledge of global warming need no sympathy. Humanity is fed up with the current practices of protecting the insanely wealthy and depreciating everything else. Only those with vested interests in the status quo and those who are woefully misinformed continue to deny the truth. Primary among the ill-informed category are those who think that life is magic. It is not. Humanity and the planet itself are mortal, and facing that fact is part of growing up. You're not going to heaven, but perhaps you could live there because heaven and hell are both manifestations that could be actualized here on Gaia. Which would you choose if you believed you could?

So now to confront time. Time is a concept firmly rooted in our physical experience but challenging to pin down. Although Einstein thought it was the fourth dimension, that idea has fallen out of favor because it seems uni-directional. You know that time marches ever forward. The physicist calls this "the arrow of time," and its direction is defined by the thermodynamic system's entropy. But for our purposes, let's just say that the arrow points in the opposite direction of our memories. You can remember the past, but not the future. Time has, seemingly, only one trajectory, unlike up/down or forward/backward. True dimensions do not limit the direction(s) of travel within them.

As time marches on, everything falls apart a bit more. People age, rocks roll downhill, the universe inches closer to potential heat death. You are denied the ability to return to the past to fix your mistakes. It is only through the birth of the next generation that we hold entropy at bay. And then, only if we can teach the next generation to avoid the mistakes we've made.

Time is unavoidable, omnipresent, and rapidly catching up with us. But as you have now learned, life swims against that current. Life surfs the tides of energy, space, and time. Life builds as it grows. And it grows over everything, it seems. We are still alive and, therefore, have the agency to choose our path. Our ability to choose is embedded in time as well. Now marks the point of departure, the fork in our path. It is our choices now that will either forestall or advance our decline.

Time is relative. Einstein discovered that space/time is the fabric of the universe. And then he proved it with calculations that show that as an object increases in speed, its relative time slows. Conversely, as an object approaches an enormous mass (a black hole for example), its relative time can virtually come to a stop. Experiment has proven these theories, repeatedly. So, time is also relative to mass or your proximity to it. Without going deeply into space/time curvature, rest assured that the farther away you are from a large mass, the faster your clock will tick relative to a clock on the surface of that mass. If you were to synchronize an extremely accurate pair of watches with a friend and then travel to the top of a skyscraper while they had a latte at the sidewalk cafe, your watch would run just a bit faster.

But perhaps, metaphorically, life's time might also be relative to the scale. A moth may emerge from its cocoon and fly, love, and die, all in a matter of days. Is its experience of life lesser than ours? A giant sequoia towering over the forest lives for centuries, and from its perspective, our human lives may seem like just another passing season. When you live for a thousand years, and each winter is just another nap, how might your experience differ from the moths or the humans? A colony of bacteria can overrun a petri dish in hours, consuming all the nourishment, and finally, die in an overcrowded bath of their waste products. In the end, after consuming each other in an ultimate effort to eat their way out of fate. These bacteria have lived entire generations, some in plenty and some in horror. Does it feel like decades or centuries to them?

And those bacteria illustrate for us what the result of a life of rampant consumption will be, must always be. Is this what humanity will choose on petri dish Earth? Will the sun, still an adolescent in its multi-billion-year life, look upon us quizzically as we squander our inheritance? How might stars experience their time?

Many a philosopher would tell you that the only actual time is now. And while it is true that the now is our point of action, the past, and future as just as real, if currently inaccessible. And these other lands are just as consequential. Our shared now is formed by our past choices (and boy, do we have a past). Those already made choices inform our future, combined with our decisions enacted in the now. That is why now we can choose to join together and survive. If our civilization refuses to change, then in the very near future all of our poor choices will come for our children and the children of the planet. And it will be real, real as a heart-attack, as real as the past and present.

We are always free to choose, although it often seems we'd prefer not to. What future would you decide upon if you believed you could? Not one of failure, decay, and extinction, for sure. Yet this is the choice so many keep making, second by second, plastic water bottle by disposable smartphone.

Time is incredibly vast. Estimates for the age of the universe are currently around 13.8 billion years. No one knows how long it has to go, but probably at least that amount of time, if not longer. Humanity is less than a million years old. Civilization is even younger, a mere few tens of thousands of years. There's not even another century left for humanity at our current rate. And if we continue to make these same terrible choices, it will be an unfortunate century indeed. Sad and slow, as we grind ourselves back down into dust, fighting and screaming as we go.

What a crime against humanity to steal the future from ourselves in this manner. Humans could thrive, potentially for billions of years. We could become the space fairing race of our science fiction. But if we're not very careful and diligent, we might not even leave much of a fossil record. Our reign will be but a fraction of the dinosaurs', a pittance of the trilobites.'

We have to change our ways now. There isn't any other time available.

This potential end to humanity seems especially sad if considering that people have the wits to survive even these catastrophes of our own making. Our training is holding us back. Our cultures have claimed to have answers but have seemingly just developed strategies to avoid the questions. Most civilizations before us, to some extent, felt that the world, and its life, were a given, or even that they were literally—given to them. We thought that the world was our oyster and would always be there to fuel our lofty desires and cradle us when we faltered. So, we treated it like past New Yorker's treated the oysters in their harbor. The world is our home, not our oyster. To ignore our precarious position would be a capital crime carried out against us by us. A crime we are capable of either perpetrating or prosecuting.

Upon our examination, it seems possible to reduce our carbon release drastically, and in a relatively short time, by continuing the shift to renewable energy technologies. But there is still the ecological damage already inflicted upon our life support systems. To repair the ecosphere, we must use life's resilience to heal the human inflicted damage. Once we've stopped adding carbon to the atmosphere, we will still need to remove the gross excess already put there. Forests are one of the best mechanisms for achieving this; letting the living trees and soil drink in the CO_2 to make cellulose, chitin, and life. Kelp forests undersea might work the same way. As you have seen, how we go about reforestation is just as important as the number of trees planted. We need to assist them towards a restored and expanded state.

The re-imagining of our place and rightful power has to align with the template nature provides to have any chance of success. Ecosystems need restoring on a planetary scale, they need our help to achieve this. Technology is our ally, but not the damaging disposable machinery we have come to rely on, requiring more raw materials ripped from the land or more pollutants poured into the atmosphere and oceans. Humanity needs to reevaluate how we grow food globally. But not by following the same path as the Green Revolution. It is time to reject wasteful agriculture as now practiced, where nutrients are consumed but never properly replenished. A system where vanishing petrochemicals are over-sprayed and washed away to pollute the wild spaces and our seas. Instead, our amplification of fertility must come in conjunction with increased soil vitality. It is time to act with humility when interacting with the life forms we grow to consume. If our agriculture can't give an animal a decent life before slaughter, then perhaps they shouldn't take their lives at all. Where do they think the torture and cruelty will go, if not directly onto the fork that their customers place in our mouths?

It is time to clean our water and stop poisoning it with our waste and excreted pharmaceuticals. And we need to stop poisoning ourselves as well. Our treatment systems are already in place; we just need to use them better. Stop pouring chemicals and random objects into our drains. They are not a portal to Hades. They just drain to the treatment plant down the road. Our actions should facilitate balance instead of continuing to disrupt it. Our new efforts need to remove the waste that we've added to the air, land, and waters. And like cleaning up the oceanic garbage patches, that task will be monumental.

Perhaps most importantly, it is time to step aside and just let nature heal herself. The seas must be allowed the chance to revive. Once cleaned and cleared, the oceans can absorb atmospheric carbon, but only so much as it balances with the biomass held within. Planting reefs and seeding oysters is a good start, but stopping the poisoning of oil spills and chemical leaks must occur, first and foremost, lest it all be for naught. Once the plankton and micro-algae rebound, they can do their part by absorbing even more carbon and returning oxygen to the atmosphere. Currently, the acidification of the ocean impedes the phytoplankton's ability to heal. Carbonic acid is a dangerous carbon store; living tissues are not. While perhaps someone, someday, may come up with a carbon sponge machine, no such thing currently exists. And the same services can be performed without needing more resources and energy for such fanciful machines. Because the life cycle itself is our carbon sponge machine. Knowing when to step aside and when to lead is of primary importance for our hopes of survival. The planet can be restored by allowing life to heal itself with the same species and mechanisms that built this living world in the first place.

This is the Green Evolution, the stewardship of humanity working hand in hand with nature's mechanisms. It is our rightful place, and this is our right time. As a newly minted student of nature, you must now look to life itself, see what it wants to do, and then help it achieve that goal.

We must walk with lighter feet and get soil under our fingernails. Human power is fantastic; consider a labor pool counted in the billions. Human energy can move mountains without the emissions and oil spills that our current love of monster machines entails. Our waste products can return fertility to the land, not poison it. Every human wants to live, and each of them wants a purpose. Now we can offer both. Everyone has something to contribute, and our collective need is enormous. Planting a trillion trees is a monumental undertaking. To reduce our agricultural footprint and temper our overconsumption of meat will be difficult at first. Still, it will repay us all in improved health, and wasted resources returned to other uses or just left alone and in place. Oil and coal, left in place, are the greatest, most stable reservoirs of sequestered carbon on the planet. Let's keep it that way. By stopping our consumption as a means to fill the void within us, the void will heal itself shut. Because now it will be filled with a goal. To survive, perchance to thrive.

We must seek the return to the world in which we evolved to survive. The world that exists in reality and not the one in our fantasies. This challenge might be something all sentient, apex species must survive or be self-selected out of existence (assuming we are rare, but not singular, in the universe).

Humanity has reached "Childhood's End."[175] We either change ourselves just this little bit or fall into the petrifying bog of the ages.

Humans have made such a mess of this planet that it would stagger the mind of any sentient being that wasn't an intimate part of this process from the beginning. It is tempting to throw up one's hands and give up, which is exactly what some would have you do. "Well, had a good run. You might as well get as many jollies as you can before the roof caves in." You are going to hear this attitude more and more as the century creeps forward, as the deniers lose the flimsy excuses they are currently fond of and have to face what they have done. The attitudes of many who have fought tooth and nail and spent billions of dollars to hide the truth from the public are, frankly, mind-boggling. The impulse may be to make them pay for crimes not only against humanity but against nature herself. And for some, this might be appropriate, but countless others are victims. They are victims of their own delusions and the comfortable lies of the status quo. They may also have technical expertise, as well as financial capacity, and both are needed. So a better tactic would be to recruit them to help because many who may have the greatest culpability also have significant means to aid in our transformations ahead. One way to do this is to show them hope; there is a path out of the murk, a new, better way to live. The way of partnership is the path to salvation. This partnership is how nature has always survived, evolving through every cell's cooperation and sacrifice.

This is the Green Evolution, and rather than being a new idea, it is how life has always operated. It's only now that we are awakening to understand those ways and adopt them.

To envision everything that awaits us is a task beyond the scope of this book. To draw out a step-by-step plan for survival was never its intention. Since that blueprint is within all of us, it will take everyone's input to realize it. All I have tried to do is show you an accurate reflection of humankind's interactions with nature and, in many respects, each other. You have seen how the systems work and how important every life and each interaction within it is. It is with hope our collective desire to survive will outweigh our inculcated greed that I

---

[175] The promised optimistic novel by: Arthur C. Clarke.

trust you to see the patterns. The patterns that our civilization has broken. And how they can be used to heal. To that end, I have endeavored to distill the wisdom of nature into a few simple observations. And so, I offer you several, not instructions, but demonstrations.

Here are the principles that I have taken from years of mucking in rivers and puddles, from days spent wading in swamps and nights spent nursing sick children and animals. These ideas have become plain hiking through mountain watersheds, digging through dirt and gravel, scuffing knees, and breaking fingernails. It is not some ultimate list of commandments. Nor is it possibly a complete or authoritative tally of the descriptors of life. There are others, but none as helpful when planning a harmonious future *with nature* as these six. These are also not the list of the attributes of life you studied in high school science. Nevertheless, these principles should seem pretty self-evident when you see them and are an excellent way to judge how well our actions might be in, or out, of synchronization with the business as usual, of life itself.

VI principles of nature

I.   All Life is One.
II. Life is a self-perpetuating cycle.
III. Life does not play favorites.
IV. Everything that is taken must be returned.
V.   Life derives its strength from diversity.
VI. All Life is Aware.

I. All Life is One

There is only one template of life on Gaia, and it is based on Deoxy-riboNucleic Acid. All that we know of as living arises from DNA, RNA, and their various forms. DNA is the self-replicating molecule that compiles and passes on the chemical recipes that make our bodies, defines our speciation, and passes on our traits to our children. This spring of nucleotides on a phosphate sugar spiral may not have consciousness or even a plan, but it certainly acts as if it does. It is the mechanism of adaptation that can be split, mutated, fragmented, and still survive. Without it, none of this would be possible or necessary, as the wag would say. It is over 4 billion years old and has many more eons to go.

Many ignore this simple and obvious tenet to divide and control. All life is one. Yet that knowledge of the mechanics of life has been wielded to debase people and animals for the benefit of a few, often while claiming it is to meet the needs of the many. Libraries are filled with volumes of justifications, philosophies, and techniques, all intended to enforce this lie. So cleverly have we rationalized our supremacy over all other life when the driving force of that life itself is not clever at all. Replication is simple, and this is why, now that humanity has entered the age of CRISPR, many think they can wield the power of DNA with the same optional wisdom that has always worked for the ruling elite for so long. No one knows what the result of editing a molecule that is the basis of *all* life might be. But to claim that you have it under control is the basest deceit of our age.

While mutations happen every day, that process is slow, and the failures outweigh the successes a thousand-fold, a million-fold to one. Gene editing is playing with fire and should only be performed with the of utmost seriousness and care. Genetic manipulation is potentially far more destructive than atomic bombs. If there's a better argument that it should not be in the hands of a profit-seeking, quarterly earnings-based mentality, I don't know what it is. Private companies are not allowed to have nuclear weapons, now are they? Gene editing is possibly even more dangerous, but many treat it like a teenager treats a case of beer and dad's shotgun. CRISPR experiments can now be conducted in a high school biology class or some wingnut's garage. Ultimately, there is no need to mess with DNA directly. The tools for better crops, healthier stock bloodlines, and longer life exist without gene-splicing. And any use of altered genetics needs to be rigorously tested for extended periods. The changes need to be very tiny and incremental, just as in nature. The release of gene drives must be stopped. I know many believe that genetic engineering may cure cancer, reverse Alzheimer's or eliminate Downs's syndrome, and someday it might. And if anyone is going to mess with any genetic map, ours is the only one with which they might have the right to experiment. Might. It is the only one where informed consent is possible. But, if you have faith that the same system that gave us Thalidomide, Fen-Phen, DDT, OxyContin, and a host of other cures that were worse than the disorder, you need to temper your expectations and err on the side of caution. Once a mutation, which an edited gene set is, finds a way into the wild, there is no putting that genie back in the bottle. A gene drive can, and would happily drive right over you.

All of this is not to say that genetic research should be outlawed. Quite the contrary, but those who do it need to move forward with ethical standards rarely demonstrated by industry. And they should proceed with a level of regulation that governments never have never imposed before. For that is what regulation is: self-control. So from now on, when anyone uses the word "regulation", replace it in your mind with the phrase "exercise self-control," and see clearly whether their argument makes sense or is just a rhetorical charade.

Our unwillingness to accept that all life is one and interconnected on an intrinsic level, and that it not our toy, is at the root of false perceptions of god. Humans keep telling ourselves they are separate and made in the image of some god. God tinkers, and "He," said[176] it's okay for us to do so as well. But the gods were created by us, in our own image, so that we could pretend we were separate and superior. All so that one day someone might usurp that godly power. A fragment of humanity appears to want to be god, this is the root of the Dominion mindset, and they just can't be that. As we commonly use the phrase, there is no such job title.

And if there were a god, maybe it is somewhere in our DNA. It has made us in its image, after all, complete with its strengths and imperfections.

II. Life is a self-perpetuating cycle

It is generally accepted that life is around four billion years old on Gaia's planet. That can be determined this by using radiometric dating on the rock

---

[176] Not: "She said" because goddess religions tend not to encourage the rape of the world for fun and profit.

deposits found on deep ocean vents and highly eroded basaltic plains. But since there's been much effort exerted assailing science's dating methods and fossils specifically (i.e., Satan did it to fool the disbelievers), let's look at it in another way. You can also date life forms using the genetic material within. This technique is a bit more complicated, but a molecular clock is apparent within our DNA. It can determine the distance of two species from their last common ancestor. It can also move back, step-wise, through the genera to get a different estimate of life's age. Here is a picture that can confirm or deny the fossil record information. And verify it; it does. Life appears to be almost 4.4 billion years old (Betts, 2018) when traced back to LUCA, the Last Universal Common Ancestor. LUCA is intriguing since it can only be traced genetically at this point, as there isn't any clear fossil evidence yet, just clues. But the evidence that exists would show that the first life forms were not eukaryotic cells, the type of cells that make up plants, animals, and fungi. We were not the first, as the tales immortalized in fossil rock continually remind us.

The aptly named Archae, the archaebacteria, Monera, and their prokaryotic friends sit at the base of our tree of life. Prokaryotes lack a cell nucleus, their DNA more or less floating around in the cell's cytoplasm. So distant are these ancestors that they might invoke a bit of disgust in you if you were to see them as blue-green algae in your fish tank or bright-pink pond scum found in salt marshes. But as pointed out in the first principle, all life is related. We are part of a chain.

So how does life continue to exist for billions of years? It does so by reproduction. No life form is immortal in and of itself. However, our evolutionary sequence of life is giving it a good go. The parents (we'll skip asexual life cycles and parthenogenesis here) mix their genetic material and give forth a copy of themselves. Clonal and asexual reproduction has the weakness of passing some defects to the child. But this sexual copy is fresh and new, free of their parent's acquired injuries and the defects of time. This newbie, the product of genetic mingling, grows up strong and becomes a parent itself, continuing the cycle. The drive to reproduce is just as strong in a sponge as it is in a teenager, and without it, we'd all potentially still be pond scum. Our sexual drive is one of our greatest strengths, despite the uncomfortable situations it can get us into sometimes. So human immortality can, and could be, achieved by the march of the generations. You will not get to live forever, but humankind might live a very long time if we don't screw this up. Who are we to make such a choice for the entirety of humankind?

It is the denial of this simple principle that causes so much grief in the world of humanity. Many continue to see themselves as separate when they most definitely are not. Some think we will find a way to make the individual immortal, or at least practically so, and they should not. People may live their lives as if life were made for them personally and exclusively. Many ignore the future generations who, if you have any rights, would wish to have the same rights as well.[177] Some think life is a special thing that only they, in the most singular, can adequately appreciate. They wish to remain in this adolescent reverie forever, never aging, keeping it all for themselves.

Were humans to achieve greatly expanded lifespans, they might break the cycle of life in a different way than they are at present. This "immortality" would take the space and resources that rightly belong to the next generation and squander it on ourselves. This greed is the basis of many of our actions at present, such as the overuse of petroleum or the misuse of aluminum. While many people may wish to live in this wasteful manner in perpetuity, the built-in stop button keeps them from inflicting such damage. They would consume with no thought for a future and no shame in pursuing more and faster. This self-delusion arises from the status quo selfishness that thinks it would be okay to live forever because the world and its wonders were made for us and us alone.

Consider now the Haudenosaunee and the concept of living your life for the future, based on the wisdom of the past. It's a tricky notion for our modern capitalistic societies because they have discarded wisdom, for the most part, and traded it for a chest of glass beads and trinkets. Life is a chain that only lasts as long as each link stays strong. Try to make your choices based on their effect seven generations hence. Don't be the weakest link.

If you want to live, join forces with life. And this means you need to take your bow and gracefully step off stage when your role is played out.

---

[177] It is because we pretend the future is not real that we can pull off this mental gymnastic. The future is real, just not completely formed yet.

III. Life does not play favorites

The future of life may now seem to rest in our hands, but if the worst were to happen, it could revert all the way back to pond scum and begin its cycle anew. It won't have to, but it could. Modern humans are the product of evolution and not evolution's master. Nor are we meant to be its master. We are far too foolish for that title. This can't be stressed enough: this world wasn't created for your benefit. Its natural forces created you and you are just as disposable as a tetradactyl. In this mandate lies the power of evolution. It can create greater and greater life forms, and if they can't survive, for whatever reason, then there is an understudy that will be happy to step up and take the stage. Survival of the fittest does not mean supremacy of the cruelest, or the triumph of the powerful. It simply means those who can adapt best will succeed.

That there are no free rides is an inviolate rule, and there is nothing that will ever do to change that. So the fact that the conditions that are now challenging our fitness for survival are self-inflicted is irrelevant. Or, more precisely, its only relevance being if we did it, we could undo it if we choose to do so.

The Francevillian biota is our parable and warning. If not for scientific research, no one would know there was another distinct tree of life on this planet, but we do now. The resulting extinction that was the Great Oxygenation Event resulted from a species growing too fast, getting too much and busting after the boom. The tree of life that is ours took root in their fouled petri dish.

It is worthwhile to ponder the ancient path of life on Earth. Life seemed to have popped out shortly after the crust settled, even before it had finished subsiding. Our best dating puts the planet's age at around 4.5 billion years old. Life appeared less than 200 million years into that span. (Schopf, 2017) That's a blink of an eye in cosmic terms. How is it that something as unique and counterintuitive as life could appear on a barren rock in space so quickly? The answer may be that life, rather than being a rare supernatural event of creation, is a factory feature of existence. The evidence that there was a different multi-cellular life system on this planet before us reinforces this idea. Science may be very close to discovering that life existed on Mars and currently exists on Venus or one of Jupiter's moons. And with that revelation will come the realization that planetary life systems can rise and crash. Will we get that memo only once it is too late to save ourselves? And why do we need an email from outer space when the writing is already on the wall? Saving the human race is the job of the human race.

Again, this is the end of our adolescence; our make it or break it point in history. If humanity can't get past our stories about ourselves and our maddening sense of privilege, humanity will burn and damage life in the process. However, life will not be destroyed; it can continue onward in our absence. And if we persist in rejecting our proper place as the stewards of this world, it will hand us a pink slip. And there will be no other jobs for us to take. Survival of the fittest doesn't care if you have military-style automatic weapons. Bullets are finite, and life is as close to infinite as imaginable. A 100-round magazine would be spent quickly when trying to shoot up a carpet of moss. The moss would then replicate from every fragment and thrive, stronger and larger than before it met Rambo.

No more deities to save us and no intergalactic aliens poised to intercede; there are, however, others who can help. Other humans who are currently called aliens. Our imaginary separations between nations and cultures must soften enough to see each other as we are, relatives and potential friends — allies. It is only together that we can save this planet and ourselves. We must abandon nativism and prejudices of all stripes to achieve this.

The desire of some is just to watch the world burn. This form of psychosis realizes that it is not the ultimate creation but thinks that it has to power to destroy it. That mindset is a testament to our reluctance to grow up and act like adults. Nations are governed with adolescent rage. They may seek conquest with pubescent lust. Our concepts of masculinity are sadly broken. People obsess about celebrities. The rich and famous are distractions that keep us from seeing that our current goals are hollow. The American Dream is pointless if you can't drink the water or breathe the air. Our political geniuses are spouting meaningless pablum when they choose to diminish our ecological crisis's severity and urgency. Our leadership and their common sense are leading us over a cliff like a bunch of Disney lemmings. No one ever said: "The lemmings will inherit the Earth." But now, they just might.

There is no dominant philosophical structure that can lead us to our survival. Instead, we must follow the wisdom of no wisdom at all. The mimicry of nature is our hope.

This hurdle is the next step in our evolution, and it is one we must clear because nothing is going to do it for us. Choosing to evolve is the evolutionary step no creature has yet made, and now we must do it. That's pretty special.

IV. Everything that is taken must be returned.

Recycling is how the entire construct of life sustains itself. When the lion eats that gazelle, it incorporates the materials that the antelope had eaten during its lifetime. The lion excretes what it can't use, which becomes food for bacteria and the grasses covering the savanna. When Simba dies, every mineral and tissue is taken back into the cycle of life and used to benefit the new life that his death will nurture. This succession is the real secret of the resurrection. Death, renewal, and birth, in that order. Life is cyclical, and the scrapping of its materials is the depreciation of life. Throwing any substance into a landfill is a crime against life itself. It is also one behavior that only humans display, that and building monuments to themselves.

Here is the crux of one of our greatest challenges. Recycling is the way of all life. Many of the compounds needed for life are expensive to synthesize, and not reusing them could be disastrous or, at best, a saddening waste. Vitamin B12 is an excellent example of this. B12 (Cobalamin) is a complex and relatively rare substance built around a cobalt atom and is synthesized only by certain bacteria and Archae. Yet every animal cell needs it to function, or it will die. Grass eaters get it from the bacteria in the soil they consume while grazing. Once inside the bunny/cow/pig, B12 accumulates and is passed on to the predator that eats them in nutrient bioaccumulation. You can only naturally get this vitamin from eating bacteria in the dirt or consuming meat. You don't need a lot because it will be put to good use and jealously guarded once you have it. This substance is vital to the continuing dance of life. Sharing what was once ours with others is imperative, and not optional: you can't take it with you; you must give it back.

There is more talk now of a circular economy, cyclical materials processing systems, and closed water treatment operations. And this represents a significant glimmer of hope. Society will see the ability to view anything as disposable as an aberration in the future. Our acceptance of waste is a by-product of our training that humans are superior supernatural beings. Most even refuse to allow their bodies to decompose and re-enter the cycle of life after death. Our cultural methods of treating and burying the dead are, sadly, also off the rails. These practices began with the mummification of kings and shamans in multiple cultures in multiple locations worldwide. They were so unique, so perfect that they needed to be removed from the base world and elevated to perfection. So again, we see that perfection is truly perpetual death. It is now expected, and in many cases, illegal not to embalm the dead and bury them in hermetically sealed caskets. Or at least post-process them in some manner that is more cultural than practical. These rituals are akin to stealing all the gifts life has given us, the nutrients, vitamins, and proteins that make up our bodies, and embezzling them from the future. The final slap in the face of nature is to try to take it with you. This action after a life of making useless plastic baubles and disgorging them about wantonly. This last act, embalming and burying the dead, takes energy from the system and wastes it. It is not a slap, as much as it is spitting in the face of life itself. It is only because of cultural training that most of us cannot see it as such.

There is already a scarcity of places to bury people. Arable lands have been taken and filled them with non-rotting corpses. The waste is staggering. Those bodies belong to the land; they could make it fertile. The crops that can't be grown on those artificial green lawns feed nothing and no one. This practice represents a ghoulish form of bioaccumulation, one where all those exotic and essential compounds are taken and locked away from the ecosystem. If, carried to its conclusion, there would be no dry land left as the tens of billions of bodies in their tens of billions of boxes consume the 29% of the planet's surface that isn't underwater. Here would be the final consumption of everything, leaving no matter or energy behind. It is one possible, albeit unlikely, outcome of our current manners of living, another face of extinction. But even if we wanted to, the human race won't survive long enough to accomplish this Lovecraftian[178] task. There's a horror film in this somewhere, though.

Cremation is no better. Remember how the severely burning forest loses all of its nitrogen-based nutrients to a re-conversion into atmospheric nitrogen? The same thing happens with the body. Cremation robs nutrients from the system and vomits them into the air. The B-12 is rendered useless, and released along with more carbon dioxide, into the atmosphere to fuel yet more global warming. Since bodies don't burn well, the additional fuel required to consume them is just more planet-killing waste. It's all appalling when you really stop to think about it.

These traditions show how far our cultures have diverged from nature and the conditions that gave us life. Most waste thoughtlessly and without care. In every manner conceivable, they refuse to follow the path of life, the way of renewal and recycling. We need to make this one of the most important tasks before us. Zero waste is not an unachievable goal. Creating waste was the original impossibility.

V. Life derives its strength from diversity

Life is a jungle. We walk upon the dust of those who fell before us. That Tiger will take a human child to feed its cub. A fungus can grow out of control from within and consume you inside out. Could a stranger's cough carry COVID-19? In a dog-eat-dog world, you may seem justified in an automatic distrust of the "other." Seen from a human perspective, our world is a ruthless competition where only the strongest survive. Evil deeds await the weak, and its perpetrator can wear the mask of innocence. So perhaps we are justified in placing refugee children in cages.[179]

But seen from the perspective of life, cooperation and not competition is the

---

[178] Howard Philips Lovecraft, a master of otherworldly horror. Also a racist, and misogynist, whose points of view led to mindbending terror. Love him or hate him, you wouldn't have wanted to live inside his head.
[179] NO, We are not. The separation of families enacted under the administration of #45 was an abomination.

hallmark of existence. The single cell that was the beginning of all of us needs to feed. All organisms are undoubtedly related, as has been seen, and they all operate together, waltzing across the visage of Gaia. The plant needs nitrogen, in the proper form, to survive. It can get this by absorbing it from the soil. How it got there is perhaps from an animal's corpse, the decay of last year's crop, or this morning's ablution. It needs minerals from the fungi that shroud its root hairs and engage in natural trade. The herbivore eats that plant and enjoys the rich mixture of nourishment involved in the, until recently, lush green grass. As it chews the tender leaves, bacteria begin the process of digestion that will ultimately transform that timothy hay into delicious bison.

When you eat that steak, you do the same thing that every part of the system has done before us. Our saliva is rich in digestive juices, but it is also a nutrient broth for the thousands of bacteria that live in our mouths. These bacteria will begin to break down the meat, and it's challenging to break down muscle tissue, so it's all-hands-on-deck. Our stomach will then attack whatever comes inside it with powerful acids, breaking chemical bonds and denaturing proteins. When it passes into our intestinal tract, the absorption begins. This time, billions of bacteria attack and render this preprocessed chyme even more. As lunch moves through our long pair of intestines, it is further churned and broken down, releasing the chemicals and nutrients needed to live. Gut fungi are there too. Everyone does a job, and if it all works out, everyone gets to live. Every meal is a banquet attended by thousands.

There are bacteria and fungi on our outer bodies as well. We are bathed in microscopic life. Shower all you want, use anti-bacterial soap, scrub until you are raw, and you will not get rid of them all. And that's a good thing because if you did, your health would suffer, and you might even die. Everyone is a community. The Human Microbiome Project found that our bodies are host to trillions of micro-organisms, outnumbering our human cells 10 to 1. There is a difference in cell sizes between us and them, making the human bodies tally: 97% human and 3% bacteria, fungi, and others by weight. This natural community is normal and healthy. All life is an act of compromise, and all life is cooperation.

So intertwined are we and the organisms that bathe us, the wrong partnering bacteria could make us ill, both physically and mentally. A study has shown that introducing an ancient gut ally, *Mycobacterium vaccae* to mice not only increased their learning abilities, but reduced anxiety as well. (Matthews, 2013) Other studies have suggested the same thing, we may "miss" our bacterial and fungal pals if something changes our microfauna. But the thought that a missing gut bacteria might cause some sort of anxiety is something to ponder deeply in our apprehensive times.

Life thrives not only because of diversity but because of redundancy. Many organisms perform the same functions at different times or in different places. This replication is because life, in the singular, is frail and tenuous. Disruptions are common, and someone (or thing) needs to be ready to take over if another, for any reason, fails. In a healthy ecosystem, many are prepared at all times. Life is like a poker game where the ante for each hand is everything you've got.

When an asteroid hit the Yucatan 66 million years ago, it triggered events that led to the extinction of much of the then-current bio-diversity on Earth. Life absorbed the blow, and as the last dinosaurs heaved into final submission, some little furry creatures scuttled about, dancing around their rotting corpses. These rat-like mammals would eventually lead to us. This rodent infestation did not disgust life, and the rats would patiently wait tens of millions of years for their first slice of pizza. Here, the mammals were the dinosaur's understudies, and they rapidly took the main stage in life, eventually becoming the apex. And they ultimately led to us. Who will take over when we're gone? It will be someone if we let it happen.

No one likes mosquito bites, so why not eliminate all mosquitos? Cockroaches, yuck, let's poison them into oblivion and not think about what that poison might do to us. "Those people: we don't like them. All of our problems would be gone if they were just no longer plaguing us." This type of thinking is an ethical and emotional trap, and it is so opposite of the truth that it staggers the healthy mind. Our old ideas are in free-fall with no end in mind or sight. We're waiting for something, and you can feel it's close. But our current myths deny us these final understandings.

An unintended side-effect of the climate crisis is that it offers us something essential that is needed to create or renew our civilization. It gives us a shared goal. It presents us with a mutual obligation. And avoiding it also requires many diverse viewpoints so that we can save the only actual race: the human race. Finally, it offers us the opportunity to write new stories. And to find new friendships, as we all hold each other's grimy hands at the end of the many days of hard work that lie ahead.

The path to avoiding our extinction is to save all of our companions so that all systems continue to function. Not only do we need to save the world we like, but we also have to keep those aspects of nature that we don't. This is the Green Evolution unbound. The world grew up the way it did because that was the way that worked. We were just too naïve to comprehend that simple fact. Life is diverse; we must nurture each other, as we must nurture the planet to ensure our survival and create our future.

Avoiding extinction is an all-or-nothing proposition.

VI. All Life is Aware

This facet of existence must be understood. For it is not just an attribute of life. It is its core characteristic. All life — from bacteria to barrister — is aware. The ability to find food, warmth, and shelter all requires some level of self-awareness. This awareness is not the same as projecting human traits on lower animals. It is, however, imperative that we come to understand its validity if we hope to work with life to save us all.

There is a reverse type of anthropomorphization that we now need to face as well. One where we refuse to recognize conscious traits in other life forms. A considerable part of our ability to depreciate life and wantonly degrade it is our ignorance of its consciousness. It's not human consciousness, but it is awareness. And it deserves acknowledgment and a modicum of respect.

We have been trained to believe that lower life forms don't feel pain. But if they react in repulsion to what are assumed negative stimuli, what would you prefer to call it? There is a long list of intellectual gyrations that have been developed to avoid this simple, essential understanding. And that's because if we allow ourselves to accept that all of our comrades are aware and feel, then an additional responsibility burdens us, one many of us might prefer to shirk.

It is still believed in the scientific literature and on many bar stools that fish don't feel pain. So, if you've had one squirm in your hand as you removed the hook from its mouth — that's not pain. Then what is it?

Robert Arlinghaus of the Leibniz Institute of Freshwater Ecology and Inland

Fisheries, and some colleagues, published a paper on this very topic in 2013. Their conclusions were based on the difference in brain formation between fishes and mammals. Fishes do lack the analogous brain structures where people and other mammals would feel pain, true. And if you were to believe that nature has only one way of doing things, then it would look like the jury had rendered their verdict on that one. But nature is full of tricks and methods of manifestation. Unfortunately, the studies' scientists seemed to miss that little understanding; and they ruled fish feel no pain.

However, there is a "but" in their study and a huge but. As the Leibniz team stated in their conclusion: "… bony fish certainly possess simple nociceptors, and they do, of course, show reactions to injuries and other interventions. But it is not known whether this is perceived as pain." (Arlinghaus, 2012) So, they weren't unequivocally stating that fish felt no pain, only that they don't feel via the same mechanisms as we do. This realization makes the title of their article: "Do fish feel pain," seem a bit over-sensational. Perhaps it could've been: "We still don't know if fish feel pain," or, "If fish feel pain, it's not the pain we feel when we fail to publish a new article this term." Call it what you want; fish are aware, even the scientists agree, and that's the important takeaway here.

We call blue jays camp robbers up here in Colorado. Because they are adept at snagging tidbits and morsels from unexpecting campers, the moment they turn their backs. Jays are members of the corvid family, a very intelligent group of birds that includes crows and Clark's nutcracker. The Clark's nutcracker is a small, 140-gram (5 oz.) bird with a prodigious memory. Scientists from the University of New Hampshire studied their food-collecting behavior and were dumbfounded by what they discovered. The Nutcracker relies on food stashes to survive the winter months, where nuts and seeds hidden all summer long are their only source of subsistence. They found that within the bird's nearly 40 km² (15 mi²) territory, it would establish up to 30,000 different food caches. This behavior allows them to maintain adequate stores and have backups in case of tragedies such as fire or squirrels. The thing that amazed the researchers was that the little bird remembers every single site. Even when the land was blanketed in deep snow and the trees had lost their leaves, these little feather headed geniuses found every snack that remained. (UNH, 2006) Now, few humans can remember a 30,000-location matrix in the course of a few months, and to the few who can: I'm sorry. The condition known as hyperthymestic syndrome seems more like a curse than a blessing to those with such persistent, detailed memory. This type of memory is impressive to witness but seems a disability to the many who work with or suffer from it. The rest of us may have to admit that our minds are somewhat less than those of these cute little birdbrains.

"The Secret Life of Plants" by Peter Tompkins and Christopher Bird was published in 1973 to great controversy. As mentioned earlier, it documented experiments that seemed to show awareness in plants. The data presented was too much for some fragile egos and was derided and challenged in nearly every possible way. Challenged in nearly every way except for the correct way: by replicating it and showing either an error or a null result. Now science is seeing a new group of experiments and more robust evidence that plants have awareness. Monica Gagliano, a research ecologist in Australia, has been experimenting with plant awareness for several years. She has uncovered evidence that plants can learn. (Gagliano, 2019) Plants are capable of learning. That's right, Vegans, and they know what your game is!
You can't eat without taking life, that's just how it is. That is the foundational deal that all life has made with itself and each other. Life is sharing, it is transient, and it is a group exercise. So we should get over it and move on to how we treat the life that we're taking.

All of these examples points to what is clear to any child who has become inseparable from their new puppy — life is aware. Puppy love is as real as any other. Doggie kisses are real kisses.
Treating your planetary roommates as objects is a common trait of us humans. It is the first step in developing the lack of respect for life that many have been toiling to perfect. It is the last stop on the road to demeaning. We've seen it before, in how we treat food, how we treat nature, and how we treat each other. Because life is all connected, this disdain affects everything and says something about those who practice it as a way of life. And the more one disconnects themselves from life, the greater the existential terror that wells up within them in a self-destructive feedback cycle. The deeper you fall down this hole, the farther you will feel removed from others, from relief. The relief that

is our re-connection with nature. And it feels good once we do it, like the wet slurp of a fresh smelling puppy.

Planaria worms can learn mazes, have their heads chopped off, and then complete the labyrinth again once a new head regrows. (Shomrat, 2013) This ability is astounding, as is every characteristic of life. Life has the answer to questions we haven't yet thought to ask. So, we need to stop treating everything that isn't human as an object. These planaria worms recoil when decapitated, as would any of us. And even though two new worms will grow from the pieces, they deserve at least a little admiration and some gentle care when working with them.

We should stop treating people as objects as well. This admonition should be obvious, but a quick look at the nightly news will confirm that it is not. If we stopped to consider other viewpoints, perhaps even other non-human perspectives, we might be on our way to wisdom. Ignorance is no longer a viable survival strategy for the most intelligent creature on the planet.

And so, we reach the crux of our situation; here is the very definition of an existential crisis. Do we choose to continue to exist, or do we go down whimpering, as T. S. Eliot[180] suggested? As I've said repeatedly, there is no Plan B for us. We can survive, and we will be all the better for it, poised for the next great human adventure. But only if we accept change at every level of our lives. If we reject diversity in any of its forms and choose our tired old ideas over the life that grows before us, we choose death. Then, the vines and beetles, the bacteria, and funga will devour what is left of us. Rust will conquer our machines, and time will wear down our towers, dams, and monuments. In the end, there may be no trace of us left behind.

Nothing.

---

[180] The poet, T.S. Elliot wrote "The Hollow Men" in 1925 concerning the horror of post- WW I Europe. "This is the way the world ends, not with a bang, but with a whimper." It was so powerful a poem that it remains relevant to this day.

Evolution is a fact, and if you disagree, you probably stopped reading and burned this book many pages ago. Burning this book would release about 1.3 kilograms (2.9 pounds) of CO_2 into the atmosphere. So, best to bury it and sequester that carbon, let it return to the greenness of the world.

Evolution is how nature improves itself and continues to survive. It is how we must renew ourselves now so that we might continue. I have often tried to impress that our key problem is our childish ways and need to grow up. This immaturity is valid as a metaphor, but perhaps it is more accurate to suggest we need to grow out.

We need to grow out of this system that keeps us at the lowest common denominator level of social development. People don't need to be paid off to do anything. And least not for anything that is really necessary for our lives, or the lives of those around us. Doing things is our most primal of drives. We don't need to punish people who make mistakes. We need to help them find their way. And we certainly don't need to kill each other over political differences, religious creeds, racial or sexual identities. We don't need to steal when we can trade.

As nations, many have often preferred to steal just to show their dominance. As a result, systems have developed that reward accidents of birth and even acts of treachery. For endless profit is just that, theft and treachery. It is the betrayal of life. And our current capitalism is often just a money-laundering scheme where all takings are deemed fair because they were captured as profit, regardless of that profits unfairness. But now we can realize that the real profit from life can, and must, be the ability to grow and thrive, to achieve and inter-relate. These are the profits denied by a system that measures value only as pieces of paper or columns in databases.

When one chooses to act in a new manner, the very structures of their brains, formed under a previous set of suppositions and actions, will resist it. We have to fight our minds to change them, but it is something we most definitely can do. And each step we take down that novel path will get easier. Our brains will alter their ways of processing data in line with the new information. If you've read this far you can probably already feel it. That is the true goal of this work, to stretch your awareness just a tiny bit. The goal is to reshape your thinking and guide behaviors that slogans and non-stop marketing have starved, to refresh our education about the world and each other.

A person who sits in a chair for years will find it hard to stretch and run. But with time and effort, they can run a marathon and even win it. Quitting smoking is a difficult and uncomfortable process, but it has a payback greater than the struggle. Millions have done it, and none of them were exceptional or unusual in any important way.

This task is achievable. And we find that the things we always wanted are possible, have always been there, just clouded by the rote and repetition that we wanted a dream of wealth and privilege instead of a life of health and harmony. It is a privilege to be alive. Yet, our current way of life threatens to erase even that original, primal aspiration.

Humanity faces a deep and multifaceted crisis that won't be ignored by even the most reluctant among us. Humans can bond together in the face of an emergency. We've done it before, and we'll do it again. It is not a matter of faith, and that is why it will happen. Faith can be broken.

We do not believe in climate change.

Belief is something you maintain despite the lack of concrete proof.

Global warming is a fact, and that it will keep getting worse until we confront it is the truth. No faith or belief is required. What has been needed is some understanding of what the problems are and how they might be addressed. Our problems can't be repaired with the presuppositions currently held. We need a new pattern, a Green Evolution.

Nature has a pattern, and it's right in front of us. We've forgotten that order because we separated ourselves from it. The pattern is that of chaos and how it fosters order in it's wake. Order that then sets the table for the next round of chaos. This pattern is the model of all growth, all movement. It is the prototype of the variety that is life. We need to see that there is a possibility of success if we adapt to that rhythm, that primal pulse. Nature has given us the model of a sustainable future. We need only follow its blueprint with a student's humility, not the arrogance of lords and masters. We are, after all, hitchhikers on this magnificent voyage.

What is required is the evolution of our ways of thought and our manners of living. We must align with the forces that gave us life or abdicate life itself. This evolution will free us from the imaginary bonds that imprison our species and set us on a path of success in the new world that has already begun. Evolution is a continuously ongoing process and will occur with or without us. In this marvelous case, we have the chance to take part actively. A species choosing to evolve is something we have yet to see.

This is our Green Evolution unbound. This is the Gordian knot untied.

This moment will be a revolution as great as any humankind has achieved, with the remarkable exception that here we understand what we are choosing and why. We will have crossed an epic threshold when we join together and save ourselves. We can't yet imagine all that this action will mean or the opportunities which will present themselves next. We only know that crossing this threshold is the choice of life over submission to oblivion. Humans don't like to submit, and therein lies our hope.

We can restore the planet using the forces of nature to heal our damage. Then nature will be using us to heal itself, and our value to Gaia will be secured. Our place onboard earned. The path of survival (life) is to choose to work on the side of the living, not on the side of ancient fables and fairy tales of power and perfection. We can heal the soil with our agriculture. We can restore and expand the forests by taking only what we need and returning the stands to their natural trajectories. That means also returning healthy fire. We will come to understand chaos as well as order, for it is the interplay of what only seems to be opposites that provides movement, growth, and transformation.
We can rid our world of the nerve gases that pass as pest control and save the bees and other pollinators. We can cleanse the waves and watch the ocean stocks return as we cease the onslaught of chemicals and nutrients with which we've burdened them. Ceased because we no longer require them. Our soils and trees will breathe in the carbon from the atmosphere and draw down the overburden. The winds will be safe, no matter from whence they blow.
The salmon will return because we came to understand that the river system was never broken and in need of our fixes. When we release the impounded rivers because we have reduced our need for imprisoned water by keeping

what we have cleaner, we will begin sharing what was never ours alone in the first place. As diversity returns, so will stability. The seasons can once again become mild and predictable, though that will take some time. The ecosystem was never broken until we broke it. Now we have to face the costs of fixing it. We have the intelligence, scientific knowledge, and the power to do this. We have the human heart to do it as well, and the promise that our hearts will grow beyond their previous bounds by saving the environment. All we'll need is patience, which is a fundamental aspect of nature. It is only our imposed will that lacks this capacity. Nature has the will of no will, so it has no reluctance to act.

So here is a meaning of life: "To shepherd every plant, funga and creature so that they can nurture us in return."

Here is our purpose, and if we can get this right, maybe we can go to other planets and improve them (I hear Mars is a "real fixer-upper"). Or at least co-exist stably with their systems. Otherwise, all attempts to expand across the galaxy are doomed to meet the same failures our lack of manners threatens to deliver us here at home.

If we were to calculate the mass of all animals by their weight and compare them to human biomass, we would discover some enlightening facts. Total human biomass is around 317 trillion kilograms (325 million standard tons; 3.25×10^8 tons). A roughly estimated weight of all other animals (not plants) is 912 million million kilograms (~1 million million standard tons; 1×10^{15} tons). Humans represent 0.034% of all animals on the globe by weight. In terms of biomass, the greatest Kingdoms of life on Gaia are the Protista and Monera, the bacteria and their relatives, weighing in at a staggering 907×10^{12} kilograms (0.9×10^{15} tons), or 99% of the total living biomass. Insects weigh in at around one gigaton, 1000 million tonnes. Spiders, which are arachnids, and not insects, eat up to 727 million tonnes (800 million tons) of prey[181] per year, over twice the weight of the entire human race. (Nyfeller, 2017)

---

[181] It should be noted that spiders eat mostly insects. A large number of these insects are what we would consider agricultural pests. Our insecticides kill spiders. Our culture is afraid of spiders and kills them indescriminantly. We might want to re-think that and let them get on with their job.

We represent ~1/3 of one percent of all animals on the globe by weight. So, our burden is not too great for Gaia to carry, if we could stop polluting and degrading and heal the ecosystem and its productivity. Suppose we only took what we needed and returned everything once we were finished with it? And we returned it in a form that was readily useable by other life forms? Then humans, still a small enough percentage of total planetary biomass, would be fine. We just need to let go a bit to let the tops' spin right itself. Life is self-healing.

If we can stop desertification and restore the lands we have broken since civilizations' beginning, we will find ourselves the recipients of plenty. When we stop throwing away the wealth we have pulled from the ground and put it to good use; then we can finally allow everyone a fair share. When we halt the accumulation of wealth by those who have no goal other than just wealth for its own sake we'll find the means to fund our salvation. When we share, when we practice human-created ideals such as fairness and justice, the weight of the world will fall from our shoulders and instead lift us up. We achieve our safety by keeping Gaia and her children safe. If we want to keep fighting her, we will lose. There is no other option.

I want to live. I want my children to live, and theirs as well. I want you to live too. I want your children to thrive, and theirs as well. This healing will take time, decades, at least, perhaps a century or longer. We will undoubtedly see things get worse before they get better. We will hear those certain folks screaming just to give up and let it be. "It is what it is," they will say. For that's what people say when they are without a plan and lack resolve. From now on out, we must react to such outbursts appropriately by laughing at them. Not everyone will be on board right away, but enough will be, and eventually, the others will follow. Humor will temper the harshness, and joy will attract the outliers. So let's join now, and don't just demand change from the status quo; let's make change happen. And despite the effort required, we can not only achieve this, but we can also do it joyfully. For it will bring joy when we begin to see the bounty that following the path of life provides. And imagine the delight when you realize you chose this way over the void.

This is our mea culpa: The only way to undo a lie is to admit it. Then the only way to restore trust is to make it right. And that's going to take some work, so roll up your sleeves.

Our Green Evolution began 4.4 billion years ago and is already in progress. Now let's go out there and join in.

Ecology AS Technology

EPILOGUE

"The day the power of love overrules the love of power, the world will know peace."
— **Mahatma Gandhi**

Our time is nearly over. Our time is only beginning. This is the choice that confronts us. One path is that of nihilism, surrender to the void. It will be bittersweet, but without rage. That rage is being spent on those who won't surrender to their cult. The consumptive personalities that presently hold the "conservative" political steering wheel have already let go of it, and released all hope as well. From their point of view, this coming annihilation will be freedom. Freedom actualized as attacks on social norms, government buildings and proceedings. Freedom from the debt the human race owes to the ecosystem. This is the arsonist, the personality who wants to burn down what they can't understand, much less control. They may unleash violence until their own release. So stay out of their way as much as possible.

The path of optimism feels good at this point. It is empowering to say you will face the coming storm. Welcome, we're glad you're here to help. There is a great relief to stepping off that fence. But it is certain that at some point, you will encounter the rage. What we're about to live through is going to be transformative, and there's going to be no one left untouched. This next century will not be comfortable, no matter what. We will see the end of oil and the collapse of any economies still connected to it. Governments and social order will collapse in some places. The faster we get on the stick, the shorter-lived this transition will be. We will lose a good percentage of our forests and sustain much coral loss. Coastlines are going to be swallowed. We are on a trajectory to the bottom and we've procrastinated too much to not at least graze it. Our job is to make sure we have the resiliency to bounce back out. We're falling in to a burning ring of fire. And in 30 -60 years, on the other side, we will be very different people. As long as we can remember the history that got us to this place, we'll be better people.

You are now a student of the living world. You should be brimming with information and armed with the tools to continue growing your arsenal of answers and ideas. And while this book may have seemed exhaustive, it is not. It is merely a jumping-off point for further investigation. The living world is so vast and fantastic that this work barely scratches the surface of the wonders you are about to discover. If we are to become stewards of this beautiful world, we must never stop learning about it. We must never stop learning about each other. We must never stop investigating and questioning. Science does not promise an ultimate answer because even to assume that one exists is just an assumption. We have found a damn good meaning, however, and that is how to thrive, not just survive. What this book does promise is the ability to live in awe, the opportunity to find significance for yourself and then to live it openly. It offers the ability to rip away the shroud of myth and habit and join your brethren and become whatever we are to be.

Now is the time to get to work.

This book is not an ultimate, anything. I conceived it to inform and draw connections that have been intentionally fragmented. An adult connect-the-dots exercise for a complex subject. It is a bit of education and a spot of humor. Because humor can help soften the blow of understanding hard truths. It is a vaccination against the misinformation that surrounds us. If you want to know more, look into the citations or begin some research of your own. Hopefully, your tools for sniffing out useful information have been sharpened. I expect that there will be a backlash from the status quo over this text. That was the idea. We needed to kick this debate in the under-gut. For this book is the Affirmative case: to affirm life and its inherent worth.

The most important understandings you need to take away are these:

Our situation is dire. Our survival is possible. Nature is our template for a meaningful life that can continue to be almost perpetual (as far as we know). If you see nature trying to do something, find a way to help it. Why have a revolution when you could have an evolution?

But You Must Engage!

It is up to you, the reader, to act on the information you gather. You're trapped now, though. You've learned too much to turn back. And that's good because our only way out now is to work through the coming changes. Your perspective is valuable; through diversity, we will grow stronger.

Keep **Learn**ing- keep investigating, find your niche, and dig in; this is a fight for existence and not a fad. If you are young, consider an education in ecology or other natural sciences. But don't discount a complete education. Schooling bereft of art, music, and literature gives tools without context on how to use them. We should strive to be like Michelangelo and not tedious hacks. Unfortunately, the current obsession with STEM education sometimes misses this point. Well-rounded individuals are mighty individuals.

> "Educating the mind without educating the heart is no education at all."
> - *Aristotle*

If you fall into an older demographic, like Aristotle in his prime, keep learning. Volunteer and find your new passion. Treat it as if it was your only career. Use the tools and wisdom you have gathered and let them loose upon the world. Take your organizational skills and help the youngsters get it done!

Always **Question**- Question everything, including your motives, before charting your actions. Never fall into the trap of thinking you have everything figured out and have every bit of information you need. Nor should you wait until you think you have it all figured out. Balance your actions, but above all, act. Natural systems are complex and deeply cross-linked, but that makes them interesting. Nature knows how to heal. We just need to figure out how to assist her in the least damaging ways possible. Nature is within us as well; feel its rhythms inside you and let them guide you.

Listen- Listen to your allies and those with whom you disagree, especially those with whom you disagree. While we must work to save the planet, we need to do so while considering many perspectives. Be aware that some people won't, or even can't, change their paths. So you should listen to them to know who you are talking to and where they are. Some people could become agitated, or worse, if you try to bend their minds too far. Please recognize the psychology at play here, and practice presenting to your audience. You might find, if you listen, that you will agree on far more than you disagree, even with those who would appear to be the opposition. And that's a power those who try to bluff and bully will never grasp. It's an opening that can be used to give people new information in a manner that they won't immediately reject. Remember that they are the opposition; they are not the enemy until they actually strike you. And even then, try to de-escalate and incorporate them into the work of survival- they want to live too. And once you've found some common ground, walk a little way together until they can see the path as well. Know when to speak and know when to shut up and listen.

Speak- Engage with others, help educate them. Speak in their language. Speak in your voice. Spread opportunities and information to anyone who'll listen. If they show resistance, show them respect and try to discuss through their doubts. But remember the warning above. We have been trained to be stubborn in our beliefs. Some folks believe that changing their mind (growing when presented with new information) is a weakness. This reluctance is a remnant of their training to follow those who are "perfect" and never to question their "wisdom." They have already accepted pat answers to all the tough questions and find peace in never having to consider them again. But the actual strength, the proper road to our personal growth, is the ability to change our minds when warranted. A healthy mind is strong and flexible. It will not break when confronted with new information. So speak with recognition, but firmly, try to impart upon your audience that belief has little to do with facts or survival. If they can't hear you, sign off with a respectful goodbye, and move on.

And finally,

Organize- The voices of the many need to be heard. But despite the power of mass demonstrations, they are not the most effective method of implementing lasting change. They are significant events to enthuse those already on board and a net to collect interested newbies, but they can also turn off those who are still undecided. Many people hid inside their homes in fear of demonstrations of inclusiveness and tolerance in 2020. Some showed up on their decks and waved their guns, thinking themselves brave and righteous. Some killed unarmed protestors and were rewarded by a segment of our society that included several courts. So not everyone present at peaceful protests is going to "get it." Mass protests can also provide cover to anarchists who love the confusion of a large crowd to perpetrate some vandalism. They also allow the opposition to infiltrate peaceful gatherings with violence, diluting the message. These disadvantages do not mean we should never use protests or mass celebrations, but we must be do them carefully, and we must understand their effectiveness. I hope the Black Lives Matter demonstrations of 2020 will lead to continued change, but using the example of the mass protests of the 1960s, there's an equal chance they won't. But that's no reason to stop working for peace and unity. It is instead just an opening for a new tactic and continued activity. The change we seek, the salvation of the ecosystem, is one of social equality and diversity as well. We need to embrace the cause of equality. And it will take more action than some demonstrations to resolve the injustices of the past.

Change happens in boardrooms and council chambers; we must be there to make ourselves heard. Laws are made in legislatures and the people belong there too, both on the floor and in the balconies. We must be able to speak clearly, respectfully, and with the power of truth. Do not go off the handle like those who have been prepped by propagandists and inauthentic politicians. The first step is to be informed, get past the rage and stand firm with knowledge before you take the podium. If you find anger overpowering yourself, step back and hand the baton to the next speaker in line.

What seems to work for change is to organize into small groups who can implement small actions- groups where every face is known and each opinion can be directly addressed. Use the psychology built in us to create tight-knit groups that can identify and complete local projects. Once that works well, reach out and join other small groups to tackle larger issues. Don't wait for larger organizations or the government to reach out to you and assign a task, although if you are successful, that may soon happen. Always start where you are and go from there. Chances are it will take some time before you (we) have everything we need, but don't wait; get started now. Build that community garden, start composting, recycle, and join with others to make it more effective. Don't forget to smile. Joy and enthusiasm are contagious and will win more hearts than facts or citations in the long run. And to finish this marathon, we need to run now.

In the 1950s, and for a period beyond, society took the threat of nuclear war very seriously. We trained our children with (albeit silly) films about survival in a nuclear holocaust. We practiced self-control in international affairs and avoided the undue escalation of hostilities. To this day, there are thousands of nuclear warheads with a combined destructive power that could wipe out most of life on Gaia. But we still treat them with care and have stepped back from the brink of nuclear posturing and threats of mutually assured destruction. In the 1980s, Russia even offered to eliminate these weapons for good. But our rearward-focused President, Ronald Reagan, refused. However, now we can begin that task anew.

Our reactions tempered, have avoided nuclear extinction so far. And that's because we took the danger seriously. And we still do. Our first task is to teach the same seriousness and awareness of the global climate crisis we recognize in the atomic menace. Once the world is on board, we will be on a fast track to implementing the frameworks outlined in this book. So let's get out there and start persuading.

This book cannot save you. It is only your starting point. This book can't even give you a big enough step up to reach the cereal boxes on the top shelf. You must save yourself. We must save ourselves. I'll be out there working, too. Look for me in the forest after the rain, grubbing for fresh mushrooms. Or in the Lab on dry days (or on the deck, take a rest now and then). We need an irresistible human tidal wave of action to overcome the immovable object of the status quo. The future is in all of our hands — so let no one be empty-handed. If we get this right, we will be known in the future as the true conservatives of the 21st century.

Let's begin!

-Ravage-

Further reading:

"Silent Spring"	Rachel Carson, 1962
"Humans versus Nature" (A Global Environmental History)	Daniel Headrick, 2020
"Silenced Rivers" (The ecology and politics of large dams)	Patrick McCully, 2001
"The Hidden Life of Trees"	Peter Wohllbern, 2016 (English)
"Entangled Life"	Merlin Sheldrake, 2021
"The Omnivore's Dilemma,"	Michael Pollan, 2006
"How to Change Your Mind"	Michael Pollan 2018

A brief list of Gaia's extinctions:

1) The Great Oxygenation Event: 2.4-2.0 Billion years ago.
2) Ordovician-Silurian Extinction: 440 Million years ago.
3) Devonian Extinction: 365 Million years ago.
4) Permian-Triassic Extinction: 250 Million years ago.
5) Triassic-Jurassic Extinction: 210 Million years ago.
6) Cretaceous-Tertiary Extinction: 65 Million years ago.
7) Anthropocene Extinction: (coming soon if we don't stop it)

CITATIONS

Agency, U. N. (2015, March). *Backgrounder on Nuclear Waste*. Retrieved December 24, 2018, from nrc.gov: https://www.nrc.gov/reading-rm/doc-collections/fact-sheets/radwaste.html

Ahmed, N. (2013, December 23). Former BP geologist: peak oil is here and it will 'break economies'. *The Guardian*.

Allison, E. (2018). *Methane Emissions in the Oil and Gas Industry*. Alexandria, VA: American Geosciences Institute.

Allred, B. (2015). Ecosystem services lost to oil and gas in North America. *Science, 348* (6233): 401.

Altieri, M. (2004). *Roundup ready soybean in Latin America: a machine of hunger, deforestation and socio-ecological devastation*. Berkley, Ca: U. Cal. Berkley.

Andrews, W. (1989, April 3). Exxon takes out spill ads in 165 publications. *United press International*.

Andrzejewski, A. (2018, August 14). Mapping The US Farm Subsidy $1M Club. *Forbes*.

Arlinghaus, R. (2012). Can fish really feel pain? *Fish and Fisheries*.

Arnason, R. (2017, January 19). *Survey reveals 'amazing' soil loss in Great Plains region*. Retrieved January 26, 2019, from the Western Producer: https://www.producer.com/2017/01/survey-reveals-amazing-soil-loss-in-great-plains-region/

ASCE. (2018). *Infrastructure report Card*. Reston, VA: American Society of Civil Engineers.

AWI. (2007). *Legal Protections for Animals on Farms*. Washington, DC: Animal Welfare Institute.

B.P. (2018). *B.P. Statistical review of World Energy 2018*. London, UK: British Petroleum.

CITATIONS

Baker, B. W. (2003). *Wild Mammals of North America: Biology, Management, and Conservation. Second Edition.* Baltimore, Maryland: The Johns Hopkins University Press.

Baker, B. W. (2005). INTERACTION OF BEAVER AND ELK HERBIVORY REDUCES STANDING CROP OF WILLOW. *Ecological Applications*, 110-118.

Battaglia, M. (2018). *A regional assessment of the ecological effects of chipping and mastication fuels reduction and forest restoration treatments.* . Fort Collins, CO: Joint Fire Science Council.

Berdick, C. (2018). *What's school without grade levels?* New York, N.Y.: The Hechinger Report .

Bergmann, M. (2019). White and wonderful? Microplastics prevail in snow from the Alps to the Arctic. *Science Advances*, 1157-1160.

Betts, H. (2018). Integrated genomic and fossil evidence illuminates life's early evolution and eukaryote origin. *Nature Ecology & Evolution 2*, 1556–1562.

Black, R. (2011, March 17). *"Reactor breach worsens prospects".* Retrieved from BBC News: https://www.bbc.com/news/science-environment-12745186

Blincoe, K. (2016, September 21). Half of UK women flush tampons away – this has to stop. *The Guardian.*

Boon, J. (2012). Evidence of sea-level acceleration at US and Canadian tide stations, Atlantic coast of North America. *Journal of Coastal Research.*

Borunda, A. (2019, September 6). How tampons and pads became so unsustainable. *National Geographic.*

Borunda, A. (2019, August 29). See how much of the Amazon is burning, how it compares to other years. *National Geographic.*

Bowd, E. (2019). Long-term impacts of wildfire and logging on forest soils. *Nature Geoscience*, 113-118.

Bruce, F. (2017, February 26). Denver Water, US Forest Service, partners embark on $33 million forest health deal . *Denver Post.*

Burchstead, D. (2010). The River Discontinuum: Applying Beaver Modifications to Baseline Conditions for Restoration of Forested Headwaters. *BioScience*, 908–922.

Burke, H. (2021, September 29). *Baby born with 'pack-a-day' smoke exposure despite the mother never touching a cigarette*. Retrieved from news.com.au: https://www.news.com.au/lifestyle/parenting/babies/baby-born-with-packaday-smoke-exposure-despite-the-mum-never-touching-a-cigarette-in-her-life/news-story/ff018cbd9c269e7cc7b92e8d9b864914

Burke, L. (2011). *Reefs at Risk Revisited.* Washington, D.C.: World resources Institute.

Calloway, A. (2017). *Demand the Supply.* Washington D.C.: Enough Project.

Census, U. (2016). Washington D.C.: U.S. Census Bureau.

Chandler, W. (1984). *The Myth of the TVA: Conservation and Development in the Tennessee Valley.* Ballinger: Cambridge, MA.

Charbonneau, J. (2010). *"Conserving America's Fisheries-An Assessment of Economic Contributions from Fisheries and Aquatic Resource Conservation".* Arlington, VA: US Fish and Wildlife Service.

Chen, M.-M. (1993). *Mushrooms: A Fine Agricultural Crop.* Berkley, CA: University of California.

Cincinnati, U. o. (2008, February 4). Plastic Bottles Release Potentially Harmful Chemicals (Bisphenol A) After Contact With Hot Liquids. *Scince News*.

Clemmonsen, K. E. (2013). Roots and Associated Fungi Drive Long-Term Carbon Sequestration in Boreal Forest. *Science*, 1615-1618.

Cody, W. (1894, June). Famous Hunting Parties of the Plains. *The Cosmopolitan: A Monthly Illustrated Magazine* .

Cohen, P. (2020, June 24). Roundup Maker to Pay $10 Billion to Settle Cancer Suits. *The New York Times*.

Comes, H. (1998). The effect of Quaternary climatic changes on plant distribution and evolution. *Trends in Plant Science*, 432-438.

Cox, K. (2019). Human Consumption of Microplastics. *Environmental Science Technology*, 7068-7074.

CSB. (2014). *Investigation Report Overview- Explosion and Fire at the Macanado Well.* Houston, TX: US Chemical Safety Board.

Curry, O. (2019). Is It Good to Cooperate?Testing the Theory of Morality-as-Cooperation in 60 Societies. *Current Anthropology*, 47-69.

Cury, P. Y. (2002). Marine Ecosystems: towards responsible and sustainable fisheries. *Les dossiers thematiques de IRD, Institue de Researche pour le developpment*, 255-265.

D'Arrigo. (2014, August 24). *nirs.org.* Retrieved December 24, 2018, from Nuclear Information and Resource Service: https://www.nirs.org/wp-content/uploads/radwaste/atreactorstorage/prvotewc82614.pdf

Datoo, M. (2021). High Efficacy of a Low Dose Candidate Malaria Vaccine, R21 in 1 Adjuvant Matrix-M™, with Seasonal Administration to Children in Burkina Faso. *The Lancet.*

DeConto, R. (2016). Contribution of Antartica to past and future sea-level rise. *Nature*, March.

Dhar, R. (2017, December 6). *Diamonds are Bullshit.* Retrieved 2 9, 2019, from Huffington Post: https://www.huffingtonpost.com/rohin-dhar/diamonds-are-bullshit_b_3708562.html

Downs, C. (2015). Toxic pathological effects of sunscreen UV filter, oxybenzone (benzophenone-3) on coral planulae and cultured primary cells and its environmental contamination in Hawaii and the U.S Virgin Ieslands. *Archives of contamination and toxicology.*

E.I.A., U. (2011). *Annual Energy Review.* Washington, DC: United States Energy Information Administration.

E.P.A. (2019). *Facts and Figures about Materials, Waste and Recycling.* Retrieved from EPA.gov: https://www.epa.gov/facts-and-figures-about-materials-waste-and-recycling/national-overview-facts-and-figures-materials

EPA (2020, January 17). *Water Conservation at EPA.* Retrieved from EPA Official Site: https://www.epa.gov/greeningepa/water-conservation-epa

E.S.R.L. (2013, May 10). *CO2 at NOAA's Mauna Loa Observatory reaches new milestone: Tops 400 ppm .* Retrieved from N.O.A.A. News: https://www.esrl.noaa.gov/gmd/news/7074.html

Eig, K. (2017, November 3). *Where was the world's first oil well? Poland!* Retrieved from Adventures in Geology- Karsten Eig: https://karsteneig.no/2017/11/where-was-the-worlds-first-oil-well-poland/

El Albani, A. (2014). *The 2.1 Ga Old Francevillian Biota: Biogenicity, Taphonomy and Biodiversity.* San Francisco: PLoS One.

Eweje, E. (2006). Environmental Costs and Responsibilities Resulting from Oil Exploitation in Developing Countries: The Case of the Niger Delta of Nigeria. *Journal of Business Ethics*, 27-56.

Falade, A. (2018). Ligninolytic enzymes: Versatile biocatalysts for the elimination of endocrine disrupting chemicals in wastewater. *Microbiology Open.*

FAO. (2009). *Fishery and Aquaculture Economics and Policy Division; Fisheries and Aquaculture Management Division.* New York, NY, USA: United Nations, F.A.O.

Filbee-Dexter, K. (2020). Substantial blue carbon in overlooked Australian kelp forests. *Scientific reports.*

Frankel, T. (2017). *Why Apple and Intel don't want to see the conflict minerals rule rolled back.* Washington, D.C.: The Washington Post.

Gagliano, M. (2019, 08 13). *What a plant learns. The curious case of Mimosa pudica.* Retrieved from Botany One Blog: https://www.botany.one/2019/08/what-a-plant-learns-the-curious-case-of-mimosa-pudica/

Galanda, D. (2014). Mycoremediation: the study of transfer factor for plutonium and americium uptake from the ground. *Journal of Radioanalytical and Nuclear Chemistry*, 1411–1416.

George, R. (2013). *Ninety Percent of Everything.* London: MacMillan.

Gerlin, A. (1994, Sept 1). A Matter of Degree: How a Jury Decided That a Coffee Spill Is Worth $2.9 Million. *Wall Street Journal.*

Giesler, C. (2017). Impediments to inland resettlement under conditions of accelerated sea level rise. *Land use Policy*, 322-330.

Giessen, V. H. (2015). Emerging pollutants in the environment: A challenge for water resource management. *International Soil and Water Conservation Research*, 57-65.

Gilbert, N. (2012). Organic farming is rarely enough. *Nature*.

Gill, J. (2009). Pleistocene Megafaunal Collapse, Novel Plant Communities, and Enhanced Fire Regimes in North America. *Science*, 1100-1103.

Gimon, E. (2019). *THE COAL COST CROSSOVER:ECONOMIC VIABILITY OF EXISTING COAL COMPARED TO NEW LOCAL WIND AND SOLAR RESOURCES*. San Francisco: Energy Innovation Policy and Technology.

Gliessman, S. (1982). Polyculture cropping has advantages. *California Agriculture*, 14-16.

GlobalWitness. (2019). *Enemies of the State?* Global Witness.

Goodnough, A. (2016, January 23). When the Water Turned Brown. *New York Times*.

Gossner, M. (2009). How many species live in the oldest pine in the Bavarian forest. *Der Wald*, 164-165.

Grand, J. (2019). The future of North American grassland birds: Incorporating persistent and emergent threats into full annual cycle conservation priorities. *Conservation Science and Practice*, pending.

Greenpeace. (2015). *The Trash Vortex*. Retrieved 2 9, 2019, from Greenpeace International: https://www.greenpeace.org/archive-international/en/campaigns/oceans/fit-for-the-future/pollution/trash-vortex/

Gregory, D. e. (2017). Transgenerational transmission of asthma risk after exposure to environmental particles during pregnancy. *American Journal of Physiology – Lung Cellular and Molecular Physiology*, L395-L405.

Hacskaylo, E. a. (1967). Inoculation of Pinus caribbea with pure cultures of mycorrhizal fungi in Puerto Rico. *Proceedings of the 14th Congress of the International Union of Forestry Research Organizations*. (pp. 139-148). Munich: Freiburg: Deutscher Verband Forslicher Forschungsanstalten.

Hall, S. (2015). Exxon Knew about Climate Change almost 40 years ago. *Scientific American*.

Harris, P. (2013, February 12). *Monsanto sued small farmers to protect seed patents, report says*. Retrieved from The Guardian: https://www.theguardian.com/environment/2013/feb/12/monsanto-sues-farmers-seed-patents

Hawken, P. (1997, March-April). Resource Waste. *Mother Jones*.

Headrick, D. (2019). *Humans versus Nature: A Global Environmental History*. Oxford, UK: Oxford University Press.

Heckman, D. S. (2001). Molecular Evidence for the Early Colonization of Land by Fungi and Plants. *SCIENCE*, 1129-1133.

Hendryx, M. (2011). Self-Reported Cancer Rates in Two Rural Areas of West Virginia. *Journal of Community Health*.

History, A. M. (1909). *The beaver man*. Retrieved from Indian mythology: http://www.indianmythology.org/assiniboin/beaver_man.htm#Buffalo

IAEA. (1999). *Inventory of radioactive waste disposals at sea*. Vienna, Austria: International Atomic Energy Agency.

IAEA. (2011). *Japanese Earthquake Update (19 March 2011, 4:30 UTC)*. Vienna, Austria: InternationalAtomic Energy Association .

IRENA. (2019). *Global Energy Transformation: A roadmap to 2050 (2019 Edition)*. Abu Dhabi: International Renewable Energy Agency.

J.H. van Oppen, M. (2015). Building coral reef resilience through assisted evolution. *PNAS*, 2307-2313.

Jackson, B. (2008, July 1). *Fuel Efficient Freight Trains?* Retrieved March 9, 2019, from Fact Check.org: https://www.factcheck.org/2008/07/fuel-efficient-freight-trains/

Jones, T. C. (2012). America, Oil, and War in the Middle East . *The Journal of American History*, Vol. 99, No. 1, pp. 208-218 .

JRC. (2008). *Urbanization: 95% Of The World's Population Lives On 10% Of The Land*. European Commission, Joint Research Centre (JRC).

Kasinof, L. (2018, April 19). An ugly truth behind 'ethical consumerism'. *Washington Post*.

Kennedy, A. (2002). Host range of a deleterious rhizobacterium for biological control of downy brome. *Weed Science*, 792-797.

Kimbrell, A. (2005). *Monsanto vs. US Farmers*. Washington, D.C.: Center for Food Saftey.

CITATIONS

Koplitz, e. (2016). Public health impacts of the severe haze in Equatorial Asia in September–October 2015: demonstration of a new framework for informing fire management strategies to reduce downwind smoke exposure. *Environmental Research Letters*.

Kraft, J. (1978). Relationship between energy and GNP. *Journal of Energy Development*, Journal Volume: 3:2.

Kulshrestha, U. C. (2001). Investigation of Alkaline Nature of Rain Water in India. In K. Sutake, *Acid Rain 2000* (pp. 1685-1690). Tsukuba, Japan: Kluwer Academic Publishers.

Laden, F. (2006). Reduction in Fine Particulate Air Pollution and Mortality. *American Journal of Respiratory and Critical Care Medicine*, 667-72.

Lin, N. (2012). Physically based assessment of hurricane surge threat under climate change. *Nature Climate Change*, 462-467.

Litman. (2006). "Changing Travel Demand: Implications for Transport. *ITE Journal*, 27-33.

Litman, T. (2018). *The Hidden Traffic Safety Solution: Public Transportation.* Washington, D.C.: American Public Transportation Association.

Lonergan, S. (1998). The role of environmental degradation in population displacement. *Environmental Change and Security Project Report, Issue 4 (Spring 1998): 5-15.*

Long, E. C. (2016). Non-cultivated plants present a season-long route of pesticide exposure for honey bees. *Nature Communications*.

Lovett, B. (2019). Transgenic Metarhizium rapidly kills mosquitoes in a malaria-endemic region of Burkina Faso. *Science*, 894-897.

Lustgarten, A. (2013, May 27). *Causes of — and potential solutions to — the Water Crisis on the Colorado River.* Pro Publica.

Lyon, D. (2016). Aerial Surveys of Elevated Hydrocarbon Emissions from Oil and Gas Production Sites. *Environ. Sci. Technol.*, 4877–4886.

Madowell, N. (2016). Multi-scale predictions of massive conifer mortality due to chronic temperature rise. *Nature Climate Change*, 295-300.

Mai, C. (1996). Pharmaceuticals and Endocrine Disruptors in Wastewater. *Water Qual. Res. J. Canada*, Volume 41, No. 4, 351–364.

Malthus, T. (1798). *An Essay on the Principle of Population.*

Mason, K. L. (2015). *Occupational Fatalities During the Oil and Gas Boom — United States, 2003–2013.* Washington, DC: US Centers for Disease Control and Prevention.

Matsuzaki, S. (2018). "Bottom-up linkages between primary production, zooplankton, and fish in a shallow, hypereutrophic lake". *Ecology*, 2025–2036.

Matthews, D. (2013). Ingestion of Mycobacterium vaccae decreases anxiety-related behavior and improves learning in mice. *Behavioural Processes*, 27-35.

McClain, C. (2019). Persistent and substantial impacts of the Deepwater Horizon oil spill on deep-sea megafauna. *Royal Society of Open Science.*

McCully, P. (1996). *Silenced Rivers; the Ecology and Politics of large Dams.* Zed Books.

McHenry, L. (2018). The Monsanto Papers: Poisoning the scientific well. *International Journal of Risk and Safety in Medicine.*, 193-205.

Md Wasim, A. (2009). Impact of pesticides use in agriculture: their benefits and hazards. *Interdisciplinary toxicology*, 1-12.

Merchant, C. (2007). *American Environmental History: An Introduction.* New York, NY: Columbia University Press.

Mikkelson, D. (1997, February 27). *Did Disney Fake Lemming Deaths for the Nature Documentary 'White Wilderness'?* Retrieved from Snopes.com: https://www.snopes.com/fact-check/white-wilderness/

Millero, F. J. (1995). Thermodynamics of the carbon dioxide system in the oceans. *Geochimica et Cosmochimica Acta*, 661–677.

Miyawaki, A. (1992). Restoration of Evergreen Broad-leaved Forests in the Pacific Region. *Ecosystem Rehabilitation.*

Moomaw, W. (2019). *Intact Forests in the United States: Proforestation Mitigates Climate Change and Serves the Greatest Good.* Oxford, UK: Front. For. Glob. Change.

CITATIONS

Moore, F. (2015, January 12). Temperature impacts on economic growth warrant stringent mitigation policy. *Nature climate change*.

Murphy, J. (Director). (2019). *Artifishal* [Motion Picture].

NASA. (2004, April 27). *NASA News*. Retrieved March 16, 2019, from nasa.gov: https://www.nasa.gov/home/hqnews/2004/apr/HQ_04140_clouds_climate.html

NASA. (2009). *Atmosphere's Energy Budget*. Retrieved from Earth Observatory: https://earthobservatory.nasa.gov/Features/EnergyBalance/page6.php

NOAA. (2019, June 12). *NOAA forecasts very large 'dead zone' for Gulf of Mexico*. Retrieved June 15, 2019, from National Oceanic and Atmospheric Administration: https://www.noaa.gov/media-release/noaa-forecasts-very-large-dead-zone-for-gulf-of-mexico

Nobel, C. (2019, June 19). Current CRISPR gene drive systems are likely to be highly invasive in wild populations. *eLife (pre-publication)*.

Norwegian Inst, B. R. (2017, January 18). *Food security threatened by sea-level rise*. Retrieved February 3, 2019, from phys.org: https://phys.org/news/2017-01-food-threatened-sea-level.html

NRCS. (2015, October 23). *NRCS*. Retrieved from http://www.nrcs.usda.gov/wps/portal/nrcs/detail/ia/newsroom/releases/?cid=NRCSEPRD410034

Nuwar, R. (2019, August 5). Environmental Activists Have Higher Death Rates Than Some Soldiers. *Scientific American*.

Nyfeller, M. (2017). An estimated 400–800 million tons of prey are annually killed by the global spider community. *Science of Nature*.

Osterling, K. (1999). *An Objective assessment of the dangers of Plutonium*.

P.S.C., W. (2011). *psc.wi.gov*. Retrieved August 25, 2019, from Public Service Commission of Wisconsin: https://psc.wi.gov/Documents/Brochures/Under%20Ground%20Transmission.pdf

Panel. (1983). *A Nation at Risk*. Washinton D.C.: U.S. Dept of Education.

Park, C. (2016). Condensation on slippery asymmetric bumps. *Nature*, 78-82.

Patricola, C. (2018). Anthropogenic influences on major tropical cyclone events. *Nature*, 339–346.

Pell, M. (2016, December 19). Thousands of US Areas Afflicted with Lead Poisoning beyond Flint's. *Scientific American*.

Petrusich, A. (2020, June 5). K-Pop Fans Defuse Racist Hashtags. *The New Yorker*.

Pew. (2015). *Religion and Views on Climate and Energy Issues.* Washington, D.C.: PEW Research Center.

Picasso, V. D. (2008). Crop Species Diversity Affects Productivity and Weed Suppression in Perennial Polycultures under Two Management Strategies. *Crop Science*.

Porter, K. (2021). *Big Profits, how big pharma takeovers destroy innovation and harm patients.* Washington D.C.: The Office of Katie Porter.

Powell, J., & Danby, G. (2003, July). Maglev: The New Mode of Transport for the 21st Century. *21st Century Science and Technology*, pp. 43-57.

Rahmstorf, S. e. (2015). Evidence for an exceptional twentieth-century slowdown in Atlantic Ocean overturning. *Nature Climate Change*.

Ravage. (2023). Fungal degradation of woody by-products of forest management activities. *TBD*.

Read, A. (2008). The Looming Crisis: interactions between Marine Mammals and Fisheries. *Journal of Mammalogy*, 541-548.

Redfearn, G. (2019, December 13). *"Australia's bushfires have emitted 250m tonnes of CO2, almost half of country's annual emissions".* Retrieved from The Guardian, AU edition: https://www.theguardian.com/environment/2019/dec/13/australias-bushfires-have-emitted-250m-tonnes-of-co2-almost-half-of-countrys-annual-emissions

Redmond, J. (2017). *DIRTY ENERGY DOMINANCE: Dependant on Denial.* Washington, D.C.: Oil Change International.

CITATIONS

Rhoades, C. (2015). Pile burning creates a fifty-year legacy of openings in regenerating lodgepole pine forests in Colorado. *Forest Ecology and Management*, 203-209.

Rigaud, K. K. (2018). *Groundswell : Preparing for Internal Climate Migration.* Washington, D.C.: World Bank.

Ross, M. (2004). Occupational respiratory disease in mining. *Occupational Medicine*, 304-310.

Schiermeier, Q. (2019). Eat less meat: UN climate change report calls for change to human diet. *Nature*.

Schopf, J. W. (2017). Evidence of Archean life: Stromatolites and microfossils. *Precambrian Research*, 141-155.

Scientists, U. o. (2018, October 11). *Union of Concerned Scientists*. Retrieved from https://www.ucsusa.org/global-warming/science-and-impacts/science/each-countrys-share-of-co2.html

Scudder, J. (2015, February 13). *The conversation-The sun won't die for 5 billion years, so why do humans have only 1 billion years left on Earth?* Retrieved December 27, 2018, from phys.org: https://phys.org/news/2015-02-sun-wont-die-billion-years.html

Seibel, R. (1970). Variables Affecting the Critical Thermal Maximum of the Leopard Frog, Rana pipens, Schreber . *Herpetologica*, 208-213.

Shogren, E. (2011, April 21). *BP: A Textbook Example Of How Not To Handle PR.* Retrieved from npr.org: https://www.npr.org/2011/04/21/135575238/bp-a-textbook-example-of-how-not-to-handle-pr

Shomrat, T. (2013). An automated training paradigm reveals long-term memory in planarians and its persistence through head regeneration. *Journal of Experimental Biology*, 3799-3810.

Simard. (1997). Net transfer of carbon between ectomycorrhizal tree species in the field. *Nature*, 579–582.

Singh. (2006). *Mycoremediation: Fungal Bioremediation.* John Wiley & Sons.

Singh, R. (2013). Oil Palm genome sequence reveals divergence of infertile species in Old and New worlds. *Nature*, 500 (7462): 335-339.

Slat, B. (2018). Interview with non-profit: Ocean Cleanup's Founder. (CNN, Interviewer)

Smith, D. (2018). *It didn't start with Trump: how America came to undervalue teachers.* Washington, D.C.: The Guardian.

Smith-Spangler, C. (2012). Are Organic Foods Safer or Healthier Than Conventional Alternatives?: A Systematic Review. . *Annals of Internal Medicine*, 157:348–366.

Solis, S. (2016, June 1). Donald Trump tells Californians there is no drought. *USA Today*.

Sovacool, B. (2009). Contextualizing avian mortality: A preliminary appraisal of bird and bat fatalities from wind, fossil-fuel, and nuclear electricity. *Energy Policy*, 2241-2248.

Spewsic, D. (2017, Jan 1`7). *Betsy DeVos is alarmingly stupid.* Retrieved from Medium.com: https://medium.com/@donspewsic/betsy-devos-is-alarmingly-stupid-9ce2639cdc98

Staff. (2012, April 15). *Recycling & the Future of Mining.* Retrieved January 28, 2018, from the business of mining: https://thebusinessofmining.com/2012/04/15/recycling-the-future-of-mining/

Stauber, S. R. (1997). *Mad Cow U.S.A.* Common Courage Press. Retrieved from http://cspinet.org/foodspeak/laws/existlaw.htm

Steele, J. (2007). *Flathead Indian Reservation Fire Management Plan.* Ronan, MT: Confederated Salish & Kootenai Tribes.

Stevens-Rumann, C. (2019). Tree regeneration following wildfires in the western US: a review. *Fire Ecology*.

Surana, K. (2019). The climate mitigation opportunity behind global power transmission and distribution. *Nature Climate Change*, 660-665.

Sutake, K. (2001). *Acid Rain 2000 Proceedings from the 6th International Conference on Acidic Deposition.* Tsukuba, Japan: Kluwer Academic Publishers.

Tanzer, J. (2015). *Living Blue Planet Report. Species, habitats and human well-being.* Gland, Switzerland: World Wildlife Fund.

Taylor. (2014, March 24). Remembering the Exxon Valdez Oil Spill. *The Atlantic*.

CITATIONS

Taylor, A., & Stamets, P. (2014). Implementing Fungal Cultivation in Biofiltration Systems –. *National Proceedings: Forest and Conservation* (pp. P-72. 23-28). Fort Collins Colorado: Proceedings Rocky Mountain Research Station.

Tero, A. (2010). Rules for Biologically Inspired Adaptive Network Design. *Science*, 327-439.

Testor, e. a. (2015). *Sustainable Energy- choosing amoung options.* Cambridge, Ma: MIT Press.

Tidwell, T. (2016). State of Forests and Forestry in the United States. *World Conservation Congress.* Honolulu, HI.

Tokonami, S. (2012). Thyroid doses for evacuees from the Fukushima nuclear accident. *Scientific reports*, 507.

Tzimas. (2005). *Enhanced Oil Recovery...* Petten, the Netherlands: European Commission Joint Research Centre.

UCS. (2015). *Deforestation causes 10% of global warming emissions.* Cambridge, Ma: Union of Concerned Scientists.

UN. (2013). *United Nations World Water Day.* Retrieved from UN Water: http://www.unwater.org/water-cooperation-2013/water-cooperation/facts-and-figures/en/

UNEP. (2019). *Thawing Arctic peatlands risk unlocking huge amounts of carbon.* Copenhagen: UN Environment Program.

UNH. (2006, October 6). *Researcher Uncovering Mysteries Of Memory By Studying Clever Bird.* Retrieved from science daily.com: www.sciencedaily.com/releases/2006/10/061012094818.htm

US EIA. (2020, March 20). *Electricity explained.* Retrieved from Independant Statistics and Analysis: US Energy Information Administration: https://www.eia.gov/energyexplained/electricity/electricity-in-the-us.php

USBLS. (2018). *Consumer Expenditures 2016.* Washington, D.C.: United States Bureau of Labor Statistics.

USDA. (2019). *2017 Census of Agriculture.* Washington D.C.: U.S. Dept. of Agriculture.

Vozzo, J. A. (1971). Inoculation of Pinus Caribaea with Ectomycorrhizal Fungi in Puerto Rico. *Forest Science*, 239–245.

Wagener, W. (1972). *Logging slash: its breakdown and decay at two forests in northern California.* Berkely, Ca: Pacific SW. Forest & Range Exp. Station. USDA Forest Service.

Wang, X. (2020). Emergent constraint on crop yield response to warmer temperature from field experiments. *Nature Sustainability*, 908-916.

Wiess, A. (2018). *ANALYSIS: In 2018, the West's top oil and gas producing states reported over 2,800 spills.* Denver, CO: Center for Western Priorities.

Williams. (2019). Observed impacts of anthropogenic climate change on wildfire in California. *Earths Future*.

Williams, M. (2018, October 25). *Rice University news and media relations.* Retrieved August 25, 2019, from news.rice.edu: https://news.rice.edu/2018/10/25/nanotubes-may-give-the-world-better-batteries-2/

Williet, M. (2019). *Mid-Pleistocene transition in glacial cycles explained by declining CO2 and regolith removal.* Potsdam, DE: Science Advances.

Wright, D. K. (2017). Humans as Agents in the Termination of the African Humid Period. *Frontiers in Earth Science*.

Yang, C. Z. (2011, July 1). Most Plastic Products Release Estrogenic Chemicals: A Potential Health Problem That Can Be Solved. *Environ Health Perspect*, pp. 989–996.

Zhang, L. (2019). Exposure to Glyphosate-Based Herbicides and Risk for Non-Hodgkin Lymphoma: A Meta-Analysis and Supporting Evidence. *Mutation Research/Reviews in Mutation Research*.

Natura hominis est Natura

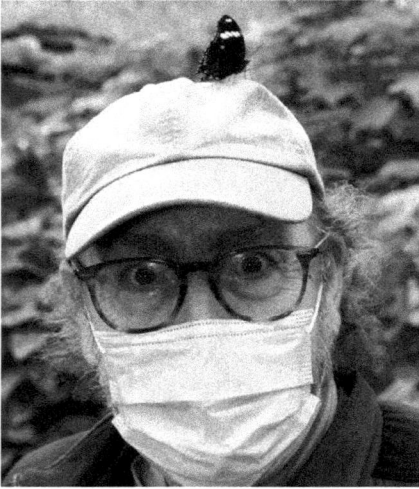

About the Author

Jeffrey Ravage is a practicing forest manager and Naturalist in central Colorado. His work includes ecological mitigation, post-fire restoration, afforestation, myco-restoration and carbon protocol work. Mr. Ravage is an Adjunct researcher for the Sam Mitchell Herbarium of Fungi at the Denver Botanical Gardens. He is a member of the Mycological Society of America and the Colorado Mycological Society. He has a lot of colorful frogs, some of which can be quite loud.

Acknowledgements:

I save my last superlative for my Editor, the incredible Dr. Janna Goodwin, without whose input this book would've been just a tad lesser. I want to thank the test readers of this book for their time and endlessly useful comments. Terry Hilgers, David Harak, Dave Brown, Lisa Patton, Dr. John Ravage, Jonathan Bruno, Adrien Vanore, Paige Fulghum and Doug Bierend. I also want to thank the many publishers and agents who assessed parts of this book- for their education on how the publishing business currently works. This book was never meant to be a happy or safe read. It's only reason for being is to be useful. I was wondering why no one was writing this book, you taught me why. To that end, I want to thank you, Dear Reader for having the courage to open this strange green book. I hope you feel better about our world and our prospects than when you began.

First Edition
Initially published July 4, 2022
Copyright 2022, Ravaged Books
ISBN: 979-8-9862664-1-1

www.ingramcontent.com/pod-product-compliance
Lightning Source LLC
Chambersburg PA
CBHW062113020426
42335CB00013B/941